S0-ABR-565

Print Unchained

Fifty Years of Digital Printing, 1950-2000 and Beyond
A Saga of Invention and Enterprise

This is St. Catherine who for many epitomizes the power of learning. She is said to have lived in 4TH Century Alexandria where she annoyed Emperor Maxentius because she was converting all she met to Christianity, including members of his court. She was condemned to be tortured and killed on a spiked wheel, but the wheel shattered instead.

Here she serenely reads by the broken wheel, a scene that might be taken to evoke the power of print over the forces of barbarism. The first print revolution flowed from the invention of movable type and the printing press that spread learning through the printed word. History names this a key force that lifted Europe out of the Dark Ages into the Enlightenment.

Digital print is the second revolution. It has brought not just print (the noun), but more importantly the ability to print—print, the verb—to industry, commerce, and the whole fabric of society. Print unchained in today's digital age has gained power yet to be fully discovered. It is a revolution still in process.

(Stained glass reproduction of a 16TH Century Albrecht Durer print, 10 x 14 inches. From the estate of Warren Sturgis. Digital photo courtesy of Jonathan Lanman, Bergmans Digital Imaging.)

Print Unchained

FiftyYears of Digital Printing, 1950-2000 and Beyond
A Saga of Invention and Enterprise

Edward Webster

with the assistance of IT Strategies, Inc., Océ Printing Systems GmbH,
The Print Unchained Advisory Board, John Webster, and scores of
industry professionals

DRA OF VERMONT, INC.,
West Dover, VT USA

Published by:

DRA of Vermont, Inc.

226 Handle Road, West Dover, VT 05356 USA

© 2000 DRA of Vermont, Inc.

All rights reserved.

Except for brief reviews, reproduction or transmission in whole or in part,

by any means, without the express written permission of the publisher is prohibited.

NOTE: Every effort has been made to ensure that the content of this book is as accurate as possible. It is based on sources believed to be reliable. However, the author, publisher, contributors, and distributors make no representation or warranties of any kind with respect to the material herein and shall not be liable for any damages claimed to result from or relate, in whole or in part, to said content. It is acknowledged and affirmed that individual trademarks cited herein are the property of their respective owners.

Cover and Book Design and Composition: Dede Cummings, DC Design

Editorial and Research Support: John S. Webster

Production Consultant: Jennifer B. Brown

Indexing: Marty Jezer

Printed in China

03 02 01 00   1 2 3 4

First Printing, November, 2000

ISBN 0-9702617-0-5

dedicated to . . .

The Infinite Network that joins us all;

and to Edie, life partner, birther of dreams;

and to Irving L. Wieselman,

who planted the seed that grew into this book.

CONTENTS

ACKNOWLEDGEMENTS

F IRST OF ALL, thanks are due the thousands of men and women around the world who have advanced human communication over the years by building the digital printer industry. My heartfelt thanks also go to the following organizations and individuals who have made this book possible, and to all who took time to contribute to its accuracy and completeness:

I T Strategies, Inc. for organizational and content support. Without the enthusiasm and wisdom of the entire staff, and especially Mark Hanley, Marco Boer and Patti Williams, this project would certainly never have come to fruition. Their insightful guidance contributed greatly toward making this book a unique strategic planning resource as well as a history.

Océ Printing Systems GmbH, an industry leader with vision, for financial and content support.

The Editorial Advisory Board for their encouragement, endorsement, and expert review: Marco Boer, IT Strategies, Inc.; Richard A. Fotland, President, Illuminare, Inc.; Larry Lorah, Concord Consulting Group; John Schneider, John M. Schneider Associates; R. Hugh Van Brimer, Founder & Chairman, I/J Printing Corporation; Vivian Walworth, Fellow, IS&T, The Society for Imaging Science & Technology; Mike Willis, Managing Director, Pivotal Resources.

–Edward Webster
West Dover, VT USA
December, 1999

NOTE: For more information on these individuals and organizations, see Publisher/Contributor Information, p.253

FOREWORD

This book, among other things, is a time capsule. It preserves the story of digital printing in a form that will still be readable decades from now: print. All that stuff recorded on magnetic or CD media will suffer in the future as Data Archeologists try to dig up the players.

Printing was invented when Johann Gutenberg broke language down into individual letters. Digital printing was born when people like Robert Howard broke letters down into individual dots. The dot-matrix printer saw the world as a mosaic of discrete particles. The character-based printer was just a souped-up typewriter. It got faster and faster but not fast enough. The belt- or chain-printer got fast enough to make, well . . . an impact, and the impact printer led to the forms market. All through the 1970s printer manufacturers struggled to keep up with the computer. Mainframe behemoths in corporate MIS centers spewed out data at a rapid rate, yet printers were incapable of keeping up.

I once commented that if IBM had been able to protect the typeface "Courier", the printer market as we know it would never have existed. Every competitor had to have a typewriter-like font. I recall winding up with a character printer for text, a dot matrix printer for charts, and an A-B box that switched between them. Then the laser printer gave me both at once. Today non-impact technology dominates. "Non-impact" – a technology described by what it does not do. Like, what do I do? I'm a non-lumberjack.

The inkjet printer is now pervasive. In the beginning the ink faded as soon as the print came out. One unit had a print that was permanent but the printer faded away. Today, their quality is superb and the future of photography lies in pixels and drops of ink. I still think they should give the printers away since inkjet ink is probably more expensive than gasoline. OIEC is the inkjet cartel – the Organization of Inkjet Exporting Countries.

All printing today is digital, no matter how you do it. Film and plates are made with laser imagers; toner and inkjet are patterns of spots. The defining devices of paper-based replication are the printer, the copier, and the press. A printer uses inkjet or toner or other technology to make marks on paper from data files, resulting in the production of first-generation originals where every one of them can be different, thus allowing the production of a collated document or individual, personalized impressions. A copier uses inkjet or toner technology to make marks on paper

from an original, resulting in the production of a second-generation copy, which, when copying multiple originals in an automatic document handler, can also produce a collated document. A press typically means a mechanical device that uses an image carrier to replicate the same image on paper, over and over again, resulting in a large quantity of the same images.

International Data Corporation reported that the number of pages output on printers in 1995 for the first time exceeded pages output on all models of copying machines. This led Hewlett-Packard to coin a new buzzword: *mopier* – a multiple original printer – prints from an original-producing printer instead of an original-copying copier. Since copiers are evolving to digital approaches – scanners on the top and printers on the bottom – they become de facto printers. Printers at the high end, like the Xerox Docutech, are challenging offset duplication at the low end of the black-and-white printing world. Low-end printers are absorbing some of the work of both offset duplicators and mid-level copiers.

The objective of replication technology over the next decade will be to build into the printing process very high levels of automation. The cost of paper-based communication comes down. Cycle time is reduced which leads to all the current buzzwords: short run, on-demand, just-in-time, distributed printing and more. When you say "digital printer" you are covering a great deal of ground, from very low-end desktop devices to high-end color devices that compete with printing presses. The latter devices are now competing against offset litho. It was a fateful day in 1993 when the Indigo and the Xeikon were first introduced. I was quoted in the Wall Street Journal and received 432 phone calls. Digital color printing struck a nerve. One third wanted to buy the machines; one third wanted to buy the product of the machines; and the last third wanted to invest in the companies. The RIT switchboard operator is still in therapy.

The reproduction of information on paper is static or dynamic. Static printing refers to traditional ink-on-paper approaches – offset lithography being the most common – where each and every sheet is reproduced from the same image carrier which is fixed with the same image. The copies look exactly the same. If they are different, you did something wrong. Digital printers, conversely, use an image carrier that is imaged each time a sheet comes in contact with it, re-imaging for each copy. The copies look the same, but each is generated individually. Dynamic printing means that the printer must re-generate the image for every page; thus, every page can be different.

This means digital printing can do things that all forms of conventional printing cannot do: variable data printing and on-demand publications. Making each sheet different lets you produce personalized or customized sheets that add information about me, to me, for me, concerning me – me, the most important person in my life. You can also output each page of a document in the proper order so as to produce a paginated publication on demand. The dream of on-demand books finally becomes a reality. As a subset of these two outcomes, we get two others: very short runs – like one – and very quick turnaround – like now.

We commend Ted Webster for preserving the history of digital printing. Now if we can only preserve Ted Webster.

Frank Romano
Rochester Institute of Technology
June, 2000

E. MAS PHOTO

COURTESY OF HEWLETT-PACKARD, INC.

Digital printing today.

First digital printing?

INTRODUCTION

This book is part celebration, part documentation, part counsel – and part recreation. My primary motive for undertaking this project, after years of involvement in the industry, was to fill a yawning gap in technology history. In the outside world – and even in the worlds of computers, communications and publishing which they serve – digital printers have always tended to be something of a footnote. Within this ill-defined industry, myriad trees are well documented, but the forest and its significance is not. If nothing else, my hope is that with this book not just the substance of digital printing is documented, but more importantly, its meaning and significance.

Here's the bottom line: with digital printing, print has been unchained.

Digital print is not simply an evolutionary enhancement of the various technologies growing out of Gutenberg's invention. It is a quantum leap. It unchains printing from the shackles of the older technologies in which it was imprisoned: those of the printing press and the photographic technologies. Lithography, even with digital makeready, is basically just an enhancement. Digital printing is a new paradigm, a major window between our increasingly digital world and the human brain.

Digital technology has liberated print from the old limitations of content, time, distance, substrate, and, to an extent, of economic constraints. We can now print on demand, when and where we need it. As for surface independence, there was the memorable Videojet ad in the 1980s showing ink jet printing on the yolk of a fried egg. Even as recently as a decade ago, it would have seemed visionary to predict that high quality color printing, including photographic images, could be done on the desktop by printers priced at under $100. Printing has been democratized. Printing has gained power yet to be fully discovered.

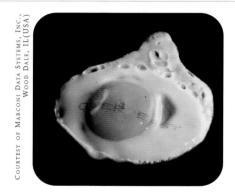

COURTESY OF MARCONI DATA SYSTEMS, INC., WOOD DALE, IL (USA)

COURTESY OF XEROX CORPORATION

COURTESY OF CASTLE GRAPHICS LTD.

Among the goals of this exploration:

- To celebrate the contribution of digital printing to commerce, science, society, and to the human adventure in general.
- To document how individuals and organizations have brought about this liberation of print. Through selected vignettes, to show the human drama, the motivators, the hurdles, and the triumphs of a few of the innovators.
- To recount history, and also to look for patterns that may crystallize lessons useful to planners and entrepreneurs now and in the future. By placing digital printing in a larger, historical context, it is hoped the way forward – the future of print – will be, in at least some small way, illuminated.
- To provide for the general reader an interesting excursion through the development of a technology and an industry that has now placed printing at our fingertips.

In one sense, this book is the fruit of a lifetime of researching and reporting about and for the digital printer industry. The process has been long, the topic vast. So part of the challenge has been to keep the resulting book manageable in size, useful, and fair. To this end, here are some of the guidelines:

Where possible, let people speak for themselves. Large portions of this book consist of oral history, the words of some key entrepreneurs and managers who helped make the digital printing revolution happen.

Minimize bias. By necessity, the content of this book is very selective. The confines of this one volume limits this exploration to a very small sampling, representative of a much larger web of stories, many no doubt of equal significance. Many companies, inventors, products and even aspects of the industry have been skimmed over more lightly than their importance deserves. The sources – individuals and companies – varied widely in their responsiveness to inquiries. One goal was not to let the varying levels of responsiveness color how companies are presented. But the reality is that the depth of coverage of various companies is at least partially gated by their level of responsiveness.

Emphasize material not available elsewhere. A torrent of information has been published over the years by the industry. This has of late increased with the Web. I have leaned toward selecting material which is original or not easily available elsewhere.

Don't try to do it all. The leaders of course are here. Other companies were chosen because they illustrate a significant pattern within the tapestry. There is Genicom, representative of programs by major industrials in the 1960s and 1970s to diversify. Adobe is here, helping to open the door to desktop page printers with its page description language. Centronics pioneered dots, but then was eclipsed by an invasion of Japanese vendors in the 1980s, represented here by Okidata. In addition, keep focused on the core of the industry – the development and commercialization of the marking engine. This book is a sampler, not the whole story. Yet I suppose this is true of any attempt to pack a history into a book.

Speak to a variety of readers. This story is first and foremost for the industry itself: not only for the engineers, but also for those whose technology IQ

may resemble my own — marketers, new executives from other industries, business partners, suppliers, investors. The second audience is students of management and high tech entrepreneurs no matter what their business or industry may be. Users are a third audience, both professional and private. Then there are the many individuals around the world drawn toward technology and technology history, and just plain lovers of print.

Keep it interesting, and light. Find some printer humor. There is some, believe it or not, which may or may not work for you. Where possible, highlight the drama. To invent and innovate and especially translate that innovation into commercially successful products is an adventure. That is where the human story — the saga — comes in. The prescription for writing this book is not a bad one for life: if it isn't interesting, if it isn't fun, don't do it.

Structure and Content. There are many ways to slice a history. This book cuts it two ways. It is not a simple, continuous narrative.

Part One, the Overture, contains background and tutorial material. It looks at the roots of digital printing, then the various technologies, applications, and broad sea changes in the products and the industry over fifty years.

Part Two, five Movements, the decades, focusing on one or several "defining companies" in each. It is a tapestry, with representative or industry-changing companies introduced and their stories normally followed through to the present.

Finally, there is the Coda, Part Three. Included here is a review of the patterns and lessons suggested by this history and a look at the future of digital printing. The usual reference aids follow as appendices: a Glossary, Bibliography, and Index.

Scope and Definitions. This story covers primarily a somewhat arbitrary subset of a larger, amorphous field commonly termed "electronic imaging" or "electronic printing." Our "digital printing" is related to, but a bit broader than computer printing,

computer print-out, or hard copy. It excludes other forms of imaging such as plotting, facsimile, analog photo processing and microfilm printers. A digital printer is not a printing press, regardless of the level of digital makeready.

Our chosen term "digital printer" can mean different things to different people. "When I hear you say 'digital printing,' I envision finger prints." That was my father-in-law, a 96-year-old retired surgeon. But come to think of it, this, indeed, may have been the first digital printing. (The first definition of "digit" in my dictionary is "a finger or toe.") For our purposes, "digital" is used to connote the use of numbers in a scale of notation to represent discretely a set of variables, as opposed to "analog."

COURTESY OF APOLLO CONSUMER PRODUCTS

"Digital printing" might be defined as printing with devices that receive and convert digital input to a visible image applied directly to the final surface. In addition, the content and/or format is subject to at least some level of programmable or computer control. The "directly" qualifier excludes various indirect digital printing devices such as digital platemaking on- or off-press, microfilm recorders, and digital cameras. The "programmable control" qualifier excludes products such as electronic typewriters with digital keyboards, copiers, facsimile, and teletypewriters in strictly (non-computer) communications networks. Although the image needs to be visible, we do not mean to exclude at least one curious application, that of printing invisible ink for security documents.

Some readers might be inclined to narrow the definition to perhaps only those machines that print dots: building the output image with binary, on-off dots. In our definition, "digital" refers to the input, not the output. Our "digital printing" stretches the

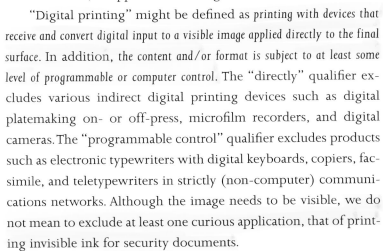

Print Unchained: (opposite page) ink jet pioneer Videojet (now Marconi Data Systems) in the 1980s demonstrated surface independence by printing on the yolk of an egg; on-demand printing by Xerox; digital printing on fabric, in this case a snappy vest modeled by Danielle C. King, President of the wide-format graphics company Castle Graphics Ltd. of Concord, MA; everyone can now be a publisher including an eight-year old, shown here with her Apollo Barbie Printer.

classical implications of "printing." Etymologically, the word "print" derives from pressure or impression. The traditional press makes an impression, or print, through pressure. (This would seem to make "non-impact printing," an oxymoron.) It furthermore implies the facility to make multiple, identical copies. A key strength of digital printing, on the other hand, is that it can make multiple copies, but each one can be different.

This is getting more complicated than it needs to be. At the Xerox Palo Alto Research Center they simplify, describing printers and scanners as the "portals" between machine and human intelligence.

. . . .

In the beginning, printing meant pressure.

The printer *industry* for the purpose of this history is defined as the companies developing and manufacturing the printing technology and the print engine itself. This narrows it down quite a bit, since for every company that develops and manufactures the basic marking engine, there are dozens of OEMs and systems integrators who use the marking engine as the basis for general purpose or specialized printing systems. (The printer industry, by the way, is not to be confused with the printing industry – the world of printers who run printing presses – which is for now still more or less distinct from the printer industry.)

Emphasis is on companies building marking engines used primarily for printing pages or documents. The many other types of specialized printers – listers, data loggers, label printers, wide format digital printer-plotters – will be touched upon but are not part of the central theme.

Then there are the related industries. These might be categorized as *enabling* industries and *derived* industries. Examples of enablers are those who develop specialized materials such as ink jet inks, page description software, or high performance microprocessors. Derived industries are those which ride on the coattails of the industry proper. Among them might be listed OEMs who take a marking engine, build it into a printing system, and affix their own label. Specialized channels including value added resellers, systems houses, and, more recently, computer retailers might also be seen as derived industries.

The stream of materials and components through development and fabrication to sales and support is (or at least should be) a seamless flow. All phases, obviously, are connected and interrelated. Enabling and derived industries and developments are covered as needed to trace the history of a technology or company. Yet the main focus will be on the innovators and companies that fit into the more specific "digital printer" definition.

Enough introduction. Enjoy.

HYPOTHESES

One value of historical inquiry is distilling broad, instructive truths from specific stories. This history suggests a number of these, which might be framed as hypotheses. Listed below are some examples. Although they are derived from the digital printing industry, many have a universal flavor.

— The era of the bold, individual inventor or entrepreneur is gone. Today innovation and success are more the realm of large organizations applying sophisticated market planning and overwhelming capital resources to product development and marketing. Teams rather than individuals make the history.

— On the other hand, although with every decade the financial resources needed to win seem to grow, expending financial resources alone isn't enough. In printers, companies such as Exxon, Texas Instruments, GE and IBM have all discovered that throwing money at a market is no recipe for success.

— It has been said the fast will win over the big. This sounds good, but doesn't quite do it. "Bigness" is needed more than ever, so maybe it should be "big and fast." But that doesn't always do it either. You can be too fast. Fielding a product too early will fail regardless of its merits. "When" and "what" are both essential.

— Without people skills — charisma, teamwork, successful marketing of ideas to top management — the best technologies or product strategies will fail.

— The product normally births the company, but then the company must take priority over the product. Put another way, corporate behavior needs to balance defending current markets and attacking new markets through innovation and risk-taking. This means a willingness to obsolete your own products at some point (see Chip Holt in Chapter 7.)

— Manufacture of the marking engine is no longer central to the success of companies now viewed as leaders in the industry. The marking engine is better viewed as just another component rather than the core of the product.

— The payoff is less and less from building hardware or even from owning proprietary technology. At the low end it is expendables, and at the high end, systems and expendables. In fact, products are beginning to become secondary to know-how and that buzzword of the 90s, "solutions." The swing is from iron to information.

— Both products and marketing are important, but marketing — getting mindshare, getting brand recognition — is the ever more important success factor. A name such as Xerox or IBM won't ensure success forever, but may nevertheless be a company's strongest asset.

— We don't need to worry about a paperless office or a paperless society. When a new medium is introduced, existing media fear obsolescence. But history shows that various media feed off one another, and that our capacity to absorb information — or have the option to access it — seems virtually limitless. Print is here to stay . . . well, at least over the near-to-medium range future. What will change is who will provide it and how.

— Digital printing is a winner that will take all. All printing — not to mention copying, photography, and all other forms of imaging — will be digital.

PART ONE

Prelude

Telegraphy proved you could send words over wires, a good start toward today's digital world. This 1877 commemorative tableau includes the relevant inventors and a photo of the earliest telegraph instrument said to have been used for the Baltimore to Washington, D.C. demonstration in 1844.

CHAPTER I: ROOTS

Streams of technology develop more like roots than individual lines, our timelines notwithstanding. The pattern is closer to a matrix in which various threads or fields develop not only chronologically, but also cross-fertilize one another. These cross-fertilizations are often under-appreciated when reviewing a history since history tends to be told in a linear format. This simplifies the presentation and makes it more easily grasped by the normal human mind which also tends to be linear.

In looking at the pre-1950s roots of digital printing, then, ideally we would need not only to chronicle the progression of each thread – typewriting, copying, photography, etc. – but also the many parallel threads which nourished print-related developments, such as chemistry, fluid dynamics, papermak-ing, coating technologies, and on and on. All of these cannot be explored in this brief overview. This chapter will sample a selection of historical threads which are seen as significant to our current variety of digital printing technologies.

"Print" is much broader than our timeline or this book might imply. Here we focus on text or images on paper documents. Looking back into the origins of print, as well as the print around us, we find a much broader assortment of applications. Today there is print in printed circuit boards, laminates, decorative ceramics, the fine arts, wall coverings, and carpets and fabrics. Applications such as these employ technologies that may or may not be also found in the main, paper applications of digital printing.

The timelines can shed a bit of light on that perennial question of time from invention to commercialization. The path can be amazingly long. In some cases, the technology building blocks exist, and the challenge is to find the right ones. Ink chemistry serves as a building block for ink jet printing. Affordable, massive digital memory capacity is needed to bitmap a digital color image. In many cases, at the time of inspiration, the technology building blocks have not yet been invented. In this case, the concept has to wait until science and technology have sufficiently advanced.

Leonardo da Vinci is a familiar example. He conceptualized designs of everything from advanced weaponry to the airplane in 1500 AD. Most of these inventions in time were realized, but in many cases it took 400 years! Charles Babbage documented many of the features of modern computers with his "Difference Engine" in 1833, more than a century before a computer was actually built.

In digital printing, the most familiar example is probably xerography. Looking at electrophotography, we will see Carlson and Kornei demonstrating the principle with their famous "Astoria" exercise in 1938. But the invention was not implemented as a really successful commercial product until the Xerox 914 a full 21 years later.

Most innovation is evolutionary, building on prior developments. In a 1965 interview with Dartmouth Professor Joseph J. Ermenc, Carlson was asked about the creative process. While acknowledging inspiration, he responded in part, "One must build on what went before and adapt or adopt as much as he can to solve specific problems. . . . I merely combined a set of facts in a new way."[*] His humility belies another ingredient, which is belief, the fuel of persistence.

Innovation needs belief, in some cases bordering on obsession. Carlson is a wonderful example, but there are many others, including some of those who pioneered ink jet over the decades and others who today continue to hone various drop-on-demand ink jet and other print technologies.

In this chapter we survey some of Carlson's "what went before," important roots from which today's digital printing sprang. We suggest 1950 as the beginning of digital printing. Yet in the case of almost every technology used in this industry, inventors – including Chester Carlson – cite antecedents going back into the 1800s and earlier. It is clear from our timeline ("Landmarks in Print Communication", pages 4-5) that as part of the entire sweep of human print communication, digital printing is only a very recent blip.

The timeline landmarks represent various threads which can be seen as related to current digital technologies. Thus, from the typewriter and telecommunications evolved the teleprinter and later word processing printers. From the adding machine evolved first the tabulating machine printers and later impact line printers. From traditional analog printing grew typography, proofing, rotary image transfer, ink science, binary tone representation (halftones), and various paper handling technologies. And from conventional photography grew electrophotography.

Analog Printing and Imaging. This root includes the development of various graphic arts technologies such as lithography, typesetting, and the printing press.

Gutenberg leaps into mind when the history of printing is mentioned. Yet he was preceded by many other reprographic de-

[*] Carlson, Chester, with Professor Ermenc of Dartmouth College, *The Invention and Development of Xerox Copying*

velopments, some known to him and some, as is often the case, no doubt unknown (such as Mr. Sheng, 400 years earlier in China.) In any case, at least in the West, Gutenberg is credited with the invention of movable type and the printing press. Print speed: perhaps 300 copies per day – a quantum jump over previous hand copying. The invention is said to have fueled the transition of Western civilization from the Dark Ages to the Renaissance.

Histories of traditional printing normally cover presses, plates and platemaking, typography and typesetting, and paper. Among the classifications of presses are flatbed and rotary, printing process (letterpress, lithography, gravure), paper handling (sheet or web), and color capability. Each of these has its own history, beyond the scope of this exploration.

Before 1950 analog printing was pretty self-contained. Processes evolved beyond letterpress and printing press speeds accelerated with automatic sheet feeding and later rotary sheet and web presses. But many basics were static. Up to 1950, according to RIT's Mike Bruno, the convergence of traditional printing with digital that we see now was not even a dream. Bruno writes, "If Gutenberg or his contemporaries had stepped into the average printing plant of 1950, they could have stood at a type case and set type by hand almost exactly as they did over 500 years earlier."*

Since 1950 all that has changed. Setting type by hand and mechanized letterpress linecasting are rare. Today the line between traditional and digital printing is increasingly porous. Convergence seems inevitable and is already well underway. Digital prepress is rapidly becoming the norm. We find Robert Howard taking it a step further. His most recent company is Presstek, which has commercialized an on-press digital platemaking system. In a recent speech he stated, "We have made every printing press a computer peripheral." To this one might respond, "Not quite yet. It doesn't really make every printing press a digital printer since it is not direct. The consumers of print don't read computer plates." But it is a significant step forward in the convergence of these two streams of technology.

How, then, has traditional printing served as a root of digital printing?

NEGATIVE THINKERS vs. TRUE BELIEVERS

Technology lore is filled with examples of vision deficit – sometimes called "realism" – in response to new ideas. At times it seems amazing anything ever gets invented let alone pushed all the way down that long path from concept to commercialization.

Some examples:

- There was computer pioneer J. Presper Eckert, as mentioned elsewhere, unable to imagine why anyone would want high speed computer printing.

- IBM considered backing Chester Carlson and went to Arthur D. Little Company for expert advice. After looking at the xerographic process (39 steps in an early implementation) their verdict was, "Too complex to be affordable."[†]

- Later, there was the finding of a "reputable New York firm", after an in-depth survey of businessmen commissioned by Haloid Company in the mid-50's, advising Haloid that "we could not look forward to placing more than 5,000 copiers in American offices .... Joe Wilson kept shaking his head and saying, 'I cannot believe it. I *simply cannot believe it.*' That was the spirit – a stubborn defiance of reports, a show of unyielding optimism or perhaps blind courage. It pervaded all those in research and development."[††] Wilson of course persisted, bringing out the Xerox 914 in 1960, a product which has been described as the most successful ever, with 200,000 units shipped over the next twelve years or so.

- "Even Mark I inventor Howard Aiken had suggested that the two men (Eckert and Mauchly) were on the wrong track, that there would never be enough work for more than one or two computers."[†††]

- "Everything that can be invented has been invented." Attributed to Charles Duell, Commissioner, U. S. Patent Office, 1899.

---

* *Pocket Pal*, published by International Paper Company, 1995 edition
† Robert Gundlach, Xerox Sr. Research Fellow, SPSE Boston Chapter lecturer, October, 1985
†† Dessauer, John H., *My Years with Xerox*, Doubleday, 1971
††† Slater, Robert, *Portraits in Silicon*, MIT Press, 1992

Photography, Facsimile, Duplicating and Office Copying, Other

1749: Nollet experiments with the interaction between static electricity and a drop stream, anticipating ink jet printing

1777: Christoph Lichtenberg experiments with electrostatics for developing patterns with powders on cakes of resins

1839: Louis Daguerre demonstrates early photography and William Talbot uses papers soaked in silver chloride with negatives to make multiple positive prints

1842: Alexander Bain in Scotland invents facsimile (remote copying)

1867: Lord Kelvin demonstrates an operational continuous ink jet device which leads to the use of ink jet for oscillographic recorders

Calculating, Accounting Machines, Computers

1820: Charles Xavier Thomas of Colmar, France, builds 1st "modern" calculator featuring the "Thomas One-Way Principle," later a key feature of Friden calculators in the 1930s

1872: Thomas A. Edison patents "electric type-wheel machine," a receiving device for printing out gold and stock prices

Communications, Typewriting, Typesetting

1714: Queen Anne grants patent to Henry Mill for a typing machine

1794: Chappe demonstrates optical telegraph in Paris

1833: Xavier Progin in France invents the first typewriter with individual typebars that converge at a common printing point

1837: Sir Charles Wheatstone in England and Samuel Morse in the USA both (independently) invent the telegraph

1844: First telegraphic system — Samuel Morse transmits "What hath God wrought" over a 40-mile telegraph line between Baltimore and Washington, DC

1856: Western Union Telegraph Company formed

1866: Single element typewriter built by John Pratt

1867: Christopher Sholes invents (actually, re-invents) the typewriter

1867: Giuseppe Ravizza develops first practical moving inked ribbon

1873: E. Remington & Sons commercialize the Sholes "TypeWriter"

1874: Baudot invents multiplexing that enabled six signals from telegraph machines to travel together over the same line

1875: Bell displays the electric telephone at the Philadelphia Trade Fair

Media and Analog Printing

4000 BC: Hieroglyphics; media: clay tablets

2000 BC: new media: papyrus

1100 BC: newer media: parchment

400: Wei Tang in China perfects ink for block printing using lamp-black

1041: Movable type, Pi Sheng, China, using fired clay

1397: Oldest known example of printing from cast bronze movable type, Korea

1450: Johannes Gutenberg converts a wine press to a printing press

1609: First regular newspaper, Germany

1690: First paper mill in North America: Rittenhouse, Philadelphia, PA

1798: Alois Senefelder of Munich invents lithography using a porous stone "plate"

1850: Andreas Hamm establishes workshop that advances printing press design, a company later renamed "Heidelberg"

1887: Albert Blake Dick awarded patent for the first true rotary mimeograph machine, leading to the establshment of A. B. Dick Co.

1923: First diazo patent issued in the USA said to have been developed by a monk named Koegel and assigned to Kalle; trade named Ozalid

1925: Domestic wirephoto (facsimile) system launched by AT&T; later sold to Associated Press

1930: Paul Selenyi, Hungarian physicist, demonstrates electrostatic facsimile using dry powder and photoconductivity, coining the term "electrography"

1938: First electrophotographic copy, "10-22-38 Astoria" by Otto Kornei working for Chester Carlson in his Astoria, NY lab

1940: xerography: Chester Carlson concept patent issued

1947: Haloid licenses xerography from Battelle Memorial Institute

1947: One-step photography demonstrated by Edwin Land

1949: First commercial xerographic copier, the Xerox Model A, introduced by Haloid Company

---

1886: William Seward Burroughs invents the first adding machine

1890: First major use of Hollerith punch card tabulating (U.S. Census)

1911: Burroughs introduces the first adding-subtracting machine

1938: IBM 405 Tab Printer, 88 reciprocating type bars, 100 lpm

1942: First computer using binary arithmetic, Iowa State University by John Atanasoff and Clifford Berry

1946: ENIAC, considered the first digital computer, Eckert and Mauchly at University of Pennsylvania

1947: The transistor is invented, Bell Laboratories

1950: IBM 407 tabulating machine printer, 150 lpm, 120 print positions

---

1886: typesetting is mechanized as The New York Tribune installs first recirculating matrix typesetting machine, the Mergenthaler Linotype

1893: Blickensdorfer single element typewriter

1896: Monotype invented by Tolbert Lanston, said to be inspired by the work of Herman Hollerith

1897: Marconi demonstrates wireless telegraph communication between Italian war ships

1904: First phototypesetting machine by René Higgonnet and Louis Moyroud

1904: Korn transmits a photograph by telephone in Germany

1914: Teletype machines begin to be installed by Kleinschmidt Electric (which later became Teletype Corp.)

1918: Morkrum Company, later Morkrum-Kleinschmidt, introduces its teleprinter

1923: Picture, broken up into dots, transmitted by wire

1925: Remington introduces first electric typewriter

1933: IBM buys Electromatic Typewriter Co., invests over $1 million in further development, resulting in the first popular electric typewriter

1937: Digital transmission by pulse code modulation

1950: Proportionally spaced typewriter (IBM Executive)

1883: Drum cylinder printing press patented by Calvert B. Cottrell and Nathan Babcock

1906: Offset lithography, invention credited to Ira A. Rubel, a New Jersey papermaker

1912: Heidelberg obsoletes manual sheetfeeding, introducing rotary grippers on its platen press for breakthrough speed of 1,000 sheets per hour; plate inking system

1923: Planeta introduces first two-color press in Germany

1930: John Webendorfer introduces first web offset press in the USA

1935: Heidelberg Cylinder (OHC) letterpress boosts speed to 3,600 sheets per hour

In short, there are first the various techniques borrowed by digital printing from traditional printing. Secondly, traditional printing created a market for print. Conversely, in recent decades, digital printing has been oozing into the graphic arts as part of the prepress workflow. It created the market and also serves as a market for digital printing.

As digital printing technologies evolve, they are becoming an increasingly realistic alternative to traditional printing in many applications.

Typewriters. Many computer printers are directly descended from the once-familiar typewriter. Typewriters and teletypewriters were adapted for use as console printers and remote printing terminals for computers. The word processing systems of the 1970s and early 1980s were computerized extensions of the typewriter. And the typewriter keyboard remains with us as a dominant data entry method.

According to one history, the typewriter – at least the idea of the typewriter – first appeared on January 7, 1714 when Queen Anne granted a patent for such a machine to an English engineer named Henry Mill, preceding the Sholes typewriter by well over a century. In 1829 a form of typewriter was granted U.S. Patent 259, and in 1833 Progin in France built a typewriter using individual typebars which converged at a common printing point.

None of the several typewriters which appeared before 1878 achieved much commercial visibility in the USA. Commercialization is generally attributed to Remington, a gunsmith turned office machine manufacturer. The Remington typewriter was invented by Christopher Sholes in 1868 and first manufactured by Remington in 1873.

This Remington machine incorporated many of the typewriter features which in time became the norm. The keyboard activated typebars, each with one letter, which struck the paper through an inked fabric ribbon. The paper was supported by a cylindrical platen. But there were differences. Carriage return was by foot pedal. It printed only upper case letters. Because the configuration was "up-strike," the user could not see the page being typed. At $125 each, it was expensive (around $2,000 in today's dollars). In five years, just 5,000 were sold. It had its enthusiasts, however. An early Remington owner was said to be Mark Twain, who bought one as soon as he saw it. But he apparently just used it as a toy, hiring a typist to convert his handwritten manuscript for a 1883 novel to a typewritten document.

In 1878 Remington introduced a second generation, the Remington Number 2, which more or less set the pattern for the next fifty years. It featured a shift key so it could write upper and lower case letters. Although still a "blind" up-strike machine, it sold well and was the first of a series of successful Remingtons. Remington Number 5 kept them ahead of their many competitors, thanks in part to its redesign as a visible format typewriter. Although several inventors in the U.S. and Europe introduced visible typewriters, the blind design was favored for quite a while because it was considered desirable that the operator not see the page being typed. This presumably was to encourage touch typing. But by around 1908 most typewriters were "visibles" and also had the four-row QWERTY keyboard we still use after all these years.

It was largely typewriters that fueled Remington's evolution over the years into a major office machine and computer vendor. In 1928 entrepreneur James Rand took the Remington Typewriter Company and merged it with nine other companies to form Remington Rand. In the 1930s they competed with IBM, manufacturing an assortment of keypunch machines, punch card tabulators (round holes), card sorters, and interpreters. Then, in the 1950s, they acquired UNIVAC and became a pioneering computer manufacturer. While rival IBM was pretty much committed to its 150 lpm computer output printer, Remington Rand Univac fielded one of the first on-the-fly line printers rated at 600 lpm.

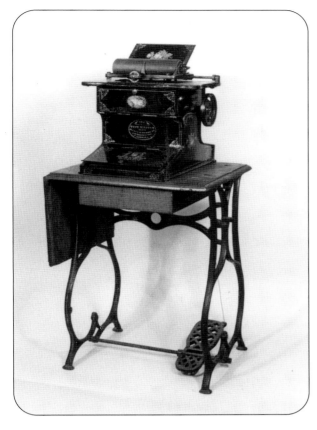

Sholes typewriter, 1868
(Courtesy of typewriter historian Darryl Rehr)

In the 1940s and 1950s three typewriter-based serial impact printing terminals were the dominant human-to-machine, input-output interface in early computers: adaptations of the IBM Model B typewriter, Teletype, and the Friden Flexowriter (which was built around IBM typewriter mechanics).

Typewriters and teletypewriters are of course still with us, widely used for document preparation, telecommunications, and data entry. Typing also merged with typesetting: Linotype and Monotype keyboards were derived from the typewriter. In the 1960s and 1970s several "strike-on" typesetting machines – in effect expensive typewriters with film ribbon and sophisticated proportional spacing – gained popularity. The two leading examples were the Varityper (Addressograph-Multigraph) and the IBM Selectric Composer.

The document preparation function of the typewriter merged with low-end computing to power the word processing industry of the late 1970s and 1980s. And for all of us who use the computer, the most prominent legacy is the QWERTY keyboard, apparently destined to be with us for the foreseeable future. In the 1950s and 1960s students on a vocational track – most often girls – were steered toward touch typing and forced to spend untold hours on finger warm-ups, drills, and typing to music. In this day of ubiquitous computers, once mundane touch typing skill has transformed into an unexpected asset.

IBM, meanwhile, had purchased the Electromatic Typewriter Company in 1933 and pioneered powered typewriters. According to IBM, the powered typewriter was made possible by a rollover cam driven by a rubber power roll. The inventor, James Fields Smathers, took his inventive inspiration for this from the rollover cam action of a hay-raking machine.[*] IBM's electric machines took time to win acceptance, but eventually, by the late 1940s, IBM had become a major typewriter power.

Telecommunications. The digital information which drives computer printers arrives in the form of an encoded binary data stream from almost anywhere in the world, or from a local computer or keyboard. This transmission technology is rooted in telegraphic communications.

Telecommunications goes back pretty far, to Samuel Morse – generally viewed as the main pioneer in the USA – born in 1791. The Samuel Morse story is another interesting glimpse of an in-

---

[*] Beattle, H. S., and Rahenkamp, R. A., in *IBM Journal of Research and Development,* September, 1981

ventive personality type. In this case, a primary goal and gift is eclipsed by a pragmatic side path taken to achieve that goal. His means to an end became an end in itself.

Morse attended Yale University and then went to Europe to study painting in pursuit of his lifelong yearning to make a mark and a living as a painter. He was never able to earn much of a living at painting, but in 1835 he gained a teaching appointment at New York University. It was teaching "the literature of the arts of design," not painting, and it didn't make him happy. The story of his conversion to an inventor is artfully recounted as follows by Steven Lubar.[*]

His painting career frustrated, Morse turned instead to inventing. Inventing a telegraph, he thought, would earn him enough money so that he would be able to return to painting and to paint what he wanted. This was not a practical scheme, but Morse was not a practical man. He was one of the best artists in the country but not particularly skilled as an inventor or a scientist. Neither was he a talented mechanic or an accomplished commercializer. His principle qualification was that he combined some of all of these talents, and was able to bring together the right people to work for him. He was also able to use his personal and political connection to gain first government and then commercial support for his work. He had no qualms about claiming more credit than was his due. Like many others of his day, he believed in the idea of the great inventor and began to see himself as one. (The telegraph would indeed make Morse wealthy, but he never returned to painting.)

Morse's key insight was said to be a flash on an ocean voyage, triggered by a lunchtime conversation in which he was told electricity flows instantly through a wire. If so, he thought, at any point in a circuit, the presence or absence of electricity could be sensed, and this could be applied as instant communication.

In 1837 Morse improvised an early sending and receiving machine which never worked very well, then hired an experienced mechanic, Alfred Vail, who made a set that would really work. More significant than the send-receive electromechanics was the coding system which may or may not have been Vail's work. The first system was simply sending sets of dots to represent the numbers. A dictionary was coded, assigning a number for each word, allowing the users to encode and decode word messages.

Along with Vail's improved instruments came a new coding system with two types of signals, a dash and dot, with combinations of them linked to the letters of the alphabet. Inspired by the Wheatstone Telegraph in England, this newer system required more data to be sent over the wire, but greatly simplified the encoding and decoding at each end. This was the origin of the Morse Code still used today.

Next, telegraph lines needed to be built and with the help of others Morse convinced the U. S. Congress to finance the first line between Washington, DC and Baltimore, which was completed in 1844. Morse saw the telegraph and its potential to instantly relay news and information across space as a wonderful, democratizing opportunity with benefits to not only government but also to the people at large. But he was unable to convince Congress to finance more lines. This was left to a multitude of telegraph companies that sprang up in the latter 1840s.

The telegraph is said to have become a popular passion, used by journalists, lottery runners, financial brokers and businesspeople. Accounts of this enthusiasm sound strangely familiar – transferred to another century, they would perfectly fit today's Internet frenzy.

The next major steps were the improvisation of machines to

[*] Lubar, Steven, *Infoculture*, Houghton Mifflin Company, 1993

print messages and parallel encoding systems. Early telegraph printers were devices that used an electromechanically controlled pencil that made short and long marks (dots and dashes) on a moving strip of paper. These were so limited that operators more often preferred to hand-write messages based on the sound of the incoming signals and these early printers fell into disuse.

Thomas Edison is said to have made his first fortune with various improvements to the telegraph. One of his accomplishments was building an advanced telegraph printer to print out gold and stock prices in 1873. It was not until around 1910 that typewriter-like teleprinters came into use, namely those developed by Charles and Howard Krum. These became widely used by the Associated Press to deliver news articles by 1914.

Two other major advances were the parallel coding system, most notably the five-bit Baudot code, and the use of punched paper tape to store messages and data. This brings us to the dawn of the computer and digital printing age, since Baudot code came to be also used for computer input-output functions. Baudot's name lives on as the apparent source of the generic term "baud rate" which is the maximum on-off rate at which signals can be transmitted (although engineers are likely to say there is a subtle but critical difference between bit rate and baud rate, so the two terms should not be used interchangeably.)

In time the five-bit system was found to be too limited for computer applications and the seven level ASCII code was standardized by the American National Standards Institute. An early advantage of the Friden Flexowriter over the Teletype terminals was that the Flexowriter used a seven-level encoding system while Teletype, at least early on, was limited to the five-level system.

The use of punched paper tape in telegraph systems was an important root adapted to printers as the computer age unfolded

in the 1950s. Punched paper tape, later joined by punched cards and finally magnetic tape, all served as storage media for digital messages and data over the years. These media allowed printers and other input-output devices to be disconnected from the processors. This let users work off-line and feed and receive data at high speed, thereby conserving hyper-expensive processor time. Use of these media also served as digital storage to offload the relatively minimal internal memories of early computers.

Adding Machines, Accounting Machines, Early Computers. Just as typewriters evolved into serial printers, the printing mechanisms of mechanical adding machines, calculators, and later electromechanical accounting machines evolved into line printers.

As usual, accounts of the development of desktop adding and accounting machines in the 1880s and 1890s are filled with examples of improvisation and persistence in the face of failure that defy today's imagination.

In the US, the men normally credited with inventing commercially successful accounting machines are Dorr E. Felt and William S. Burroughs. Felt is described as building his first prototype from a macaroni box and meat skewers around Thanksgiving day in 1884. Burroughs manufactured a run of fifty of his accounting machines in 1889 but hardly anyone could use them productively. He soon brought out a more successful redesign and one day on impulse opened the door of the room where his first fifty machines were stored and one by one destroyed them by tossing them out a window.[*]

After this rough beginning, by the 1920s printing calculators and accounting machines from Burroughs, NCR and lesser competitors were used worldwide. A set of type bars or type wheels functioned as miniature, numeric-only line printers, printing totals onto a paper roll with impact through an inked ribbon. Later larger, moving-carriage accounting machines boasted much more

---

[*] Eames, Charles and Ray, *A Computer Perspective*, Harvard University Press

sophisticated paper handling. By the 1950s such machines could handle a journal roll, action document such as an invoice, and ledger card under simple program control. On the ledger card, totals could be carried forward encoded on magnetic stripes on the back. Although there may still be a few die-hards using such machines, most have been long since replaced by computers.

In parallel with the development of desktop adding and accounting machines, Herman Hollerith was working on his statistical tabulating machine that operated from punched cards rather than a keyboard. This ground-breaking level of automation was inspired by the card system that had been used for years to control patterns on the Jacquard loom.

The US census crisis of 1880 was the break that put Hollerith on the map. With manual tabulating, as of 1887 the 1880 census had not yet been fully compiled. There had to be a better way, so the US Census Office sponsored a competition. There were just three entries, and the Hollerith machine won easily. It was used for the 1890 US census (counting a population of 62,979,766), and within the next few years for national censuses in Canada, Austria and Russia. In 1896 Hollerith formed the Tabulating Machine Company, destined to become a unit of IBM Corporation.

The early Hollerith machines did not print, but rather sensed the position of holes in manually fed cards and indicated the totals for each position by a set of clock-like dials. At the end of a day of reading cards, the total on each dial was manually recorded and then set back to zero.

It was not until 1919 that IBM came up with a printing tabulating machine, responding to a competitor, Powers Company, which was renting their printing tabulating machine for less than IBM's non-printing model. The classic 80-column IBM card was introduced in 1928 and was used more or less unchanged well into the 1960s. Later IBM printing accounting machines included the Type 285 and the 405, introduced in 1934. The latter machine

was considered the IBM accounting machine flagship until the mid-1940s. The final pre-1950 evolution of IBM's printers included the 150 lpm 407 Accounting Machine, announced in 1949, and the impact wire matrix printer developed for the Type 26 keypunch the same year.

IBM was the leader in punched card-oriented accounting machines, but not without competition. In parallel, Remington Rand developed their own line of card punches and accounting machines using round-hole cards.

NCR, Burroughs and IBM all made the leap from calculators and electromechanical accounting machines to computers. The early prototype computers were conceived in the late 1930s and 1940s, among them the ENIAC at the University of Pennsylvania, the Whirlwind at MIT, and the ASCC (Automatic Sequence-Controlled Calculator) at Harvard (later, wisely, renamed the Mark I).

IBM was a major contributor to the ASCC. The electromechanical monster was 51 feet long, weighed around five tons, and was described as the "high tide" of the technology invented by Herman Hollerith. It demonstrated IBM's commitment to advancing the art, but had minimal technical relevance to the company's later generations of true electronic computers. The ENIAC, on the other hand, embodied concepts carried forward by the inventors to Eckert-Mouchly Corporation which became part of UNIVAC in 1950.

Computers, which are seen as birthing the real digital printer industry, actually did not become commercial until the 1950s. This means the computer story needs to wait for the next chapter. (But don't expect a great deal. Lots of computer history is readily available elsewhere.) For now, as a pre-1950s root, two seminal inventions deserve mention since they made commercial computers as we know them today possible.

One is the core memory. It was on the Whirlwind that the first random access main memory came into use based on the

HOLLERITH CODES

IBM 80-column punch card, introduced in 1928.

IBM 405 Punch Card Accounting Machine was introduced in 1934 (Courtesy of IBM Archives)

invention of solid state, magnetic core memory by Jay Forrester. Core memory was made up of a grid of tiny ferrite cores strung on wires in a way that each could be accessed directly. Under a coincident current scheme each could be polled to read or change their on-off state. In the Whirlwind, this technology was used for the main memory which had a capacity of 2,048 words. Although the principle was demonstrated in the late 40s, it was not utilized in the Whirlwind until 1953. The expense of the panels was not the materials, but the tedious hand labor required to assemble them.

Forrester co-directed the Whirlwind project at MIT which was begun in 1947 and completed in 1957. The huge machine – 15,000 vacuum tubes consuming 150,000 watts of power – is considered the first large-scale, real-time control system. Early work was tracking aircraft. Later it controlled a network of radar sites, an application that led to the later SAGE air defense system.

While Forrester was working on core memory, another development was unfolding at AT&T's Bell Laboratories in New Jersey. This was the invention of the transistor in the late 1940s by W. Shockley, W. H. Brattain, and J. Bardeen, which earned them a Nobel Prize in 1956. In 1948 a power gain of 100 times to 50 milliwatts was demonstrated, leading in time, of course, to the replacement of vacuum tubes by solid state, semiconductor circuitry.

Surprisingly, as mentioned earlier, in most descriptions of these early computers, input-output is not emphasized. Printers are mentioned only as a footnote, if at all. Operators might read results from flashing lights or dials and write them down. On-line output might be punched in a card which would then be carried to a tabulating machine for printing. The reason for this latency between computing and printing was not necessarily a gap in the technology

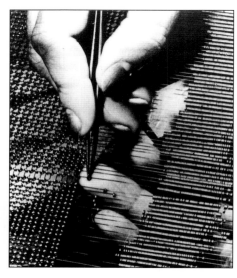

In the 1940s IBM began building the ASCC proto-computer at Harvard University. Fifty-one feet long, it contained over 2200 electromechanical counter wheels for storage and summation. (Courtesy of IBM Archives). Random access memory came a bit later with the invention of magnetic core memory by Jay Forrester at MIT. Forrester is shown here holding a core memory panel. Core memory cost was not so much materials but rather the tedious hand labor needed to assemble the panels. (Both photos courtesy of Compaq Computer Corporation, Digital Photo Library)

needed to build printers. It was rather in the applications, which at first were not input-output intensive. It was not until computers came to be used in commercial settings that printers came into demand, responding to what in the 1950s came to be popularly described as the "input-output bottleneck."

To get ahead of our story a bit, it might be added here that from the standpoint of printers, computers have become server as well as the served. They were the application which drove the development and market for printers. But with each step forward, printers required more memory and intelligence. So microprocessors and full-blown computers have became an ever more important part of the digital printing system.

Photography, Copying, Duplicating. Printing or imaging might be thought of as a marriage of two techniques. First, there is image capture. Second, there is preserving and perhaps reproducing that image. This becomes clear considering the nature of photography. This technology — writing with light — has fed many aspects of business communications including various forms of non-impact printing, photo-offset lithography, and copying.

Photography dates back into the Middle Ages when artists began playing with crystals and lenses to acquire and manipulate images. Even earlier, Aristotle and Euclid noted that a vivid but inverted image could be projected on a wall through a small hole in a darkened enclosure. Some artists in the following centuries are believed to have used such projected images to map out their paintings until the practice was superseded in the 19th century by photography as we know it today. Photography gave humanity the ability to fix and reproduce a lens-cast image. It was originally light and chemicals — electronics did not enter the picture until later.

The replacement of rigid plates with roll film in 1888 led to the popularization of photography by George Eastman, beginning the story of the Eastman Kodak Company. "You press the button, we do the rest," was their slogan in those early days. Later the frontier was color, and then "one step" instant photography. More recently, the role of photography has shifted into high tech, industrial applications. Finer and finer microlithography makes possible ever more compact printed circuits, a basis for today's computers and most other electronic gear.

Duplication with light – i.e. making copies from translucent originals onto coated copy papers – came into use in the late 1800s. First came the familiar blueprint process, used for engineering drawings, and later the diazo process, commercialized by Kalle in 1923 under the trade name Ozalid. Diazo in time more or less supplanted blueprints.

That was the engineering and design world. The office world certainly had its how-to-get-copies problem, too. Up to a point, carbon paper did the job. First it was manually interleaved, multistrike sheets. Later came one-time carbon forms, both continuous and unit sets ("Snap-Out" was a trade name that became generic). But the maximum number of carbon copies was perhaps six to twelve, depending on the power of the impact printer (or typist), the thickness of the paper parts, and the copy quality needed. The routing of the various copies disclosed a certain pecking order in an organization, with the person receiving the fuzzy, marginally legible last carbon copy usually at the bottom.

When more copies were needed, a master could be made for any of several duplicating technologies that became available after the turn of the century, namely mimeograph, spirit duplicating, and hectograph. These processes worked well for reproducing anywhere from a dozen to a hundred or more copies. Image quality, however, would be scorned by most office workers today. With both carbon sets and duplicating masters, the correction process was tedious, although that did encourage accurate typing. The motivation for a copying process that could take "plain paper" originals and make one or multiple copies – also on normal business papers – was strong. But it took a while.

Office copying didn't emerge until after World War II, in the late 40s. The main technologies were two-step wet processes,

Photography pioneers: Daguerreotype print of Louis-Jacques-Mandé Daguerre by early photographer J. B. Sabatier-Blot, 1844 (above, right). Creating photographs in the early days was a fine art for the few. George Eastman did for photography what low cost ink jet did for printing: he brought the power to the people. He democratized photography. Eastman is shown above (on the left) with fellow inventor Thomas Edison demonstrating his early movie camera. Eastman's basic patent, awarded in 1888, was the inexpensive box camera with roll film. (Images from the collection of the George Eastman House, Rochester, NY)

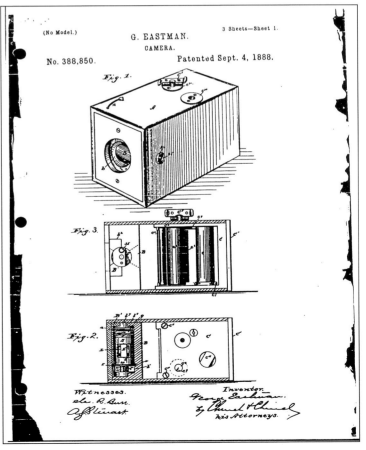

termed diffusion transfer reversal, or DTR. Quite a few vendors fielded such machines including Agfa and Gavaert (separate companies until 1964), Kodak (Verifax) and Eisbein (private labeled by various companies including A. B. Dick). In parallel, 3M Company introduced their line of thermographic office copiers (Thermofax). These machines didn't really win a strong presence in offices until the 1950s and 1960s. They can be viewed as a digital printing root, perhaps, in that they were sort of negative motivators. They required liquids and/or special papers, and copies were not particularly archival.

So in the office of the 1940s and 1950s, it was a choice between carbon copies and cumbersome office copiers requiring chemicals and/or special papers. For most office workers these limitations were taken for granted. They were happy to get copies any way they could.

Yet some people felt there had to be a better way.

Which brings us to Chester Carlson.

Electrophotography and the Long, Lonely Road of Chester Carlson. As one of today's two major digital printing technologies, electrophotography is covered in some depth in the next chapter. For now, here's a look at the human story behind this seminal development. Electrophotography offers an interesting contrast with ink jet and other more recently developed technologies in that one person played such a predominant a role.

The Chester Carlson saga is a classic story of a certain type of inventive personality. Although it has perhaps been overtold, a quick version is certainly appropriate at this point.

Psychologist Scott Peck opened his classic book, *The Road Less Traveled*, with the following simple but somehow comforting observation: "Life is difficult." It is more difficult for some than for others. Yet the upside of difficulty is that it can serve as an anvil on which positive attributes can be forged: seeing things in a different way, hanging on to a vision come what may, and having the intestinal fortitude to follow a vision no matter how difficult the path.

Carlson seems to have been a somewhat modest and even self-effacing man, quite different from some of the entrepreneurial figures profiled later in this book. He was not given to talking about his life and gifts. For this reason, an in-depth interview between Dartmouth Professor Joseph J. Ermenc and Carlson in 1965, three years before his death, is particularly remarkable.[*]

Carlson's early life is commonly characterized as "difficult." But just how difficult comes through as he related it to Professor Ermenc:

> Shortly after I was born my father contracted TB and arthritis or rheumatism. To find a suitable dry climate, which my father thought would help his TB and rheumatism, the family left Seattle and moved to various places in California, Arizona, and Mexico while I was still very young.
>
> But the family fortunes finally gave out. We settled in San Bernadino, California in 1912, when I was six years old.
>
> I went through high school in San Bernadino. We were very poor and I had to work to help support the family as soon as I could. This was between school hours. I did odd jobs, washed store windows, did janitor work, etc. By the time I was in high school I was the main support of the family, but I continued with school.
>
> My mother died in 1923 when I was seventeen.
>
> My father's health had not improved; he was practically a total invalid. He could get around some, but not very much. My father and I continued to live together until I finished high school.

This set the stage. He worked his way through college earning a BS in Physics in 1930 and found work first with Bell Labs in New York City and then the electrical equipment firm (best

* Carlson and Ermenc, op. cit.

known for batteries) P. R. Mallory. In time he found himself managing their patent department by day and by night pursuing a law degree to become a patent lawyer. It was at Mallory that he saw the need for in-house copying since getting copies of patents at the time was expensive and time-consuming.

In the mid-1930s he began his search. He was clear he was looking for a copying, not a duplicating process requiring an intermediary. First he thought there should be a chemical way to transfer the original image to perhaps some sort of chemically treated copy sheet. But when he considered the great variety of originals that people might want to copy it became clear no single chemical process could work with all of them. What do all originals have in common? Differences in how the image area and background reflect light. The process had to somehow use light.

Conventional photography was out. It had been around so long he was sure if it were an answer that some one else would have already adapted it to office copying. He centered in on electrophotography inspired in part by work being published around the same time by the Hungarian physicist Paul Selenyi. His main focus was facsimile, but observers note that had not Selenyi's work been cut short by World War Two, he, instead of (or in addition to) Carlson, might have ended up the central figure in xerography.

Carlson first looked into using electric current to create a chemical reaction. He began experimenting in his spare time in his apartment kitchen and after two years decided an electrochemical process was not the answer. For one thing, such high voltage would be needed that the paper would burn. Selenyi's work prompted him to consider using electricity to create an electrostatic charge and he formulated his basic electrophotography concept in 1937.

It apparently came as a crystal clear insight. In his words, "It was so clear to me at the very beginning that here was a wonder-ful idea. I was convinced, even before testing it, that it was pretty sure to work and that if it did it would be a tremendous thing."*

That was the beginning of the long road to commercialization. At the time his full-time patent job was getting increasingly demanding, and there was law school at night. He began experimenting without significant results. In 1938 he decided he needed to get more serious and needed help. He hired Otto Kornei, an unemployed expatriate Austrian physicist, to work with him and they set up a lab in a rented room in Astoria on Long Island. Within just a month Kornei figured out a combination of plate, charging method, and developing powder that held promise. On a Saturday when Carlson had time to visit with him they made that famous first image, "10/22/38 – Astoria."

Kornei is sometimes credited as the co-inventor of xerography along with Carlson. Actually, by most accounts, his work on xerography was just a brief, pragmatic interlude because he was desperate for money. Soon after the "Astoria" message was created, he quit and got what he saw as a real job at IBM.

But the concept had been demonstrated. From 1939 until 1944 Carlson tried to get a substantial company to back development of the technology into a product. The lore says he was turned down by over twenty corporations including IBM, Kodak, GE and RCA. Xerography earned the distinction of being yet another invention no one seemed to want.

The break was the deal with Battelle: giving them rights to the process in exchange for 40% of any future profits. Battelle developed better plates and dry inks and early in 1947 licensed Haloid in Rochester to bring it to market. Haloid came out with their first xerographic copier in 1949, but it was only marginally practical. They also came up with the more catchy name, xerography (from the Greek, "dry writing"). This didn't help at first, but Haloid eventually changed its name to Xerox, a name that has worn well over the succeeding decades. Later, in the 1950s, the process met some success as a way to image Multilith

---

* Carlson and Ermenc, op. cit.

Chester Carlson with his early xerographic copier prototype (early 1960s?). Photo courtesy of Xerox Historical Archives, Webster, NY USA.

paper offset plates – ironically, a duplicating process which Carlson was not interested in.

It was not really until the Xerox 914 came out in 1959 that xerography took off. Unlike the competing 3M and Kodak office copiers, here at last was true plain paper copying for the office. It was much more expensive than these low-end copiers. Speed, the ability to make multiple copies automatically, and especially the capability to copy onto many grades of paper were all major breakthroughs. This machine was the foundation upon which the Xerox of today was built. And it was finally then that Chester Carlson began to see significant royalties, a full 21 years after the "Astoria" demonstration.

Carlson did not have a lot of time to enjoy the fruits of his long labor. His wife had left him earlier (tired of having a kitchen filled with chemicals, according to one commentary). He died at the age of 62, nine years after the Xerox 914 was born. Of the estimated $150 million he reportedly finally earned on his invention, he had given away around $100 million to charity.[*]

By the time of the 914 the digital printer industry was well beyond its "roots" phase, growing in parallel with the young plain-paper office copying industry. In succeeding decades copying and digital printing tended to remain distinct industries, although there was crossover by leading vendors such as Xerox and IBM. In the 1970s xerography became the basis for high volume digital printing systems and from there evolved to become one of today's two dominant technologies. Now digital printing technology is making a significant impact on copiers. The development of more powerful raster image processors (RIPs) for digital printers also birthed the digital copier, leading to the convergence between copiers and printers that we see happening today.

Before going into the decade-by-decade chapters of Part Two, an Overture is in order: a broad overview of the technologies and the evolution of the products and the industry.

[*] Silverman, Steve, "The Invention that No One Ever Wanted" 1997, posted to his lively, authoritative Web site "Useless Information."

Using a stationary LED array rather than a laser for an electrophotographic printer makes sense. A lot of moving parts are eliminated and proponents say dots can be more precisely shaped and placed. The trade-off is a lot of precision manufacturing. Shown here are the fine (25 μm diameter) wires used to electrically connect the LED chips to the driver chips in the Océ/Siemens first generation LED print bars (1980s). The wires are bonded by micro-welding with 40,000 such connections needed for a 600 dpi pint bar. (Courtesy of Océ Printing Systems GmbH)

CHAPTER 2: OVERTURE

The Technologies

As the decade of the 1950s opened, many of the inventions that served as roots to the digital printing revolution were in place. Telegraph had long since proven electrical signals could be converted to printed text. Herman Hollerith's punched cards and electronic tabulators made the 1890 U. S. Census possible and tabulating machines had filtered out into business and commerce. Liquid processes for facsimile had been demonstrated. And Chester Carlson, having demonstrated electrophotography in 1938, was about half way down his long road to commercializing xerography.

The needs that have driven digital printer developments over the decades were only dimly recognized. Even those who pioneered computers seemed to consider printing out the data

as something of an afterthought. The ENIAC, generally viewed as the world's first large scale electronic computer, had no on-line printer. Output was by IBM card punch at 100 cards per minute (80 characters per card). The cards could then be carried to a tabulating machine and printed out off-line. J. Presper Eckert, Jr., the father of the ENIAC, in his oft-quoted Moore School Lecture of 1946, talked about printout. He speculated on how high speed printed output might be developed, but concluded, "It has been difficult for us, however, in thinking of these higher speed machines, to picture of what practical use such great printer speeds can be. . ."*

Common Technology Classifications
- Impact vs. Non-Impact
- Non-Impact subclassified by image formation technology, e.g. ink jet, electrophotography, thermal transfer, etc.
- Impact subclassified as dot matrix vs. full character (impact printers)
- Serial or Line (impact or non-impact printers) or Page (non-impact only)
- Impact line printers can be subclassified by their type element, e.g. drum, chain, train, etc.
- Monochrome or Color
- Media Requirements
    - "Plain" paper
    - Special paper
        - Treated for image quality
        - Imaging chemistry built-in

One reason computers did not initially drive demand for faster print was the applications. Most of the early applications were scientific rather than business. Calculating artillery ballistic tables, for instance, is computation intensive. But the output may be just one number.

As time went on, computers found their way into business, finance and commerce, applications which tended to be input-output intensive. Each solution illuminated new needs until the technologies snowballed into an explosion of printer-related invention and innovation. The needs fueled innovation, step by step toward the usual goals —

more speed;
better output quality;
lower cost machines;
more reliability;
lower operating costs.

On the other side has been the spectrum of technology enablers. These are the developments upon which the printer technology itself rests. All human progress, the cliché goes, is built upon the shoulders of those who have gone before us. The countless shoulders upon which printer technology rests might even include the development of paper and the invention of electricity. The more specific enablers through the decades include —
computers and microprocessors (which are both enablers and drivers);
exponential chip and data transmission advances;
materials and chemistry;
software, PDLs, scalable fonts;
color sciences.

Some of these enablers were covered as "roots" and others will be touched upon in the "decades" chapters of Part Two. To keep this overview focused, the emphasis will be primarily on the marking engines themselves — the imaging process — the technologies unique to digital printing. How to actually get marks on paper or other substrates? The Technology Map and Timeline provides a quick overview.

Of necessity the products included in the chart are very selective, with emphasis on the first products in a given sub-technology. The chart does hint of the broad tides of technology history: impact firmly entrenched in the 50s and 60s; non-impact dominating the 80s and 90s. But perhaps surprisingly, there was both impact and non-impact activity in every decade. And this will no

* Weiselman, Irving and Tomash, Erwin, "Marks on Paper," *Annals of the History of Computing*, Vol. 13, Nr. 2, 1991

TECHNOLOGY MAP / TIMELINE
Time/Technology Overview with First Printers in Class (plus Representative Printers)

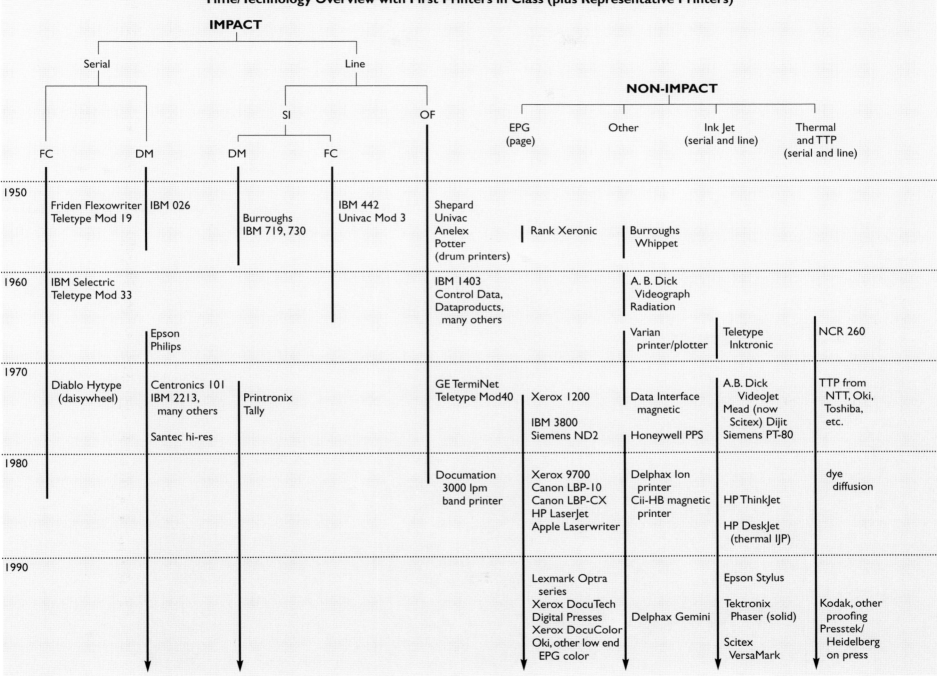

IMPACT

Serial

Line

NON-IMPACT

		SI		OF		EPG (page)	Other	Ink Jet (serial and line)	Thermal and TTP (serial and line)
FC	DM	DM	FC						
1950									
Friden Flexowriter Teletype Mod 19	IBM 026	Burroughs IBM 719, 730	IBM 442 Univac Mod 3		Shepard Univac Anelex Potter (drum printers)	Rank Xeronic	Burroughs Whippet		
1960									
IBM Selectric Teletype Mod 33					IBM 1403 Control Data, Dataproducts, many others	A. B. Dick Videograph Radiation			
	Epson Philips					Varian printer/plotter	Teletype Inktronic	NCR 260	
1970									
Diablo Hytype (daisywheel)	Centronics 101 IBM 2213, many others Santec hi-res	Printronix Tally			GE TermiNet Teletype Mod40	Xerox 1200 IBM 3800 Siemens ND2	Data Interface magnetic Honeywell PPS	A.B. Dick VideoJet Mead (now Scitex) Dijit Siemens PT-80	TTP from NTT, Oki, Toshiba, etc.
1980									
					Documation 3000 lpm band printer	Xerox 9700 Canon LBP-10 Canon LBP-CX HP LaserJet Apple Laserwriter	Delphax Ion printer Cii-HB magnetic printer	HP ThinkJet HP DeskJet (thermal IJP)	dye diffusion
1990									
						Lexmark Optra series Xerox DocuTech Digital Presses Xerox DocuColor Oki, other low end EPG color	Delphax Gemini	Epson Stylus Tektronix Phaser (solid) Scitex VersaMark	Kodak, other proofing Presstek/ Heidelberg on press

Key:
FC = full character
DM = dot matrix
SI = stationary impact

OF = on-the-fly impact
EPG = electrophotography
TTP = thermal transfer

Source: Digiprint Research Associates

doubt be the case in the decades to come. Some technologies gained acceptance in only specialized applications such as bar code labels while others dominated general purpose applications. Others just simmered away on the back burner for many years before achieving widespread commercial visibility. Still others are still simmering away, undergoing further refinement, hoping their day will come.

In this section of the Overture, the development and attributes of each of the basic sub-technologies is explored.

The Short Story. The essence of each of the technologies can be distilled into a few sentences. Here is a the short version, linked to the more complete exploration that follows.

You can adapt a typewriter to whack type slugs through an inked ribbon onto the paper. You can speed this up with various other type elements including type cylinders, the so-called golf ball, or daisywheel (teletypewriters and impact serial fully-formed character printers; for more, go directly to A1).

Rather than move a print head across the paper or the paper past a stationary print head, you can have a series of print heads whacking the ribbon against the paper more or less in parallel to print a full line or portion of a line more or less simultaneously.

You can speed it up more if you have the raised type on a continuously moving drum, chain, train, or band that spans the entire width of the document you want to print. But you can't strike it against the paper through an inked ribbon – it's too heavy, and all the characters would print at once. So you strike the paper from the back against the moving type element with a bank of hammers, each timed to make impact at the exact instant the desired character comes whizzing by (an on-the-fly back impact line printer); for more on line printers, go to A2).

You can unchain your character repertoire and create graphics and almost unlimited characters (Kanji, for example), lower the printer cost, and raise the speed by driving selected wires

against the paper through an inked ribbon to form characters or graphics built up with dots. This can be a serial dot matrix printer with the print head or heads moving back and forth across the print width, or it can be a line matrix printer (for lots more, go to A3).

A variation is to press an inked, film ribbon against the paper with a heated type element to melt or diffuse the ink onto the paper (thermal transfer or dye sublimation; see B4).

You can make a really fast and cheap printer by putting the imaging capability (and much of the cost) into the paper. Use a paper that turns dark when heated and run it by a set of selectively turned-on resistors. Take a paper with a dark substrate covered by a thin white layer, then use sparks to selectively blast off the white surface coating to reveal the dark layer beneath. Another way is to encapsulate two chemicals which when mixed turn dark, coat a paper with the beads, then crush them selectively so they can mix (as in NCR Paper). A newer alternative is to encapsulate a chemical that when crushed creates a visible pigment only when it has been subjected to an electrical charge (Mead Cycolor).

Or, why bother with hammers, moving heavy metal type elements, or expensive, treated papers? Just selectively squirt ink onto the paper. Ink and paper weigh very little. Devise a multitude of techniques to squirt the ink, and many configurations from small serial print heads moving across the page to incrementing the paper past wide, stationery arrays (continuous or impulse ink jet, serial, line, or wide format; for lots more, B2).

Finally, why not selectively attract dry ink powder or pigment particles suspended in a liquid to the paper and bond them. Here human ingenuity knows almost no bounds. The paper can be engineered to hold a charge, to serve as a capacitor, with electrodes selectively charging the spots on the paper to be imaged (electrography). Alternatively, magnetically charge the paper (magnetography). Or, instead of attracting toner to selectively charged paper, do it indirectly. Charge a photoconductive surface and se-

Comparison of Printer Technologies, 1983

Source: Adapted from *New Technology Printers*, Datek Information Services, 1983

performance criteria	impact	ink jet	electro-photographic	dielectric	electric discharge	thermal (direct)	thermal transfer
Speed	3	3	4	3-4	3-4	2	2
Hardware Cost	4	3	1	2	3	4	3-4
Print Quality	4 (FC) 2-3 (DM)	3	4	3	1	2	3
Acoustic Noise Level	1	4	3	3	3	4	3
Supplies Costs	4	3	3	2	2	2	2-3
Paper Restrictions/ Tolerance	4	3	3	2	1	1	3-4
Reliability	3	2	2-3	3-4	3-4	4	3
Color Potential	3	4	2	3	1	1-2	3-4

KEY: 1 – serious restriction or problem
2 – often a negative in some applications
3 – normally seen as an advantage
4 – considered a major advantage of the technology

lectively discharge it with light, then attract oppositely charged dry or wet toner to the photoconductor, and then bring plain paper into contact with the imaged photoconductor so that the ink transfers to the paper, and then fuse the image with heat and/or pressure (indirect electrophotography or xerography). Then there is the simpler variation which a dielectric coated drum or belt instead of a photoconductor. Projecting ions onto the dielectric surface creates the latent image which is developed with magnetic toner and transferred to the paper and fused in one step. (Xerox/Delphax ionography; details, sections B1 and B3).

That's about it. These are the basic technologies. Each has its strengths and weaknesses.

Technology Attributes. Today, for most printer users, impact printing is ancient history. Some of the attributes of impact have been forgotten. The chart on page 21 is a comparison grid dating from 1983. It is a reminder of how it was and also of what has changed and what has not.

Many of the ratings today would remain the same. However, breathtaking progress in ink jet and electrophotography since 1983 has radically changed parts of the picture. Ink jet printers at today's bargain basement prices would certainly rate a "4" in hardware cost and electrophotography perhaps a "3." Subsequent development of electrophotography has also proven the 1983 "color potential" rating for that technology wrong.

Two major characteristics of impact printers have generally been forgotten. The first relates to workplace quality of life. In the workplace of the 60s and 70s there was always the high pitched whine of serial matrix printers and the clatter of line printers — noise is an inherent attribute of impact and a serious negative in the office environment.

Second, impact printers tend to be underrated in terms of speed. Rather than taking a page of printed output to a copier to get additional copies, or printing out successive originals, why not create them as a by-product of printing the original by using carbon interleaved, multipart forms or carbonless paper. What a great idea! The effective speed can be tripled or even increased six or eight times for a job where multiple copies of the same document are needed. But there is the downside of higher paper costs, the task of decollating the multiple parts and getting rid of the carbon paper, and image quality that today would not be acceptable.

Changing Technology Mix. Surprisingly, almost all of these technologies are still being used for at least some applications somewhere in the world. But looking at the big picture, one of the most significant changes over the past twenty years has been the radical change in the dominant technologies and the technology mix (see charts, 1980 and 2000).

Over the years from the 1950s to the early 1980s, there was a diverse mix of technologies, dominated by impact. Today, ink jet and electrophotography dominate in terms of both unit shipments and value of shipments. (Note that these pies cover only equipment for page-size documents in general applications. Were they to include, say, the myriad specialized listers used for receipt printing, the mix would be significantly different. The dollar volume mix is for hardware alone; supplies are not included. Were supplies costs to be also included, the year 2000 revenues would more than double and the mix would be tilted even more toward the non-impact technologies.)

These charts do overlook the fact that there are many markets. For high quality graphic arts proofing thermal may be the dominant technology. However, in terms of dollar volume of shipments, this sort of specialized market is dwarfed by the general purpose printer market which is the main focus of this book.

So at this point in history, one wonders how long these two leading technologies will continue to dominate and which, if any, of the existing alternatives may in time eclipse them. Various alternative technologies continue to be developed. Many demand-side currents — such as the Internet, on-demand, and color — are now changing the face of printing. Considering these changes, it would seem ink jet and electrophotography might be dethroned at some point before virtual documents dethrone print in general. This will be explored in Part Three.

So much for the short story. More detailed excursions into these technologies follow which will prove interesting, hopefully, and perhaps even useful perspective from which to look ahead.

TECHNOLOGY MIX, 1980 TO 2000
World Wide Market

UNITS

1980

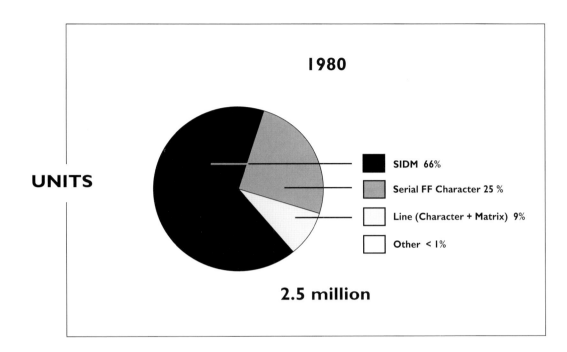

- SIDM 66%
- Serial FF Character 25 %
- Line (Character + Matrix) 9%
- Other < 1%

2.5 million

2000

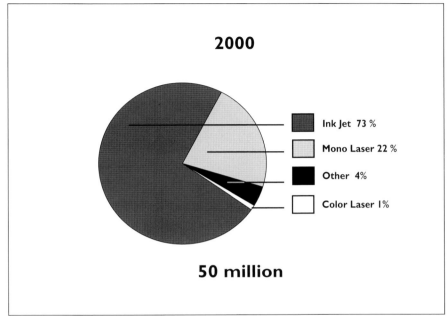

- Ink Jet 73 %
- Mono Laser 22 %
- Other 4%
- Color Laser 1%

50 million

HARDWARE REVENUE

1980

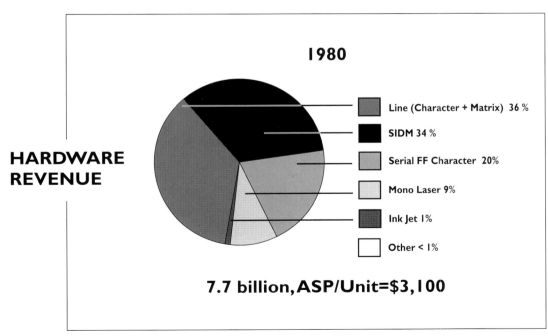

- Line (Character + Matrix) 36 %
- SIDM 34 %
- Serial FF Character 20%
- Mono Laser 9%
- Ink Jet 1%
- Other < 1%

7.7 billion, ASP/Unit=$3,100

2000

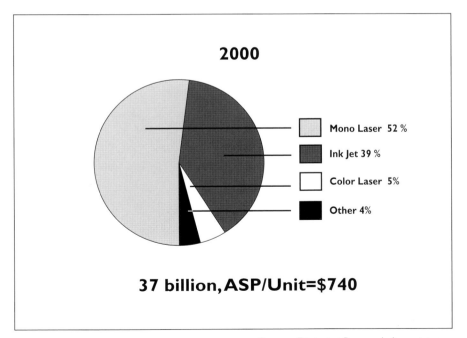

- Mono Laser 52 %
- Ink Jet 39 %
- Color Laser 5%
- Other 4%

37 billion, ASP/Unit=$740

Key:
SIDM = Serial Impact Dot Matrix
FF Character = Fully Formed Character
ASP = As If Sold Price

Source: Digiprint Research Associates
and IT Strategies, Inc.

A. IMPACT PRINTING

In the office desktop impact printing has been almost forgotten. But it is more than history. Impact printing continues alive and well all around us in various retail settings, the supermarket checkout counter and the ATM. It is also widely used in diverse industrial settings for printing bar-coded labels and a range of multi-part documents.

A1. Serial full character impact printers

This class of printer evolved from the typewriter and teleprinter. IBM and Remington Rand typewriters and Teletype teleprinters were adapted to serve as computer output printers: RO (receive-only) or KSR (keyboard send-receive) console printers, and remote terminals. One's level of familiarity with these technologies tends to be generational. Today's older generations were brought up with typewriters. But many members of the younger generation have never seen let alone used a typewriter. We're talking ancient history here. But the mechanical solutions embodied in typewriters shaped what was to follow down even to this day.

The keyboard is the most obvious. Amazingly, after all these years, the same basic QWERTY keyboard that became standard almost a century ago is still with us. The story is a bit hackneyed; writers always feel a need to comment on it and this one is no exception

The original typewriter printing mechanism was type slugs mounted on a set of levers which together made up the type basket. The basket was typically stationary, with key-actuated levers striking the paper which is then incremented to the left on a moving carriage. Character selection is multiplied by a choice of levels or shifts for each key. When the shift key is depressed, either the type basket is depressed, or the paper carriage is raised so that a second or third character on the selected type bar strikes the paper.

This type lever mechanism is subject to key clash. If typing is too fast, the lever going up can easily clash with the lever which has just printed and is falling back down. So the lore is that the arrangement of the QWERTY keyboard was intentionally planned to challenge the anatomy of the human hand, to make it difficult to type too fast. The early single element used by Teletype, the later IBM "golf ball" variation, and the still later daisywheel all avoided this key clash problem. So the deliberately inefficient keyboard design was no longer relevant. And it is certainly not relevant to today's PC keyboards.

We are still using the keyboard introduced with the Remington typewriter of 1874. There have been alternatives introduced and promoted from time to time. They have proven to offer significantly higher data entry speeds with only a few weeks of practice. But none has succeeded in becoming anything close to standard. A bit strange.

Probably the most radical departure from the standard keyboard is the Datahand minimum motion keyboard. Closer to a glove than a keyboard (see sidebar), it has been shown to increase productivity and also reduce the increasing incidence of work-related musculo-skeletal injury of the hand, wrist, arm, shoulders, back and neck, all of which are believed aggravated by the traditional keyboard. It has also been found to be a wonderful tool for the blind. We're told the DataHand speed record is held by a blind person who learned the concept and achieved 90 words per minute in less than an hour. Sighted operators, patterned to the dominant keyboard, need about a month to adapt to the DataHand's lighter touch and feel.

Yes, this is a diversion from serial impact printers. Yet it is a nice case example of the momentum of human patterning. It is

DataHand – The Better Way meets Resistance to Change

The DataHand minimum motion keyboard presents an interesting case study of new technology adoption. This alternative to the traditional flat keyboard was developed initially to address productivity. Its effectiveness in addressing the ergonomic hazard inherent in the traditional keyboard was secondary. It has been found that DataHand has allowed many people disabled with wrist injuries to return to work. Where the promise of greater productivity fails to motivate, pain will. According to proponents, DataHand addresses all the stresses associated with the flat keyboard while other alternative keyboards are likely to address only 10% or so.

Here are a few quotes from DataHand's 1999 user testimonial document.

"If aliens came to earth after humans were extinct and found a traditional, flat keyboard, they'd imagine we had thirteen fingers laid out in a straight line, like piano keys."

> – Clifford Lasser of Thinking Machines, Cambridge, MA USA, a DataHand user.

Regarding resistance to change:

"The flat keyboard is a paradigm from another century intended to meet the work requirements of mechanical typewriters. Workers had to be slowed down to avoid the clash and snarling of the mechanism. Neither the productivity shortcomings nor ergonomic limitations can be fixed until the paradigm is changed. . . .

"Human limitations have deferred the acceptance of many innovative, forward looking ideas. . . . The telephone floundered for a time under the perception that it was no more than an interesting novelty without practical application. Acceptance of steamships was slowed while the makers of sailing ships tried every conceivable means to improve the efficiency of their sail designs and deployment. The zipper, invented at the turn of the last century, was not widely accepted until almost fifty years later – when the government brought it to the attention of many people by using it on military clothing."

DataHand Minimum Motion Keyboard in Action (Courtesy of DataHand Systems, Inc., Phoenix, AZ USA.)

Datahand backer Don Patterson's words of encouragement to Datahand prospects, users and also, no doubt, investors: "Introducing an innovative, new paradigm to the world is never easy or fast – even when the concept being replaced is entirely deficient. Paradigm change is always hard for people. Perseverance in the face of skepticism and uncertainty is a necessity. Long-standing habits of an entire culture are not relinquished any more quickly than the harmful substance dependencies afflicting individual citizens."

Change, it seems, often has to be literally forced onto people. A student once said to his professor, who happened to be Howard Aiken, inventor of the Mark I computer at Harvard, that he was afraid someone might steal his (the student's) ideas. Aiken replied, "Don't worry about people stealing an idea. If it's original, you will have to ram it down their throats."[*]

[*] Slater, Robert, *Portraits in Silicon,* MIT Press, 1987

also notable as an ubiquitous legacy of the typewriter – one that may be with us at least until voice response or perhaps brain-to-machine telepathy become viable, which could be quite a while. In the words of Darryl Rehr, a typewriter historian, "It's the keyboard people love to hate, but it is as much a cultural standard as the Roman alphabet, the steering wheel, and the 4x3 TV screen . . . . a paragon of inefficiency, intriguingly bizarre."*

Single Element Full Character Serial Printers, Golf Ball Type Element. In 1961 IBM introduced its Selectric "golf ball" typewriter. It was apparently not the first single element typewriter. "We have an old Blickensderfer typewriter here from 1892," NCR archivist Bill West mentioned a while back. "It used a round drum type element, a single element which preceeded the IBM Selectric by sixty years!"

But it was IBM that made it a modern commercial success. In parallel with its introduction as a manual electric typewriter, IBM adapted the Selectric mechanism as a console printer for its large scale STRETCH computer system. A measure of the amazing success of the basic concept is the fact that twenty years later, in the early 1980s, IBM was still introducing new Selectric golf ball models.

It is no accident that this leap forward came from IBM. In the early 1930s a company in Rochester, NY called the Electromatic Typewriter Company was developing an electric typewriter. They were small and despite valuable patents, were struggling. IBM bought the company in 1933 and invested over a million dollars in design improvements and development of specialized applications. Among their early advances were special forms handling attachments, multiple print cycles from a single keystroke, and automatic carriage return.

Later typewriter enhancements by IBM and various competitors demonstrated resourcefulness in extending the life of a technology in the face of competition from newer technologies. Proportional character spacing was introduced in 1941. Film "total release" ribbons appeared in the early 1960s. It took many years for digital printers to match the print quality of an early 1960s-era proportionally spaced typewriter with a film ribbon. More innovation – cover-up and later lift off correction – was a boon to manual typists.

According to IBM, the Selectric grew from work in the late 1940s on a single element high speed printer for accounting machines.† By the mid-1950s the lightweight golf ball print element appeared with its many positive attributes. As with the existing print lever typewriters, type struck the paper through an inked ribbon. But instead of shuttling the paper back and forth, the single element print head moved horizontally across the print line. The paper was not yanked back and forth, so feeding continuous paper was more reliable. As an automatic typewriter, it increased output from the previous industry standard of 10 cps to 15 cps. The type element was interchangeable so typefaces could be manually interchanged. Incoming character selection codes were simplified. Typebar I/O typewriters were said to require a unique electrical signal for each printable character. The Selectric needed only a seven-bit code for character selection – six bits for rotate and tilt control, and one for shift control. In addition there were four functional codes: carrier return, backspace, tab, and paper advance.

The type ball itself at the time was described as a plastic molding which was nickel plated by electrolysis. The result was a very accurate, rigid, and lightweight component. According to

---

* Rehr, Darryl, http://home.earthlink.net/~dcrehr/
† Beattie, H. S., and Rahenkamp, R. A., in *IBM Journal of R&D*, Sept. 1981

IBM it would take five million impressions by a single character before the print quality would visibly degrade.

For each impression, whether under manual or electronic code input, the ball is rotated, tilted, locked into the selected character position, driven against the ribbon and paper, and then returned to a home position while the print head assembly advances to the next horizontal character position.

These and other attributes led to the adaptation of the Selectric to many IBM products including systems that helped establish the booming word processing market of the 1970s. It was also sold to many other OEMs for word processing and computer I/O applications. By the late 1970s one estimate placed the number of Selectric mechanisms being shipped for computer I/O applications at up to 40,000 per year – big numbers for those days.

The Selectric was not the only single element impact serial printer. Teletype introduced its Model 33/35 teletypewriter using a small, metal type cylinder in 1962. This was one of the most successful printer series of all time with 750,000 reportedly shipped by the time it was finally discontinued in 1981. The small metal cylinder did not tilt, but rather was rotated and moved vertically to one of four positions and then driven into the ribbon and paper. These printers were slower than the Selectric-based printers, but also less expensive and in fact used by IBM in at least one low-end output application. Still another earlier serial impact full character configuration was the type pallet mechanism used in the Teletype Model 28. The single element was actually a box-like assembly holding individual slugs for each character. Performance was 10 cps. As the box moved across the page it was positioned in the X and Y axes to position the selected slug in front of a hammer, which drove the appropriate slug into the ribbon and paper.

The Daisywheel Type Element. Serial impact printing accelerated once again with the introduction of the daisywheel printer by Diablo Systems in 1972. Xerox acquired Diablo that same year and David Lee, described as the prime mover behind the daisywheel, moved to Qume, which rapidly became the major competitor to Diablo. These two companies controlled the market until around 1980, after which a host of competitors eroded their position.

The daisywheel printers employed pretty much the same basic architecture as the IBM Selectric. The type ball was replaced by a flat or cupped plastic or metal type element with an embossed character at the end of each of an array of "petals" radiating out from the hub. The daisywheel, the wheel motor, the ribbon cassette, and a hammer mechanism ride back and forth across the print line.

The IBM Selectric typewriter was one of the most widely used printers for word processing. Later models offered extended yield, multi-strike film cartridge ribbons. (IBM publicity photo, 1970s, DRA Archives)

To print, the wheel is rotated so the appropriate character is aligned with the solenoid actuated hammer. The hammer is then fired, driving the embossed character image into the ribbon and paper. The wheel then rotates so the next incoming character is in place as the carrier moves laterally to the next print position. Wheel rotation is in either direction, taking the shortest distance to the selected character. Hammer force is varied according to the surface area of the selected character – i.e. greater for an "m" than for a period.

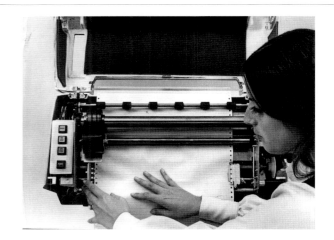

Daisywheel type element (from Dataproducts Today, July, 1981, courtesy of Hitachi Koki Imaging Solutions, Inc.); Dataproducts DP55 55 cps daisywheel printer, 1984 (from Dataproducts Dataprintout, April/May, 1984 issue); Pertec helical drum printer, 1980s. (from Pertec publicity release, DRA Archives)

The speed of the first daisywheel printer, the Diablo HyType I, was 30 cps, twice that of the IBM Selectric which had been improved in many ways, but not speeded up. Later daisywheel printer variations boasted speeds to 100 cps. Higher end versions offered horizontal increments of 1/60th inch and vertical increments of 1/48th inch for rudimentary plotting.

IBM's Wheelprinter (manufactured until 1985) offered 10, 12, and 15-pitch monospacing and proportional character spacing at a "burst speed" of 25 cps. Sheetfeeding could be manual or automatic. An important innovation with this product was a long life multi-strike film ribbon cartridge which was specified to yield a minimum of 500,000 characters.

Enablers included better stepper motors to move and position the daisywheel, and of course rapid advances in solid state microprocessors. These permitted much greater use of electronics so that the mechanical parts count was reduced by hundreds. There was a microprocessor to control character selection, hammer impact, print carrier assembly movement and paper advance. A major driver of this leap forward in full character serial printing was word processing, which was booming in the late 1970s and early 1980s.

Other advantages of the daisywheel over the golf ball included cheaper type elements ($3 to $4) and more characters available, 96 or more compared with the 88 of the golf ball. On the other hand the printers were somewhat more expensive and print quality was not so high. Xerox and others later developed metallized versions of the type element to improve life and print quality.

As the years went by the Qume-Diablo daisywheel near-duopoly eroded as perhaps a dozen other companies jumped on the bandwagon. Among them were IBM, Exxon/Qwx, Plessey (acquired by Dataproducts), NEC ("thimble" type element), C. Itoh, Ricoh, and Brother Industries.

There were still other variations. One of the more resourceful (and less successful) was the "multiple split helix." The main example was the Printec-100 from Printer Technology, Inc. This was a serial printer which used the on-the-fly technology common to the line printers of the time. The type element was a short type cylinder with embossed characters arranged around the wheel in the split helix, a two-part spiral pattern. This was to accommodate the continuous rotationof the wheel as it progressed horizontally across the print line. The axis was horizontal (so that the characters moved vertically as the wheel turned). Instead of a ribbon, an

ink roll applied a film of ink to the surface of the characters. Opposing the roll were six hammers which, as in an on-the-fly line printer, struck the paper from the back against the inked split helix drum, timed so as to print the desired character. Printer Technology's initial product offered a 64-character set, could print at 100 cps, and was list priced at $2,200.

A2. Full Character Line Printers.

The earliest line printers evolved from accounting machines. IBM adapted its 402 punch card accounting machine, and later the 407 as the basis for its early computer printers. These were front, full character printers. The 402 printed more or less like an adding machine. The available characters were embossed on type bars oriented vertically in a row that spanned the width of the paper. To initiate a print cycle, the individual bars were indexed vertically so the selected characters are positioned on the print line and driven against the paper through an inked ribbon, to print a full line of characters.

The next generation IBM 407, somewhat faster at 150 lpm, instead of type bars used individually type wheels which were rotated into position and then driven against the paper through an inked ribbon. Univac and several European companies had similar machines.

Beginning in the 1950s a number of companies introduced much faster line printers which achieved their performance by on-the-fly printing. Instead of driving a stationary raised character against the paper through the inked ribbon, a continuously moving type element was used. No start-stop mechanics or timing was needed for the type element. Continuous rotary movement is inherently faster than reciprocating or start-stop rotary movement.

But with a continuously moving type element spanning the full print width of the page, individual characters could not be

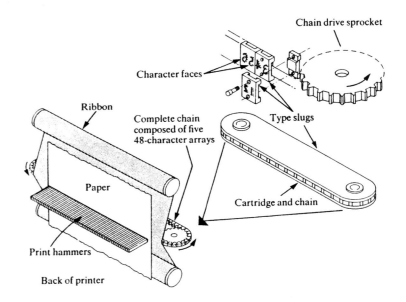

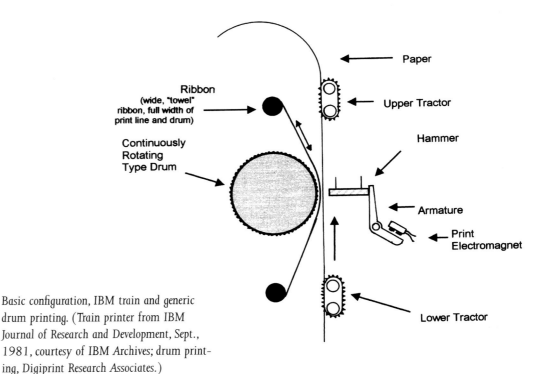

Basic configuration, IBM train and generic drum printing. (Train printer from IBM Journal of Research and Development, Sept., 1981, courtesy of IBM Archives; drum printing, Digiprint Research Associates.)

driven against the paper from the front. Instead, on-the-fly printers were most commonly "back printers:" i.e. the bank of hammers was behind the paper. Each hammer was fired timed to strike the paper against the proper character as it flew by. Fast. But as always, there are trade offs.

On-the-fly printers, especially early generations, produced mediocre print quality. They could not handle as many carbon copies. Their performance was subject to a greater number of variables. Data handling was more complicated.

Handling the incoming data stream was a challenge. A serial printer could print a serial stream of incoming digital data, if properly metered, as it arrived. In a line printer, there needs to be a buffer memory for at least one print line. This means one buffer position for each character that can be printed on a line. Line printers are all monospaced, with a fixed number of print "positions" per line. Through the years line printer line lengths were 80, 120, 132, 136 (Europe), 144 and, with later wide-format impact printers, 160 positions spaced at ten per inch. Most line printers have one hammer for each print position, also spaced 10 per inch.

Any full character impact printer is limited to the character set which can be presented to the hammer at each print position. Speed therefore tends to be proportional to the available character set and which characters the printer is being asked to print in a given application. Even if the print element contains the commonly used 64 character set, if a series of printed lines are confined to numbers, the print speed can be significantly faster. The paper can be advanced while the unused characters are flying by, so a line can be printed with every revolution of the print element. Because of these and other variables, the print

Representative drum printer, Mohawk Data Sciences, 1970s. (Source: MDS publicity release, DRA Archives)

speed specified by the vendor would normally be an "up to" speed.

All line printers are built to feed continuous, pinfeed paper. Pinfeed tractors or sprocket wheels engage the marginal punching along each side of the flat pack paper, on half-inch centers. This positive engagement with the paper means the printer always knows where it is printing on the page, for positive vertical format control.

Vertical format also affects print speed. Most impact line printers offer accelerated paper movement over vertical white space, including the top and bottom margins of the page. This was commonly controlled by a paper tape loop with start and stop skip locations encoded with punched holes. This feature, along with the aforementioned multiple part capability, if used, means the effective throughput might be significantly more than the vendor's spec. Programmable high speed skipping is a feature that continues to make impact line printers the choice for some applications with a lot of white space, such as label printing.

With an impact printer, we're talking about hammering the impression onto the paper. Hammer technology, therefore, is a key competitive feature. As will be seen in Part Two, Dataproducts attributes their competitive success to their patented hammer design more than any other factor. In the words of founder Irwin Tomash, "The hammer saved the company." Consistent timing, programmable impact energy, and minimum dwell time are key criteria. One of the all too frequent maintenance chores with many line printers is manually adjusting the flight time and penetration of each individual hammer.

Even at its best, with on-the-fly printing, image quality is well below that of both earlier printers and the non-impact printers yet to come. Unless hammer timing is perfectly consistent, there can be significant displacement of characters, vertically or horizontally. Also, with a continuously moving type element, charac-

SE PER CENT IS PERCENT OF TOTAL USE
OWN PERCENT IS PERCENT OF TOTAL DOV
PERFORMANCE PERCENT IS PERCENT OF
TOTAL TIME MACHINE WAS AVAILABLE

Output sample, on-the-fly drum printer (Shepard, 1961). When operating at its best, vertical character misalignment was imperceptible; this sample called for a bit of hammer flight time adjustment.

ter strokes perpendicular to the movement of the type will tend to blur. Type is normally designed to compensate for this. In the case of a drum printer where the type is moving vertically, the horizontal character strokes will be thinner than the vertical strokes. With a printer where the type moves horizontally, the opposite will be true.

On-the-Fly type elements included drums, chains, trains, bands and belts. Drums came first. In the early 1950s Univac, Potter Instrument, Anelex and Shepard Laboratories all introduced drum printers offering speeds of 600 lpm and higher.

Although the concept is basically simple, drum printers needed to deal with a number of subtle dynamics. For adequate paper control two sets of forms tractors, one above and one below the hammer bank, was normal (except for low-end models, some of which had just a single set of forms tractors above the print line). Paper, inadequately controlled and tensioned, could result in poor print quality or spectacular paper jams. The spinning type drum would tend to pull the paper and ribbon vertically. This

meant the ribbon needed to move parallel with the continuous paper and the direction of the moving type. This meant using a wide "towel" ribbon the width of the horizontal print area, fed back and forth between two spindles. Engineers also had to deal with strange vibration and resonance patterns. If too many

Line printer print drums are photo-etched, hardened steel. Other, smaller, lighter weight drums are magnesium. Industrial Engraving of Easton, PA is one of the few current suppliers of these components. Their volume for these products is way below what it was in the heyday of on-the-fly impact printing when they supplied a variety of major manufacturers including Dataproducts and IBM. According to Ron Squibb at Industrial Engraving, they manufacture a variety of impact type elements. A small market for replacement type elements still exists overseas where refurbished impact printers in use for twenty years or more are still alive. There is a larger North American market for type elements used in industrial, ticketing and financial applications. According to Squibb, their manufacturing technique is photo-etching. Character patterns from a photographic negative are translated into raised steel or magnesium mirror images by etching out all the background. It is a precision process. In the case of a print drum, a key challenge is to make it seamless where the edges of the stencil meet. The cost for a large, page-width steel drum is in the range of $3,000 to $5,000. (Photo Courtesy of Industrial Engraving Inc., Easton, PA USA)

hammers fire in synch, the printer could vibrate too much, maybe even walk across the floor. One solution was to configure the characters on the drum in a checkerboard or spiral pattern so hammer firing would be less synchronous.

In some drum printers cost is saved by having one hammer cover two or more print positions. The hammer bank may be shifted back and forth in the course of printing one line. In at least one clever configuration, the hammer faces were wide, covering two print positions. The raised font on the print drum was split. The set of characters that served all the even numbered print positions occupied just half the circumference of the drum. A second set of characters, for the odd numbered print positions, was on the other half of the drum, offset laterally 1/10th inch from the even position character set. Another hammer sharing technique was to shuttle the paper back and forth so one hammer could service two or more print positions.

Horizontal Font Printers. On-the-fly drum printers dominated high speed printing for the decade of the 50s. In 1960 IBM pioneered with their horizontal font 1403 train printer, introduced with the 1401 computer. It was an incredible success. Twenty years later the 1403 was still the mostly widely used high speed computer printer in the world. As a landmark product, it is in a class with the Xerox 914. According to one source, one reason IBM's sales of the 1401 computer exceeded expectations was because of the 1403 printer. It was a case of the tail selling the dog.

On early models of the 1403 the type element was a chain, somewhat resembling a bicycle chain, which rotated in a track. Later this was replaced with a train of type slugs which pushed each other along in a track.

Among the merits of the IBM 1403:

- This was the first on-the-fly printer to use a horizontally moving type element. One benefit is print quality. Horizon-

tally displaced characters are not as noticeable as vertically displaced characters.

- Unlike drum printers, the 1403 type cartridge in later models was operator-interchangeable, permitting users some variety of character sets. This permitted OCR, technical, or foreign language fonts. Or enhanced speed with a reduced character set. (The feature was not inexpensive, however. There was an upcharge for the option and each interchangeable train cartridge was priced in the neighborhood of $3,000 – around $6,000 in today's U.S. dollars).
- Print quality was also better because the horizontal spacing of the type did not need to conform to the ten per inch character format (although hammers needed to). The wider spacing meant less likelihood of shadow images from adjacent characters.
- The chain or train could hold a set of up to 240 characters, much larger than was possible on a drum.

Besides IBM, Control Data introduced their Model 512 train printer in 1967 which had attributes similar to the IBM 1403. This later evolved into the Computer Peripherals "Fastrain" train printer series which was supplied to Control Data, NCR and ICL.

The architecture of both drum printers and the IBM 1403 and later horizontal font, on-the-fly printers tended to be similar. The hammer bank was normally deep within the machine and the type element and ribbon assembly toward the operator. To load the continuous paper, the "yoke" would be unlatched and swung outward, for easy access to a twin set of tractors. Instead of a swing-out, gate yoke, some later, lower speed printers had the type element and ribbon assembly permanently attached at both sides. This assembly would drop forward and/or down to allow for threading the continuous paper between the hammer bank and the yoke.

A sea of type elements manu-factured by Industrial Engraving for a diverse assortment of impact printers including drums, bands, and daisywheels. (Courtesy of Industrial Engraving, Inc.)

As the 60s wore on, drum printers tended to fade out, re-placed with a variety of horizontally moving type configurations. Most, but not all, remained back printers.

Band printers used lightweight steel bands, typically around ¾- inch wide and several mils thick. One enabler was the devel-opment of advanced electron welding which allowed the bands to be strong and functionally seamless. Unlike the chain and train printers that generated character positioning data from a code wheel, band printers included a timing track as part of the band itself (visible in photo above). The raised vertical marks would be

sensed by a proximity sensing device, normally a reluctance pickup which generated a timing pulse.

Band printers appeared first as low end desktop printers from Nortec in 1969. Later IBM, UNIVAC and others came out with much higher speed band printers. Around 1980 Documation introduced a 3000 lpm band printer which was the fastest impact line printer to date. Later, in 1985 Documation claimed 5000 lpm for its Model 5500. Fast, but not fast enough to save impact from growing competition from non-impact page printers.

Belt printers came along around the same time in many variations. These included low end belt printers such as the GE TermiNet (a front printer) and Teletype Model 40 (a back printer) (both of which functioned as a serial printer), the Dataproducts "Charaband" printer family, the Data Printer "ChainTrain," and low end belt lprinters from Odec and other vendors. Most of these were back printers. What they had in common was the use of a flexible carrier of some sort which carried individual type slugs or metal fingers each bearing a raised character. The belts might not be interchangeable, but individual type slugs usually were. Print speeds ranged from 10 to 30 cps in the low end teleprinter applications up to 1850 lpm for Honeywell's top-of-the-line printers engineered and built in France by Bull.

Finally, there are the reciprocating stick printers, most notably the 1443 introduced by IBM in 1962 with its smaller 1440 computer. IBM's type stick, or bar as they called it, consisted of a rod holding a line of flexible type fingers. At one end was the flag, a projection from which photosensors tracked bar motion to control hammer timing. In IBM's original iteration, the type stick was halted before firing hammers, for a speed of 150 lpm. But when an engineer named Frank Furnam came in to manage the program, he found the technology worked just as well on-the-fly which increased performance to 240 lpm. Print quality at these

speeds, according to IBM, was good enough to print MICR bank checks.[*]

At the higher-end, IBM solved many of the print quality limitations of on-the-fly, back printers with another front printer, the IBM 3211, introduced in 1970. This printer had a print train with 108 type carriers. Each carrier carried four pivoted levers holding one character. As they moved along the print line, appropriate characters were struck against the ribbon and paper, which was supported by a flat metal platen. Print quality was improved, but the cost was higher. One interchangeable print cartridge alone was priced at around $10,000.

A3. Dot matrix printers: Serial and Line

Matrix printers form characters with a pattern of printed dots. Actually, besides impact dot matrix printers, almost all non-impact printers are also dot matrix printers. But in the case of most non-impact printers, the dots are small and programmed to overlap so that they are not individually noticeable. In normal usage, "dot matrix" refers only to impact printers.

The earliest dot matrix printers had a stationary head. Patents relating to impact dot matrix printers can be found dating back to the 1930's. In 1937 J. N. Loop patented a 30-wire print head that came to be used in some high speed telegraphic printers. Then came the Burroughs Model G printer built in the late 1940s for an electric utility. In the 1950s Eastman Kodak developed what they termed their Multiple Stylus Electronic Printer for addressing. Around the same time IBM introduced its Model 026 Interpreting Card Punch which incorporated a small wire matrix printer to print legible characters along the top of the card as a by-product of the keypunching. Soon thereafter IBM fielded their 1000 lpm impact matrix line printers to compete (unsuccessfully) with the

---

[*] Bashe et al, *IBM's Early Computers*, op. cit.

on-the-fly drum printers used by competitors at the time. In 1954 the Burroughs subsidiary Control Instruments announced a 1,000 lpm line matrix printer which also had only marginal market success.

Serial Dot Matrix. Twenty years later it was small, moving-head, serial dot matrix printers that revolutionized the low end of the printer market. The basic configuration was similar to that of the single element typewriter. Instead of a golf ball, a print head made up of solenoids driving print wires against the paper through an inked ribbon was scanned back and forth across the print line. The first compact, moving head matrix printer is believed to be the Philips Mosaic Printer, developed for their printing calculator in the mid-1960s. It wasn't a computer printer, but as many as 40,000 were reportedly shipped, helping to popularize the notion of dot matrix printing. Around the same time Seiko Epson introduced small dot matrix printers used to tabulate results at the 1964 Tokyo Olympics, followed by the EP-101 printer mechanism in 1968. Some IBM and ICL serial impact dot matrix patents issued in the 1959-1960 timeframe.

It took Centronics to spearhead the dot matrix revolution that swept the computer industry in the 1970s. Observers continue to debate whether Centronics "invented" matrix printing as they sometimes claimed. The courts debated the issue as well, in the form of patent disputes between Centronics and Mannesmann, a European user of similar technology. The outcome of the patent war, as is often the case, was inconclusive. But there is no doubt that Centronics implemented dot matrix printing in a market that was crying for more versatile, faster, and cheaper low-end printers. The Centronics Model 101 matrix printer, introduced at the 1970 National Computer Conference, catapulted the company to a leading position in digital printing within just a few years. In addition, there is truth in Centronics founder Robert Howard's contention (in Chapter 5) that availability of fast and affordable matrix printers was a gating technology that supported the later PC revolution.

IBM, as is often the case in printers, also pioneered, introducing its Model 2213 serial dot matrix printer a few months before Centronics. The device was part of the IBM 2770 Data Communications System. It did not use the controversial spring that was the

Inside the Centronics 101: (left to right) full frontal view, right side showing vertical format loop; left end showing control panel, fabric ribbon spool, and motors. In it's day, the printer was no doubt considered electronics-intensive. By today's standards, it is heavy metal. (Photos courtesy of Robert Howard)

subject of the Centronics-Mannesmann dispute. So IBM was in a world of its own and is not believed to have participated in this particular patent fracas.

Wire and character configurations varied. Early dot matrix printers, and later low end versions, had just seven wires in a single, vertical column which could print a variety of characters in a 5 x 7 dot format. This works for the 26 letters of the Roman alphabet in all caps, and a variety of symbols. A close look at the dot structure on receipts from today's POS and ATM terminals reveals that most of these machines still use that same, minimal format.

Why dot matrix? Part of the appeal is that printing with dots is inherently compatible with digital input. The printing itself reflects the binary, on-off nature of the incoming data codes. More importantly, the character repertoire is unchained from the limitations of a fixed selection of characters. In fact, it is unchained from characters, opening the potential to print graphics and bar codes, plot, and draw. This is one reason for the early development and acceptance of matrix printing – both impact and non-impact – in Japan and other countries using non-Roman alphabets.

Higher speed and lower cost were major attributes. Print quality, on the other hand, was the downside, at least in the beginning. Early impact dot matrix printing may have been the print quality low point in this history.

But with each passing year that changed for the better. There was the normal cheaper-faster pattern, but also diversifying applications which uncovered demand for better print quality. More sophisticated wire configurations appeared. With a single column of nine wires lower case characters could be printed with "real" descenders (as opposed to "y"s and "g"s with their tails squeezed up above the base line). The rise of word processing coincided with the rise of the impact dot matrix printer, so dot matrix developers worked to find ways to achieve that holy grail of

"letter quality" defined as what you could get with a daisywheel printer.

For dot matrix, this meant overlapping dots and smaller dot sizes. For vertical dot overlap print heads were designed with a double column of wires vertically staggered a half dot increment. Both 18-wire and 24-wire heads appeared. Horizontally dots could be overlapped by increasing the print wire firing rate and/or slowing down the horizontal scan rate. An example of this type of printer was the IBM "Quickwriter" which appeared in 1988, billed as producing true letter-quality text, which was something of a stretch. The Quickwriter delivered 330 cps in draft quality and 110 cps at 10 cpi in high quality mode. The single unit, end user price was $1,699.

Dot size is basic, which relates mostly to print wire diameter. Wire diameters from 8 to 22 mils have been used, with the average around 10 to 14 mils. The smaller the dot, the better the print quality, assuming there is a lot of overlap and the thin print wires do not punch through the ribbon or paper. But this also means printing is slowed down to near that of the competing daisywheels.

The print quality high water mark for dot matrix printers was the Sanders Technology Media 12/7 printer, which was fielded in the early 1980s. Print quality was the primary design driver. Royden C. Sanders, Jr. was a New Hampshire engineer/inventor/entrepreneur who founded and ran Sanders Associates, a large defense electronics manufacturer, for many years after earlier working for more than a dozen years with RCA. Soon after leaving Sanders Associates he was bitten by the digital printer bug and became determined to overcome the barriers that stood in the way of true "letter quality" dot matrix printing. The innovative print head developed by SDI formed the basis for the dot matrix printer that arguably offered the world's best impact dot matrix print quality. Vertical dot resolution was doubled by two pass

printing. For the second pass, the paper was incremented one half dot increment.

Despite such efforts, the Sanders printer achieved what might be judged "near letter quality" rather than the true letter quality of the daisywheels. Also, the inherent dot matrix advantage of wide font flexibility and graphics potential did not compensate

for the price of the printers and they did not achieve wide market acceptance. The Sanders Technology print head had basic attributes that advanced the art and were licensed to manufacturers in Japan and to Seitz in the U.S. However, by that time the sun was setting for dot matrix printers in the face of the rising non-impact tide.

Addressable vs. Printable Dot Positions

The diagram below shows the addressable increments for a 16-wire print head (Olivetti DM80/180). This printer could be programmed to place dots horizontally in 0.0529 inch increments. But it could not place adjacent dots centered on those increments since a given print wire could not be fired in such quick succession. It could only place dots on every fourth addressable horizontal position. This distinction between addressable dot increments and actual printable increments can relate to both impact and non-impact serial printers.

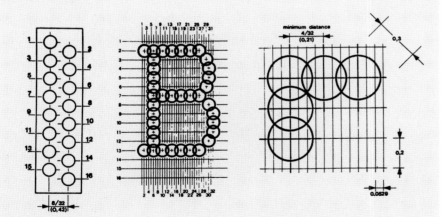

(Source: New Technology Printers for Tomorrow's Office, Datek Information Services, Inc. 1983)

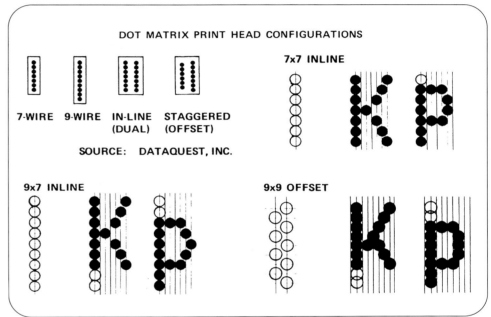

Representative print wire configurations and character dot structure. (Source: article by David Glidewell, Mannesmann Tally Corp., 1981 Printout Annual, Datek of New England, 1981)

PRINT HEAD DESIGNS

The explosion of demand for serial dot matrix printers in the 1970s and early 1980s yielded an incredible variety of variations on the theme.

The general architecture of the printers tended to be basically the same for the hundreds of models introduced. The competitive edge was in the print head itself. Among the leading designs:

Controlled flight solenoid actuators, curved print wires – this was the approach used by Philips, Centronics and others. The print wire was attached to the solenoid plunger which drove the wire forward. The spring to retract it was the subject of the above-mentioned patent dispute between Centronics and Mannesmann. At the point of contact with the ribbon and paper, the wires need to be more or less perpendicular. The diameter of the solenoids makes this possible only if the wires are long and curved. Consequently these heads are characterized by a fan or funnel shaped solenoid configuration.

Clapper Head Design – this variation works something like a baseball bat striking the ball, or, in this case, the back end of the print wire. Sub-variations – free flight, controlled flight, crush mode, ballistic mode – are moderately esoteric although at the time could arouse a bit of passion among print head engineers. A main advantage of clapper head design over the tubular solenoid is that the wire path could be more or less straight since the solenoids were off-line from the wires.

Magnetic Stored Energy – this concept was used in some earlier line printers to fire and control the hammers. Here the print "hammer" is actually a spring, normally held back by a permanent magnet. Energizing an associated bucking coil interrupts the magnetic field that holds the print wire back, allowing the spring to drive it into the ribbon, paper and platen. As the wire is moving forward, the firing pulse is turned off, allowing the magnet to retract the wire after it bounces off the paper.

Departures from print wires – several other ingenious head designs did not use print wires at all, including the Facit "Flexhammer" head, the GE "stacked blade" print head, and the Quint "Collinear 7" head.

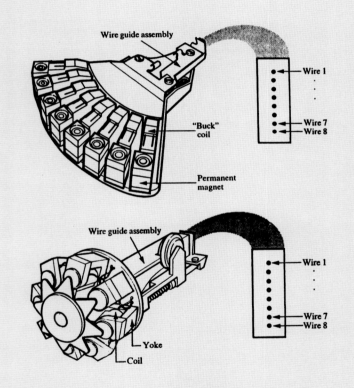

The fan-shaped print head is typical of early dot matrix printers and cost-reduced later printers with "no-work" magnets and a curved print wire path. In clapper heads with "work" magnets, the drive coil assemblies are offset so that the wire paths can be relatively straight. (Source: IBM Journal of Research and Development, Sept., 1981, courtesy of IBM Archives)

Enablers. At times a small group of suppliers play key supporting roles as little-known enablers of given printer technologies. Industrial Engraving was one of the leading independent developers and suppliers of precision print elements for on-the-fly line printers. In serial dot matrix, a key challenge was developing precision materials for wire guides that could cope with heat and friction. For these components the industry came to rely on a small group of ceramic suppliers for synthetic ruby, heat treated glass and synthetic sapphire: Maret S.A. and Seitz in Switzerland, Corning Glass in the USA, and Kyoto Ceramic (Kyocera) in Japan.

Ink science also contributed. Inks were formulated with an eye toward their lubricating function as well as print quality. Heat build-up was also controlled by algorithms to slow down printing as needed to maximize head life.

Over the years dot matrix printer costs continued to go down and performance continued to go up. New head designs had faster repeat rates, up to 3000 Hz. Printers were fielded with multiple heads. Bi-directional printing was standard.

But it takes more than ingenuity to save a technology whose time has come. Dot matrix printers live on. But in general computer output and office documents, low cost laser printers — especially the HP LaserJet and later much cheaper ink jet printers — sealed their fate. They also sealed the fate of Centronics and the larger competitors that shoved Centronics aside in the mid-1980s. Epson has managed an impressive transition to new technologies. Oki Data and several others may succeed as well. But a number of other leaders have disappeared from the printer scene, at least in North America, among them Citizen, Star, and Panasonic.

Line matrix printing. In theory, combining matrix printing with line printing should give you the best of both worlds: unchained image flexibility combined with the speed of the line printer. Several companies took the bait of this promise in the early 1970s, invested in creative solutions, and met with modest success.

But like their serial brethren, for general purpose printing the advent of affordable non-impact technologies stole away broad market acceptance over the long run. Yet they still fill an important role in specialized applications because they have something non-impact printers do not: impact. Wherever large volumes of multiple part forms need to be printed, or for high speed bar code printing, the line matrix printer remains often the best solution.

Printronix claims to have invented and been the leader in line matrix printing. For their story, in the words of company founder Robert Kleist, see Chapter 5. Printronix and Tally still manufacture line matrix printers. Other vendors included Potter Instrument, Dataproducts, Okidata (now Oki Data) and Centronics.

In the Tally and Printronix implementations the imaging component is a steel "print comb" which spans the length of the print line. Each tooth bears a small projection which forms the dot when struck against the ribbon and paper. The Printronix configuration is a front printer, Tally, back. The hammerbank shuttles back and forth as the paper is incremented vertically to build up dot matrix characters or graphics a dot line at a time. The Okidata/22 was a line matrix printer introduced in the late 1970s. Their approach was to build a reciprocating assembly with 22 solenoids which drove small hammers to make the dots.

Printing bar code labels continues to be an important application of line matrix printers. Shown here in action is the Printronix late-1980s Model P9000. (Courtesy of Printronix, Inc.)

A4. Paper and Paper Handling

Much could be written about paper and impact printing, but for now, here's just a little.

The earliest line printers and teleprinters kept it simple: they just used blank rolls of paper which were friction fed. However, in the 1950s with the increasing popularity of multi-part paper with one-time carbon (manifold forms), pinfeeding became standard. The forms, pre-printed or stock, are best described as pinfeed, flat pack, continuous forms. These forms also tend to be referred to as "fanfold." Forms old-timers, if they are really with it, will object to this usage. For the purists, "Fanfold" was the trade name of a specialized construction introduced by Uarco, probably back in the 1960s. Uarco Fanfold was zig-zag folded not just one way , but two – the typical cross web folds and unusual longitudinal folds running parallel with the web as a way to form multiple parts without needing fastenings to keep things in register. As with many trade names, especially if they are good ones, the term "fanfold" tended to become generic for all flat-pack continuous forms.

Positive control of pinfeed paper through the printer is by one or two sets of pinfeed tractors which could be tensioned horizontally and vertically. Some high end line printers also had electro-mechanical paper clamping to secure the paper between paper advance cycles while being battered by the type and hammers.

Traditional tractors were rather complex assemblies consisting of precision machined metal pins set in a gear-driven chain

This IBM publicity photo shows the 1403 printer configured with an in-line roll feed and burster-stacker. (DRA Archives)

loop. Lower speed printers managed with molded, plastic pinwheels. A little known company, Precision Handling Devices, developed a much cheaper pinfeed tractor which became popular. Their tractors replaced expensive and heavy metal chains and pins with plastic pins set in a mylar loop which cost only a fraction as much. This was one of the many technology enhancements that together were responsible for the continuing price/performance progress of impact line printers through the 1970s and 1980s.

As mentioned above, multiple part forms multiplied the effective speed of impact printers. The downside is that they had to be decollated, the carbon paper disposed of, and the resulting web burst into individual sheets, a labor-intensive post printing process.

In the 1960s and 1970s a number of companies developed forms handling accessories to speed post printing forms handling. The leading participants were those who provided the forms, Moore, Standard Register, and Uarco. The main products were decollators and bursters. Each of these powered accessories embodied their own technology and could be quite expensive.

The next step up was the in-line roll feed and burster. Cartons of flat pack paper could be heavy, yet used up quickly by a 2000 lpm printer. One solution was the roll feed. The paper roll was handled by a specialized dolly and contained several times more paper than cartons of flat pack.

Why not avoid the hassle by feeding sheets? Increasingly as word processing gained momentum in the 1970s and into the 1980s, more and more office users of low end printers called for sheet feeding. A small industry grew up developing and marketing attachments to feed sheets into desktop, impact printers, primarily daisywheels for word processing.

Convenience, as always, was a double-edged sword, however. Feeding sheets to this day is less reliable than pinfeeding continuous paper, and no doubt always will be.

B. NON-IMPACT PRINTING

Among the sea of competing non-impact technologies emerging during the 1970s it was far from clear which, if any, would dominate the general purpose applications of the future. The backers of the many variations of ink jet were the most passionate, almost a cult of true believers who at times seemed to be pouring R&D dollars into a bottomless pit of wishful thinking. Electrophotography was well accepted for copying, but early printer implementations were for only very high-end production printing. Shipments numbered only in the hundreds of units. The dozen or more competing technologies in these early days seemed on equal footing with the two which now dominate.

The emphasis in this overview will be upon technologies in use today in general-purpose and other page printing applications: the origins and pioneering companies, how they work, and their performance attributes. Major attention is given to electrophotography and ink jet. Other current page printing technologies will also be more briefly covered, including some used today in applications other than office or document printing.

B1. Electrophotographic Printing

Today's dominant page printing process is indirect electrophotography (EPG), otherwise known as xerographic printing. Although many vendors of EPG printers are inclined to distance themselves from the technology developed and commercialized by Chester Carlson and Xerox, all printers in this class are some variation of this seminal invention.

Non-impact printing had been around for a long time, but until the 1970s, all such printers required somewhat expensive, special purpose papers. The market was lusting at the time for "plain paper" non-impact printing, just as earlier the hunger for plain paper copying drove the eventual success of xerography.

As might be expected, Xerox fielded the first high speed EPG printer, the Xerox 1200, in 1973. This was followed by the IBM 3800, first delivered in 1976, the Canon Laser Beam Printer a bit later (4,000 lpm, never marketed in the U.S.), the Siemens ND2, first delivered in early 1978, and the Xerox 9700, first delivered late the same year. These high speed production printers ran at up to 200 ppm and were priced in the range of $200,000 to $350,000. A bit later, in 1980, HP fielded its 2680, priced at around $100,000. This was the first result of what has turned out to be the incredibly fruitful relationship between Canon and HP, a colorful story told by those who lived it in Chapter 6.

It is hard to believe that just ten years after these high end machines appeared, Hewlett-Packard and several other printer OEMs introduced EPG page printers priced well under $10,000. These desktop printers used the same basic technology as the production printers. They were based on the Canon LBP-CX. Canon's disposable cartridge concept was the major development behind this remarkable breakthrough, but far from the only one.

An anti-monopoly consent decree in 1974 forced Xerox to share its technology. This accelerated competition in EPG copying and printing. However, even without this, resourceful competitors managed to come up with variations in materials, chemicals and configurations that circumvented even the most formidable of patent fortresses built by Xerox and other competitors.

The main imaging steps common to all EPG printers –

1. Charging the photoconductor;
2. Exposure: selectively exposing the photoconductive surface to light to create a latent image;
3. Development: creating a visible image on the photoconductor;
4. Transfer: offsetting the image from the photoconductor to paper (to date, paper is the only substrate that works well with this technology);
5. Fusing: bonding the image to the paper, and
6. Cleaning: removing excess ink/toner from the photoconductor in preparation for receiving the next image.

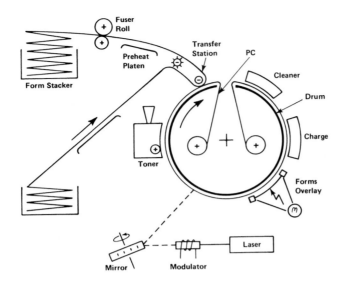

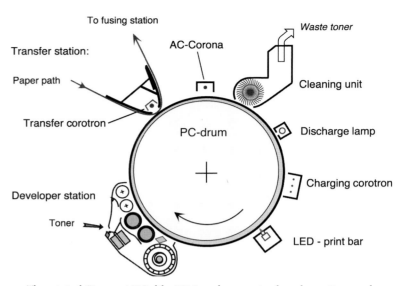

This diagram of IBM's pioneering 3800 page printer illustrates the basic elements of xerographic printing. (From IBM's April, 1975 press release announcing the 3800. Courtesy of IBM Archives.)

The original Siemens ND2, like IBM, used a scanning laser beam. Later and current models replace the laser with an LED array as shown here. (Courtesy of Océ Printing Systems GmbH)

Considering this succession of steps, it is understandable that the early implementations were expensive, high performance production printers. And it seems miraculous that the same basic technology can now be crammed into a desktop printer priced as low as $199. Diagrammed, the technology – for those with long memories – looks like something out of a Rube Goldberg cartoon, the master of portraying ways to make simple tasks hilariously complex.

All EPG printers have all these steps in common. What differentiates them is how each is accomplished. Here are some of the variations that have emerged over the past twenty years –

• Image generation: variable and semi-variable;
• Various light sources and modulation methods;
• Photoconductor materials;
• Toners and developers: wet or dry, single or dual component;
• Fusing technology: hot and cold pressure, radiant, chemical and flash;
• Paper handling: continuous or cut sheet; in-line pre/post-processing;
• Monochrome or color;
• Simplex or duplex.

1. Image Generation. Several early EPG printers were unusual in that like full-character impact printers the printing was not dot matrix. The Haloid Xerox Xeronic printer of the 1950s, Stromberg Carlson microfilm printers, the Xerox 1200, and the Uppster/Hi-

tachi page printer all generated character images by shining light through an optical mask. In addition, the IBM 3800 and some of its competitors had a "form overlay" feature for projecting semi-variable "forms" over the computer-generated data. These overlays were a kind of film negative with an image that could be projected over the variable data by flashing an intense lamp behind it.

2. Light Source. The original and most commonly used light source is a laser. Other light sources include LED arrays, liquid crystal shutters, and special purpose optical fiber faceplate CRTs (used in the pioneering A.B. Dick Videograph and, in the late 1970s, with the Fuji Xerox 1660).

The laser beam is modulated (turned on and off) under command of the printer controller which converts the buffered page data into on-off signals. The laser-generated light fragments are then directed to a rotating, faceted mirror which fans them out across the width of the photoconductor, building up the latent page image with successive rows of tiny, discharged spots on the continuously moving photoconductor surface.

This arrangement presents an interesting electro-geometric challenge. As the angle between the laser beam and the photoconductor surface increases (i.e. the further if moves from the center), the more oblique it will be so that optical and/or electronic measures need to be taken to compensate for the potential image distortion. And there are limits. For this reason, one limitation of most laser printers compared with, say, ink jet, is document width.

There (as always) are more subtleties. Spinning a faceted mirror sounds simple. But accounts from Xerox, IBM and others highlight the complications. Fine tuning the polygon mirror mechanism to provide accurate scans of the laser beam at the required speeds was incredibly demanding. The rpm of the mirror depends on the number of facets in the mirror, the resolution of the printer, and its print speed. One example cited by Gary Starkweather, who led development of the Xerox 9700 at Xerox, is a 36 facet polygon with 6.67 mm facets and 3 inches in diameter. Doing the job can call for rotational speeds of more than 30,000 rpm. At these speeds, both Starkweather and others mention an awesome menu of variables that need to be accommodated, among them,

- vibration from other parts of the system;
- ambient and machine generated thermal distortions;
- bearing wear;
- contamination from ambient air, including lubrication vapors;
- and stresses from the tendency of the mirror facets to "paddle" the air, thereby causing resistance affecting the power source.

All these variables are in addition to microscopic deviations in the manufacture of the mirror itself. Xerox and other laser printer developers have come up with various ingenious schemes to compensate for such variables, including optical correction lenses and acousto-optical deflectors to offset both vertical and horizontal aberrations. The more resourceful the compensation scheme, the less stringent the requirements for the mirror and driving mechanics.

We can go still further. What kind of lasers? Early high-end laser printers used gas lasers, cadmium or argon, and later helium neon, at least in the case of the Xerox 9700. The laser needs to be linked to the spectral sensitivity of the photoconductor. Around 1984, according to Starkweather, when infrared sensitive PCs became available, solid state laser diodes began to be used in all except the high-end printers, most of which continue to use gas

lasers.[*] The introduction of laser diodes greatly reduced cost. Gas lasers cannot be modulated directly by being turned on and off at anywhere near the speed needed. Xerox, according to Starkweather, used acousto-optic or electro-optic techniques to modulate the gas laser light source of the 9700. The complete modulated laser system cost around $4,000. In contrast, the laser diode is not only cheaper, but can be modulated directly by the drive current, for similar performance. The cost? By the 1990s, only about $4.

LEDs are an important alternative light source. The LED — light-emitting diode — is a semiconductor material which emits light when an electrical voltage is applied. Typically each LED is responsible for one horizontal spot position across the printable width. This means a 300 dpi LED printer with an 8-inch print line will need an array of 2400 LEDs. On the surface, it would appear to be a better (i.e. cheaper and more reliable) solution since with a stationary LED array a lot of mechanical clutter is eliminated. Also, compared with a scanned laser light source, the LED spot size can be smaller and the shape is inherently circular rather than elliptical. On the other hand, tooling is more expensive and LED arrays have proven difficult to fabricate. Some EPG vendors who switched to LEDs have since returned to the laser light source.

Some believe a trend which works against continued use of LEDs is ever greater resolution. Early EPG printers achieved 180, 240 and later 300 dpi. Today 600 dpi is becoming virtually a standard. Fabricating an LED array at these densities has proven challenging. Siemens (now Océ Printing Systems) succeeded, however, with their LED Plus technology introduced in 1988, which lets them achieve a print line of 17 inches at up to 600 dpi. Still higher density arrays have recently appeared offering up to 720 dpi, from Agfa, Okidata, Kentek and Océ.

LED developments by Océ and other vendors as of this writing indicate this could become the technology that will allow electrophotographic digital printing to more directly challenge the traditional printing press. Such arrays are said to hold the potential for printing 600 ft./min and up at 600 dpi.

The Liquid crystal shutter alternative was used for a time by Casio, but has not proven to be viable. More recently, in 1995, Minolta developed a variation called the "digital microshutter." The microshutter consists of two panes of polarizing material which can be rendered opaque or transparent by the application of the appropriate electrical pulse. Switching was said to be fast enough to support printing at up to 100 ppm. Resourceful, but so far not widely utilized.

In short, the laser continues to dominate, which is one reason many people refer to all electrophotographic printers — if not all page printers — "laser" printers. In addition, as Hewlett-Packard and Apple have proven, the word "laser" continues to have some sort of magic appeal to users.

3. Photoconductors. Many materials have photoconductive properties and have been used successfully. Chester Carlson's original selenium soon evolved into compounds of selenium, cadmium or arsenic. These offered excellent photoconductivity and long life, typically 100,000 pages or more. The downside is that they are toxic. Today's organic photoconductors are cheaper and are not considered toxic, but have a much shorter life span. One vendor, Kyocera, a few years ago launched a line of EPG printers using amorphous silicon, a substance then rated at around 300,000 pages. Today silicon is rated at up to a million pages and organic PCs up to 100,000.

The photoreceptor substrate is most commonly either a metal cylinder (usually aluminum) or a flexible web. Besides critical photoelectric properties, the photoconductive material must have

---

[*] Starkweather, Gary in *OE Reports*, November 1997 (SPIE)

mechanical properties that will allow it to stand up to the forces of the development, cleaning, and transfer processes.

4. Toners and Developers. Traditional xerography developed the image with a dry, two-component mix consisting of from 1% to 10% toner by weight. As described by L. O. Jones, the toner is the "ink," "generally a low melting point polymer compounded with approximately 10% carbon black and ground to a particle size of about 12 μm."[*] The rest of the developer is the carrier, a much coarser material, 3 to 50 times larger than the toner particles. Functions of the carrier are first, to disperse, or carry the toner to the electrostatically charged latent image on the drum and second, to impart a static charge to the toner particles. This is done by the rubbing action of the developer against the toner particles, a phenomenon termed "triboelectrification." Physically, the carrier consists of coated metal beads engineered to provide the ideal level of charge to the toner.

What causes the toner to adhere to the virtual image on the photoconductor is the attraction of opposites. Traditionally, with the earlier EPG printers, the selenium photoreceptor was positively charged. Organic photoconductors are usually given a negative charge. This means that with selenium the developer system needs to apply a negative charge to the toner particles, and with organic photoconductors, a positive charge. The toner is metered onto the latent image by a combination of magnetic and electrostatic forces: magnetic forces hold the carrier beads on the roll that meters out the developer mix, and electrostatic forces pull the toner to the imaged areas of the photoconductor.

Methods used to bring dry toner into contact with the photoreceptor include primarily cascade development or magnetic brush mechanisms.

* Jones, Lewis O., *Handbook of Imaging Materials*, Edited by Arthur S. Diamond, Marcel Dekker, Inc., 1991.

More recently vendors have moved to single component technology in which magnetite is added to the toner and the relatively bulky and heavy developer beads are not needed. This simplifies the development unit and gives product engineers more leeway in threading their way through the patent maze.

Alternatively, liquids can be used to disperse and meter the toner particles onto the latent image. Early applications of liquid toner appeared in the 1950s for the Electrofax copying system and later copiers from Savin and Ricoh in the early 1970s. In the late 60s Stromberg Datagraphix launched a series of high speed CRT-microfilm printers that printed on zinc oxide coated paper using liquid toner. The pioneering (but apparently short-lived) 8,000 lpm Oki Data Electro Printer of 1978 was described as the first liquid-toned ink mist development, plain paper, laser beam printer. The following year Canon had better luck with its LPB-10, the first semiconductor laser printer. The liquid toner machine is also believed to be the first table top electrophotographic printer. Recent and current applications of liquid toners include electrostatic plotters (Versatec, Océ, and Calcomp) and color proofing in the graphic arts. Liquid toners tend to be out of favor in non-industrial settings in part be-

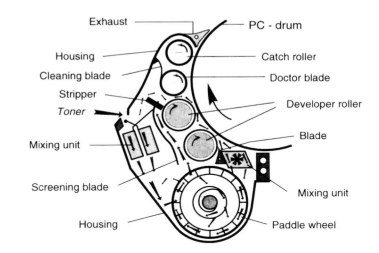

EPG subsystems can be complexity within complexity. Diagrammed here is the two-roller developer station used by Océ Printing Systems in high-end production printers. Arrows indicate the circulation of the carrier and toner particles through mixing, metering onto the developer rollers, and making contact with the PC where the toner is "drawn" to the photoconductor by the electrostatic forces of the latent image. (Courtesy of Océ Printing Systems GmbH.)

cause of environmental concerns. However a number of EPG printers and some digital presses are liquid. Process color is said to be easier to implement with liquid toner than with dry.

5. Fusing. The toner ingredient that effects fusing the colorant into the paper is any of a variety of resins. The most common fusing method is to pass the imaged pages between a heated roll and cold backup roll under fairly high pressure. Non-contact alternatives include radiant heating with a quartz lamp or heated resistance coil, or flash fusing with a short, high-intensity blast of light. An early disadvantage of fusing with heat was the chance of fire. Should the paper jam at the fusing station, it could ignite. The early Xerox 914 copiers were shipped with a small fire extinguisher (called a "scorch eliminator" by Xerox) for just this eventuality.[*] Copy paper engineered to have a low kindling point is said to have helped win acceptance for this seminal, market-making device.

Besides limiting the range of substrates, high temperatures can also compromise the kinds of inks used with pre-printed forms. The high fuser temperatures of the early IBM 3800 printer – in the range of 204°C – combined with pressure caused ink problems such as build-up on the fuser rollers. Forms suppliers were advised to avoid inks with volatile components. Flash fusing exposes the printed substrate and inks to less thermal stress. Use of impulse heat in the range of 170°C and no pressure allows use of thicker substrates and also adhesives, normally without problems.

Cold pressure fusing avoids this pitfall, and has other advantages. The power requirement is reduced in operation, and in standby there is no power draw and no wait for warm-up. On the down side, pressure fusing tends to impart a sheen to the paper, the pressure tends to degrade the edge definition of the printed characters, and image adhesion is less secure. Creasing the paper through an image is an easy way to spot a cold pressure fused document: the toner tends to crack off at the crease.

Siemens for its ND3 developed a fusing system which avoided the paper and ink restrictions imposed by hot roll fusing, and also provided image quality and durability that far exceeded that of cold pressure fusing. This was a solvent vapor technology which significantly broadened the range of papers and inks which could be used. It allowed the ND3 to be used for pressure sensitive labels. However it also had an environmental downside which prompted Siemens to return to heat in later models.

Variation: Océ Copy Press. A variation in electrophotographic printing technology is used by Océ Printing Systems in its 9200 laser printer series. OPS claims its Copy Press System is today's only real alternative to conventional xerography and that it offers high mechanical reliability and print quality that almost matches that of offset presses.

---

[*] Jacobson, Gary, & Hillkirk, John, *Xerox American Samurai*, Collier Books, 1986

Liquid vs. Dry Toners

Swiss toner expert Andre Gigon of Specialty Toner Corporation believes liquid toners have a bright future, especially for digital color presses. Here are some of the considerations on which he builds his case.

The mission of toner is to reflect light, in varying portions of the spectrum, back to the human eye. Only the top portion of the fused toner image performs this function, which implies with a thick "pile height," much of the toner is wasted. The pile height of dry toner is around 30 microns while liquid toners, like printed offset press ink, can range from 7 down to maybe 2-3 microns. This means a given quantity of liquid toner can print 5-10 times more pages than dry. Flatness has advantages.

Print quality is in part related to toner particle size. Here dry toners are faced with a difficult Hobson's choice. Large particles create irregular edge definition, partly because they are large and irregularly shaped, and partly because under pressure fusing they are flattened out. Complex driver software is needed to compensate for this dot gain problem.

On the other hand, small particles can create misting or scattering around character or image element edges, are hard to control inside the printer because of dusting, and present a health hazard since they can reside permanently in the lungs if inhaled. Cooling fans can stir up toner under the best of circumstances, but the finer the toner, the more difficult the problem of dusting.

Even if a system is able to overcome the handling problems associated with very fine powders, there is the consideration that small-

The photoconductor is nontoxic zinc oxide coated onto belt. But the material is neither as hard nor as electrophotographically stable as the commonly used As2Se3. To compensate for this, to achieve the targeted lifetime of 100,000 sheets, the belt needs to be amazingly long. The latent image is developed with monocomponent conductive toner which simplifies the system and minimizes backgrounding. However, the toner and the paper have like charges which means the usual corona transfer system cannot be used. Instead there is an intermediary, silicone rubber transfer belt. The toned image transfers to the belt and is then heated prior to transfer to the paper which is also pre-heated. This double transfer system is said by OPS to actually increase the optical density of the image.

Paper handling in EPG printers may be continuous or sheet, depending on the application. Some high-end production printers are configured as duplexing systems. In twin systems the continuous form is fed directly from one printer to a second printer for automatic duplexed (two-sided) printing.

Canon Cartridge Concept. Probably the most dramatic conceptual innovation in the history of electrophotographic printing was Canon's throwaway laser cartridge. As emphasized by HP's Dick Hackborn (Part Two), this was only in part a strategy to reduce the hardware cost. Reliability was a major goal. Both were achieved. The main trade-offs were low duty cycle and expensive per-page cost. The specified duty cycle for the CX print engine was 3,000 pages a month and the original paper cartridge held just 100 sheets. A major technical innovation was their use of monocomponent toner in a reversal development system. With the printer, light is used to discharge the image instead of the background, the opposite of the Canon Personal Copier which introduced the disposable cartridge concept.

To meet their price point, Canon gambled on high volume, mass production, and won. However their design had to be modified to meet this demand. One innovation was a compensator lens interposed between the polygon mirror and photoconductor which allowed for production variables in the mirror-based scanning system and the use of a mass produced, metal mirror.

Changing the cartridge in effect is like replacing the engine in a car. The Canon cartridge contains the PC drum, toner and development subsystem, charge corona assembly, cleaner blade, and toner supply. Cartridges can be swapped in just a few seconds. Upon insertion, the top shutter of the cartridge automatically opens to let the laser beam in, and a bottom shutter opens for the transfer of the toned image to the paper. .

Attributes. Electrophotography has proven to be an extremely flexible technology with its various implementations ranging from small desktop printers to high speed production printers to digital presses. Pricewise, EPG printers now start as low as $200 and go up to $300,000 or more for a

particle sizes are more expensive to manufacture. Mechanical grinding can produce dry toner particle sizes down to around 7 microns. There are chemical processes for producing nicely spherical toner particles down to 3-5 microns, but they are at this point expensive and the problem of controlling these very fine powders in the printer remain.

Liquid toner particles are typically much smaller. For example, in the Indigo digital press they measure 1-2 microns. Their small size facilitates very high resolution images and transparent color capability for better color rendering. Liquid toners with particle sizes down to .1 micron have been successfully developed. In any case, liquid toner is said to produce a thinner and flatter deposit than dry toner, matching the roughly 7 micron deposit of offset ink.

Fusing considerations remain. The heat and pressure typically required in EPG printing augments dot gain and can damage the substrate. It would seem that liquid would also need high or higher drying temperatures, or reduced throughput. Work is being done on liquid toners with lower Isopar content to deal with this negative. Liquid toners with up to 80% reduced Isopar content have been achieved, and it is expected such toners will be appearing in digital presses which will not require powerful dryers between each color print station.

In short, looking ahead, Mr. Gigon argues that liquid toner systems are more likely to compete effectively against traditional presses than dry toner electrophotography or ink jet.

high-end production printer or digital press. Speeds range from 6 to over 200 ppm.

Compared with ink jet, EPG printing is limited in application because to date it handles only paper. This bars the technology from the wealth of high margin industrial and other specialized applications that ink jet and other technologies are beginning to penetrate.

The complexity of the technology has made color expensive. Electrophotography was primarily a monochrome technology until the early 1990s when color copiers slowly found their way first into print-for-pay outlets and later, as prices dropped below $10,000, into office settings. Digital presses – by definition color – have been slow to penetrate the mainstream printing industry. The technology in terms of print quality has been proven to be comparable to traditional printing processes and in time the attributes of digital printing will more than offset the current unfavorable economics.

One of the major enablers upon which the popularization of electrophotographic and other page printers depends is the controller and related software. Compared with the earlier serial printers, page printer controllers posed a universe of data handling challenges. A key difference between a serial printer and a page printer is that with the latter the entire page image needs to be in place and interpreted for the printer by the controller. Once a page begins to be printed, it needs to be a continuous process. It was controller know-how in part that enabled HP to be the leading commercializer of the laser printer rather than Canon.

The classic controller needs to handle at least five functional elements: I/O, data processing, image processing, image storage, and marking engine interface. There are software decisions that affect the forward and backward compatibility of the controller as new printer models are introduced. Easy implementation of new printer applications is important since most printer users are not programmers. All these considerations become more complex as applications proliferate, as print quality standards evolve upward to 600 dpi and above, and, especially, when color is introduced.

B2. Ink Jet Printing

Back in 1979, in a brief study of the potential of ink jet printing, an observer concluded, in part, "Ink jet printing today, after more than fifteen years of experimentation and development, remains very limited in application. . . . In light of its long gestation period, the future of ink jet printing remains cloudy."

What a difference a decade makes! By 1989 ink jet for general purpose digital printing was beginning to sweep through the low end of the market and establish its place as the only technology with the promise of affordable color.

Ink jet is really a cluster of technologies, descended from an assortment of inventors and entrepreneurs, operating insofar as possible behind a protective wall of patents. It is the one non-impact technology where the names of the original inventors are often used to identify the sub-technologies, names such as Sweet, Hertz, Ascoli, and Elmqvist. Some have had their day, some have found their way into niche markets, and a several have blossomed into one of the two current dominant digital printing technologies (see Ink Jet Technology Tree, page 50).

Electrostatic Pull. In the 1960s several ink jet printers using this technology were introduced which traced their ancestry to patents by Winston and Ascoli. These included the Teletype Inktronic page printer, the Hermes HR-3 calculator printer, the Casio Typuter, and a Toshiba drum printer.

The basic common denominator of these printers is that ink droplets are pulled electronically through relatively large jet nozzles. The drop size tended to be smaller than the nozzle diameter.

Announced in 1965, Teletype's Inktronic page printing system was one of the first ink jet printers to be commercialized for office applications. The electrostatic pull system performed at 120 cps. (Photo from publicity release by Fensholt Public Relations, Inc. for Teletype, 1965. DRA Archives)

This helped overcome one of the congenital problems of ink jet — nozzle clogging. One version ejects drops on demand. Others are semi-continuous, generating drops in bursts which are electrostatically deflected to a programmed position on the document, or, if not needed, into a catcher for recycling.

A high, constant voltage is applied between the ink reservoir nozzle and a platen behind the paper. Then a smaller voltage is applied from a "valving electrode" to pull off the drops. In the Teletype Inktronic, the drops in flight are deflected both horizontally and vertically to position the drop to its assigned position in the character being printed.

Disadvantages of electrostatic pull ink jet included low drop frequency and the high voltage needed to pull the drops to the paper — up to 20,000 volts according to one source. Teletype used an array of 40 heads to achieve a print speed of 120 cps. The speed of the Casio machine was 33 or 50 cps.

None of the electrostatic pull machines of the 1960s survived for long. The basic concept, however, lives on as a development program by former IBM scientist Ross Mills and several colleagues. Their "Shadow Pulse Method" is said to overcome the problem of high voltage requirement and offers the advantage of variable spot sizes that can range from 20 to 130 μm. Trade marked the iTi ESIJET, the technology has attracted some attention and development funding. Most observers, however, remain skeptical.

Valve Jet. This form of ink jet printing has been widely used commercially for quite a few years. Valve jet has been described as not truly ink jet, but rather an array of small spray guns. Solenoid valves control the emission of ink under pressure on demand. The relatively large drop sizes range from .040 to .080 inch. The main applications are industrial, for large character printing on cartons and products, production color printing of carpets, fabrics and wall coverings, and sometimes billboards.

Valve jet has been around for a long time. Zimmer in Austria introduced their Chromojet line of valve jet printers in 1975 and has refined them over the years. They make their own heads. For their major application, carpet printing, their goal is an interesting contrast to text printing of office documents where tiny droplet sizes are sought. Valve jet printers need large volume drops because they want to saturate a three-dimensional substrate. The resolution and rep rate, at 400 drops per second, are

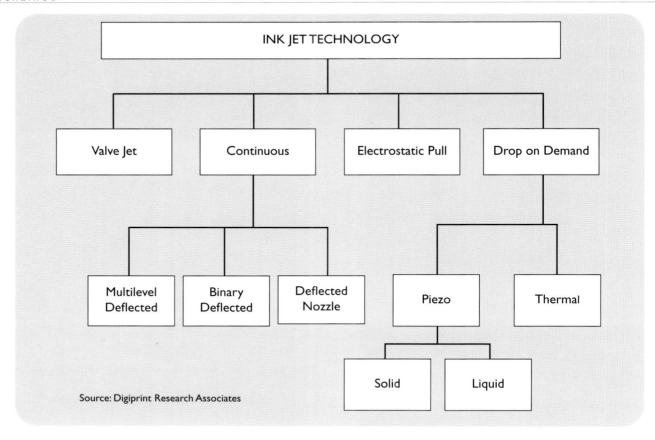

Source: Digiprint Research Associates

low. But volume can be high. Each valve can eject up to 300 grams per minute, said to be about a tenth of a gallon.

Continuous Ink Jet. The world of continuous ink jet today includes wide format printing, industrial printing, and high volume production printing. Most products using the technology are based on the Sweet patent. This is the 1964 invention by R. G. Sweet entitled "synchronous fluid drop recorder," developed at Stanford University (California). Closely related is the Sweet and Cumming patent for a "linear array fluid jet recorder with a plurality of jets."

Another key subtechnology is Hertz, developed by C. H. Hertz working at the Lund Institute in Sweden.

Less well-known is the work of Elmqvist at Siemens Elema, the Swedish subsidiary of Siemens A.G. This specialized continuous ink jet technique was implemented by Siemens in the Mingograf oscilloscopic recorder which was fielded in 1952 and successfully marketed for many years. Instead of the more common electrostatic deflection of the ink droplet stream, in the Mingograf the ink stream is deflected by an oscillating nozzle. Neither this device nor the Elmqvist concept is believed to have survived into recent decades.

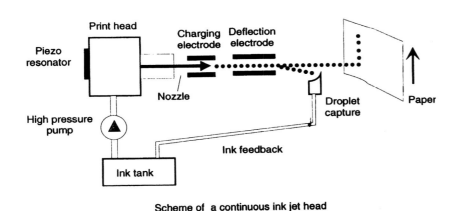

Scheme of a continuous ink jet head

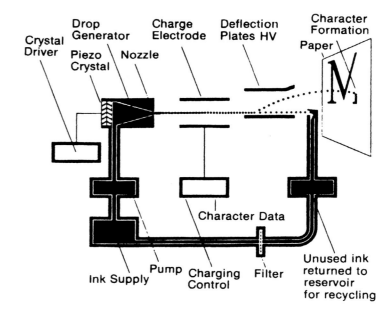

The two main approaches are multi-level deflection and binary deflection. Hertz is a variation of the latter.

Most multi-level deflection continuous inkjet printers are serial printers. A single print head may scan across the width of a document. Or an item to be coded may be moved along under a stationary print head. In either case, the continuous jet of ink is deflected vertically to create the vertical dimension of codes or characters, with the horizontal dimension provided by the horizontal movement of the head or substrate. As with all continuous inkjet systems, drops are either allowed to fly onto the substrate, or are deflected into a catcher for recycling.

A. B. Dick Company, the major licensee of the Sweet patent, implemented the technology in a number of products. They have been a leader in ink jet for industrial marking ever since introducing their Model 9600 in 1969. They claim this was the world's first commercial ink jet printer. (A.B. Dick spun off their ink jet business as VideoJet International, which later acquired Cheshire and Elmjet Ltd, and recently was renamed Marconi Data Systems as part

of Marconi plc of London, England). A. B. Dick also licensed several other companies for non-competing applications including IBM (cross licensing), Sharp, Hitachi, and Recognition Equipment. Other users of multi-level deflection technologies include Gould, Kodak, Domino, Willett, Linx and Imaje.

IBM pioneered ink jet in the office with its Model 6640, a continuous, single nozzle, deflected printer introduced in 1976. Model 2, announced in 1979, hiked the speed in draft quality from 92 cps to 184 cps. However it was fairly expensive and by 1980 sales began to slow with around 10,000 believed to have been shipped.

In binary deflection, ink drops ejected from an array of jets are either allowed to reach the substrate or are deflected into a catcher. Normally the array is stationary, spanning the width of the printable area, with the substrate moving past it. The major early user of this technology was Mead, with the development of its high speed DIJIT (for Direct Imaging by Jet Ink Transfer) system. Mead based this work on licensing Cumming's interest in the

In binary CIJ (left), ink is either allowed to reach the moving substrate or is pulled down into a gutter and re-circulated. In multi-level deflection, the stream of ink droplets is guided to various positions to more or less paint the image onto the substrate. (Binary CIJ diagram courtesy of Océ Printing Systems GmbH; Multi-level diagram adapted from IBM 6640 publicity announcement. DRA Archives)

original Sweet & Cummings continuous ink jet patent, which was soon contested by A. B. Dick. The Mead ink jet business and facilities were acquired by Kodak in 1983 and renamed Diconix. In 1993 Kodak sold the business to Scitex but kept the name Diconix for its line of portable ink jet printers.

Advantages of continuous ink jet include reliability and high droplet production rates, typically 50,000 to 100,000 per second per nozzle. A greater variety of inks can be used, including solvent inks. Disadvantages are complexity demands placed on the hardware by the need to recirculate ink. Typically, much more ink is recirculated than reaches the paper, as much as 95%. Due to solvent loss to evaporation, to maintain specified viscosity, there needs to be a solvent make-up system. The problem of nozzle clogging is avoided, but contamination of the recirculated ink is always possible, requiring a high-tech pump and filter system.

Hertz is a variation of binary deflection. Essentially, instead of deflecting a formed ink drop, the drop, if needed for the image, passes through a small hole in a buffer plate between the ink nozzle and the substrate. For an "off" cycle, a voltage is applied which breaks the ink up into a fine spray. The dispersed spray cannot reach the substrate because of the buffer plate, and is recycled. Actually, it is not binary, in that a lesser charge can be applied which allows some of the ink spray to pass through the hole. This means that Hertz technology is well suited for printing grayscale images, one reason it has found application in wide format photographic and art reproductions.

Continuous ink jet has been implemented in a remarkable variety of applications. Industrial applications have included Boeing Aircraft and American Can for wire marking systems, R.R. Donnelley and Videojet for addressing, Sharp for office printers, and coding documents in sorting operations including financial and postal systems. Moore Business Forms for a time licensed A.B. Dick technology for its on-press Compurite variable imaging system. IBM's pioneering 6640 ink jet office printer was based on a cross-licensing agreement with A. B. Dick.

Major current users include:

- Marconi Data Systems (formerly Videojet International and including Elmjet) and Domino for addressing and industrial marking;
- Scitex/Iris and other wide format vendors for high quality color reproductions; and
- Scitex Digital Printing for high speed document printing, book printing, and on-press variable imaging.

Another stream of continuous ink jet development flowed from Cambridge Consultants Ltd. in the U.K. CCL, owned by Arthur D. Little since 1970, was heavily involved in continuous ink jet from around 1975 until they spun off Elmjet in 1985. Mike Willis, now an ink jet consultant with Pivotal Resources, was

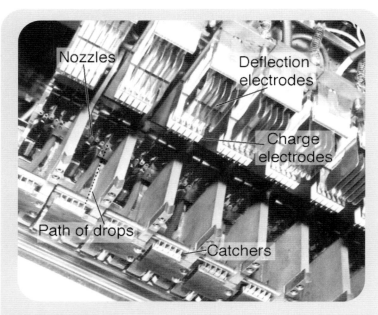

This close-up shows the workings of a multi-level deflected continuous ink jet printer built for a client by Cambridge Consultants Ltd. in 1983. The machine was used in an industrial application and could print at 80 dots/inch across a 20-inch print line at 360 ft./min. According to Mike Willis of Pivotal Resources, that performance was good for its day. Visible here are groups of 5-nozzle modules which jet downwards to sets of electroformed gutters mounted below them. Swung away and to the top right are the deflection electrodes and charge electrodes. Each module of five deflectable jets prints a swath one inch wide. Photo courtesy of Cambridge Consultants Ltd.

a key manager at CCL during the ink jet years. According to Willis, CCL has been a major seedbed of ink jet technology and enterprises. Spin-offs include Domino, Elmjet (now part of Marconi Data Systems), and Xaar, the drop-on-demand vendor which Willis co-founded. In addition, CCL indirectly contributed to the development of ink jet by Willet and Imaje.

Drop-on-Demand Ink Jet. Instead of dealing with a continuous stream of ink droplets, why not squirt ink only when you know what to do with it? Sounds logical, but myriad challenges stood in the way of commercializing such technology for many years. Today one of the main technology battles is between the two major drop-on-demand techniques – thermal vs. piezo. The leading practitioners of thermal printing in office applications currently are Hewlett-Packard, Canon and Lexmark. Leading piezo development are Epson, Brother and Xerox/Tektronix.

We will begin with piezo since historically it predates thermal. Early commercialization efforts of piezo technology included Silonics in 1975, the Siemens PT-80i in 1977, and Epson and Exxon in the early 1980s. Of these printers, Siemens had the most early success commercially. The PT-80i was a drop-on-demand ink jet teleprinter with a 12-channel print head using an ink described as oil-based. By 1982 as many as 20,000 of these units were believed to have been shipped.

Piezo printers are based on various adaptations of the piezoelectric electromechanical phenomenon. Certain crystalline materials twitch or flex under an electrical charge, thereby functioning as a transducer. This conversion of electrical to mechanical energy seems like an easy way to build a miniature pump to squirt ink on demand. Over the years inventors have come up with a remarkable variety of configurations to exploit this concept.

Two main approaches are the flat ejector and the cylindrical ejector. In the flat ejector, one wall of the ink chamber is the piezoelectric transducer which flexes when energized, reducing the volume of the chamber enough to eject some ink. In effect, the transducer is analogous to the hammer in an impact printer.

In the cylindrical ejector, the ink chamber is in a small cylinder made of a piezoelectric material. Under charge, it twitches enough to eject ink through a nozzle at one end. This concept is sometimes described as analogous to a sphincter. This approach has been used primarily for plotters and oscillographic recorders rather than general purpose printers.

Simple concept; challenging implementation. Ink jet engineers can get excited about different ejector designs, such as bender, shear mode, shared wall and piston. They can get frustrated by the challenge of uniform drop sizes. It took Silonics four years to come up with a manufacturable impulse ink jet print head, and after all that time, the product never quite made it. There is the problem of blockages from ink drying in the channel. Or impurities in the ink, or even bacterial growth in the ink reservoir or the channels clogging the tiny nozzles. Or external contamination from ambient dust. Purging cycles and mechanical capping systems have been used.

Piezo and Thermal printers are similar in their basic simplicity; the main difference is how the ink drops are actuated. (Source: Pocket Guide to Digital Printing, 1st edition, by F. Cost, © 1997. Reprinted with permission of Delmar, a division of Thomson Learning. Fax 800 730-2215.)

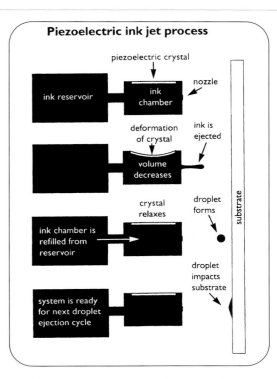

Piezoelectric ink jet process

piezoelectric crystal
ink reservoir — ink chamber — nozzle
deformation of crystal — volume decreases — ink is ejected
crystal relaxes — droplet forms — substrate
ink chamber is refilled from reservoir — droplet impacts substrate
system is ready for next droplet ejection cycle

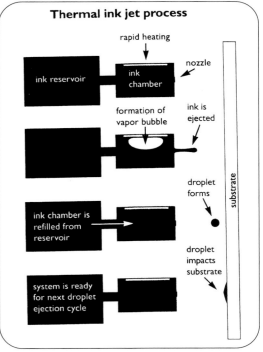

Thermal ink jet process

rapid heating
ink reservoir — ink chamber — nozzle
formation of vapor bubble — ink is ejected
ink chamber is refilled from reservoir — droplet forms — substrate
system is ready for next droplet ejection cycle — droplet impacts substrate

Ink formulations need to take into account the fluid dynamics and drying considerations within the print head as much as image quality and durability on the printed document.

The first wave of impulse ink jet development never succeeded in displacing the dominant impact dot matrix printer.

It took thermal on-demand ink jet to dethrone impact dot matrix as the low end print technology of choice. Thermal ink jet has swept the field and is now by far the most widely used ink jet subtechnology. Rather than an electromechanical "hammer" to eject the ink, heat is used. This approach was developed in parallel by Canon and Hewlett-Packard, independently at the start, and later cross-licensed. This concept, termed "bubble jet" by Canon, is demanding to manufacture, but once the process is established, in high volume it is cheap enough to allow implementation as a disposable cartridge. Like the disposable LaserJet print cartridge, for low volume applications, the throwaway cartridge has been the key to reliability and machine affordability.

The thermal ink jet head contains an ink reservoir and a matrix of ink chambers, one for each nozzle. Each chamber contains a thin-film resistor which when pulsed creates a small but intense blast of heat which vaporizes some of the ink, creating enough pressure to eject a droplet. This cycle can be surprisingly fast. The rep rate varies with the drop volume: the smaller the volume, the faster the potential speed. According to Pivotal's Willis, thermal printers as of this writing commonly fire in the range of 8 to 12 khz per chamber, although one of the latest HP thermal printers runs at 18 khz with a 5-picoliter drop volume.

This looks like a big improvement over the earliest thermal heads which industry consultant Hugh Van Brimer remembers as running at just 600-700 drops per second. "Our Exxon-type piezo heads in the late 70s and early 80s were in the 10 khz range. We saw speed as one of the big advantages of our technology. In the lab we demonstrated 20 khz per channel, but for our products we didn't need that kind of speed."

Thermal print heads do not have the power of either piezo impulse or continuous ink jet, so the gap between the face of the head and the paper is critical. Also, the required thermodynamic qualities of the ink places constraints on its formulation which can work at cross purposes to various image quality considerations.

The piezo ink jet print head is more expensive to manufacture so does not lend itself to throwaway design. However, the ink cartridges can be separate from the head itself and individually replaced as needed. This contrasts with thermal ink jet in which a lot of ink usually gets tossed when one color of a multi-color cartridge is emptied. Piezo, then, has the potential for lower running costs. Higher volume and wide format piezo printers normally offer refillable or at least independently reloadable ink cartridges, a major piezo plus.

It currently looks like a second wave of piezo development may over the next decade take significant market share from thermal. Piezo heads are said to achieve firing rates that are faster than thermal heads at comparable drop volumes. Epson has been gaining ground, touting the superior color rendering of its piezo printers. Xaar in the U.K. has developed another piezo-based technology which looks promising. Print heads with up to 500 channels have been fabricated and they claim more than a dozen licensees. Among the applications are product coding, packaging, commerical printing and hi-resolution ID cards. Among the announced licensees are Brother and Konica. MIT, a subsidiary of Nu-Kote/Pelikan, as of 1998 was a major licensee claiming among its OEMs Olympus and Xerox (through Olympus), Pitney Bowes, Raster Graphics, and Willsman. Piezo is not quite on a roll, but it's where the action is.

Solid Ink Jet. The most recent sub-subtechnology to reach commercialization is solid ink jet. Beginning in the early 1980s, Exxon, Creare, and Howtek began developing piezo ink jet printers using an ink which at room temperature is solid. Colored ink sticks are heated to their melt point and then ejected on demand

by piezoelectric transducers. The ejected ink "freezes" upon hitting the paper. The printed dots do not bleed into the paper, the colors are vibrant and water-fast. Solid ink prints tend to have a somewhat shiny, beaded look and feel, which are an aesthetic negative for some people and applications. The image is more vulnerable to abrasion and areas of solid color can flake off if creased. In addition, unlike liquid ink jet, the inks are opaque which complicates color mixing and also compromises the quality of transparencies for overhead projection. Several vendors offer optional attachments to reheat the printed ink and flatten it out to improve transparency performance. Solid ink printers require a warm-up time before operation (or are left in standby all the time), and power requirements are therefore higher.

The genealogy of solid ink jet is somewhat incestuous. As with the other ink jet technologies, some key patents have proven enforceable, resulting in a complex network of licensing among the small group of companies dedicated to this sub-technology. Original Exxon development seeded later products from Dataproducts and Trident. Dataproducts also licensed Spectra and Tektronix under certain patents. Howtek was the first out with a product, but was sued for patent infringement by Dataproducts and eventually abandoned its solid ink jet program. The major general purpose implementation has been by Tektronix (cum Xerox) with its well-received Phaser series of medium-performance color printers. Dataproducts (now Hitachi Koki Imaging Solutions) currently leans toward licensing and components. Trident and Spectra license and market solid ink printers for labeling and other industrial applications.

Ink Jet Inks, Substrates. Ink chemistry has been as important to the success of ink jet as the hardware itself. In the early days engineers tended to come up with various ejector concepts and their manufacturing processes, and then go to ink suppliers. Success more often was achieved when inks and hardware were co-developed.

Each subtechnology is linked intrinsically to one fundamental ink requirement. Thus,

> continuous ink jet needs conductivity (for deflection);
> thermal needs bubble formation; and
> piezo needs non-compressability.

Beyond this, ink jet ink science can get pretty complex. It looks like there will certainly never be an all-purpose ink jet ink. The many subtle performance requirements also imply that cloning inks for the aftermarket is challenging. The economics, however, have fueled aftermarket ink activity by numerous vendors. Many early ink jet printers were not actually "plain paper" devices, since a closely controlled sheet was needed to achieve what was considered true letter quality. It was subsequent ink development that enabled ink jet to be seen as a plain paper technology.

In any technology, materials tend to be a compromise among contradictory requirements. Ink jet inks seems to take this to an extreme. Consider drying. Ink jet ink needs to dry fast on the paper, but not in the print head. This "crusting" clogs nozzles. If drying is to be by absorption, nonporous surfaces are off limits. If drying is to be by evaporation, inks need to contain volatile solvent, which raises issues of toxicity, flammability, and corrosiveness which would shorten head life. Small nozzles mean small drops which means the potential for sharper image quality. Yet small nozzles are easily clogged, especially with inks containing solid pigment particles. But if you are looking for light-fastness, pigments have advantages over dyes. And so forth.

These and other ink considerations are summarized below (see chart, page 57, adapted from a presentation by Hewlett-Packard). Considering this matrix of requirements, it is little wonder that ink jet, despite its basic simplicity, after thirty years remains in fairly rapid development.

Ink jet copies have tended to lack water-fastness because most inks are water-based. In wide format, newer solvent inks have become available so that output can stand up better in demanding environments without laminating.

Materials also contribute to the ink/substrate bond, image permanence, and image quality. Champion's "Image Grip" paper, developed for use with Scitex continuous ink jet systems, is said to yield waterfast images at very moderate cost. Epson claims to have developed a new ink/media combination that makes output from its Stylus Photo printers as lightfast and water resistant as conventional silver halide photographic prints.

Attributes. The successful commercialization of ink jet technology has been a long and convoluted journey. Luckily, the theoretical advantages were so compelling that developers persisted. "Simplicity" was seen as a major advantage. But billions of invested dollars tell of an ocean of complexity that lies just below the surface.

These attributes motivate continued high investment in ink jet and continuing advances in functionality. Inks have been a major frontier but rapid progress continues toward better water- and lightfastness. Ink jet has already replaced both impact printing and most desktop lasers in the home and small office. Now it is becoming common for higher volume work group office printers as well. The end is nowhere in sight.

Among Ink jet's Strengths:

-Relatively high print quality with low hardware cost, moderate supplies cost (no ink is wasted, there are no intermediary chemicals or other expendables), and reliability.

-Adaptable to color more readily than perhaps any other digital technology.

-As a non-contact process, surface independent. Ink jet printing is used for wire marking, coding pipes, on hot steel, on a variety of container surfaces including glass and corrugated cartons, and on foods themselves.

-Noise level and energy requirement is low.

-Format flexibility: unlike most electrophotographic and other page printers, the image is usually formatted serially rather than page by page, which means page length as well as page width is at least theoretically unlimited.

B3. Other Non-Impact Technologies

Electrographic Printing. Also termed direct electrostatic, dielectric, or direct charge image deposition printing, this technology uses a paper which holds an electrical charge. The paper, in effect, is a capacitor. The paper is charged and then under program control selectively discharged to form a latent image directly on the paper. The image is then developed by passing the paper through toner which is attracted to the latent image and then fused for a degree of permanence.

One of the first successful electrographic printers was the Burroughs Whippet, developed for the military. Around 100 of these machines were reportedly shipped, beginning in 1955. Two years later A. B. Dick Company introduced the Videograph for high speed label printing. The Videograph used a specialized pin face CRT to direct charges to the dielectric paper. This was one of the more successful implementations of this technology, used by a number of large volume magazine mailers for many years.

Varian Associates pioneered this technology for a printer/plotter introduced in 1967, followed by Gould and Versatec, also for printer/plotters. Electroprint Corporation, later acquired by Markem Corporation, worked to develop a direct electrostatic printer between 1968 and 1973. In 1974 the major commercial manifestation appeared, the high speed Honeywell PPS.

Because some of the technology is in the paper, the hardware is much simpler than EPG. Dielectric paper has a conductive base layer covered with an insulating top coating. It has a more bond-like feel and appearance than some non-impact recording papers. It accepts handwriting and copies well. But it is more expensive than plain bond and there are only a few suppliers. In the Honeywell PPS, continuous paper was fed across a recording head consisting of a row of styli or "recording nibs" and a backup electrode. The styli were selectively pulsed to apply the latent image to the paper, which was developed by liquid toner, squeegeed,

and dried. A "form" could be overlaid from an etched metal cylinder, or the roll paper pre-printed.

The Honeywell PPS was rated at up to 18,000 lpm (30 ips) and the printer/plotters at several inches per second plotting or 300 to 3600 lpm in text mode. Direct electrostatic technology, largely because of the special paper requirement, is no longer used for general purpose document printing.

Ionography. Today's leading developer and proponent of ionography is Delphax, a 100% Xerox-owned company since 1998. Delphax and other vendors have variously termed this technology Electron Beam Imaging (EBI) or, more recently, Indirect Charge Image Deposition (CID).

The roots of ion printing go back 100 years to Ronalds who in 1842 demonstrated a form of this imaging technique in his "Recorder of Atmospheric Electrical Events." Much later commercial implementations were fielded. One of the critical aspects of these early direct electrostatic printers was the air gap between the stylus and the paper. This hard-to- control feature was avoided by an indirect system, originally termed ion printing, which was invented by Fotland and Carrish in the late 1970s at Dennison Manufacturing (later Avery-Dennison). To commercialize the technology Dennison formed Delphax as a joint partnership program with Canada Development Corporation.

In the basic ion printer engine the critical air gap between the print cartridge and the transfer dielectric cylinder can be closely controlled. The latent image is formed on the dielectric cylinder by a resourceful electronic design termed "time division multiplexing," said to greatly reduce the number of driver circuits.

Compared with electrophotographic printing, a great deal of mechanical complexity is avoided in this original Delphax design

Representative Ink Jet Ink Requirements

Ink Properties	Drop Ejection	Materials Compatibility	Image Properties
Stability	Uniform Drop Size	Noncorrosive	Optical Density
Low Viscosity	High Drop Velocity	Plastics Compatible	Color Quality
High Surface Tension	High Drop Frequency	Adhesives Compatible	Fade Resistance
Conductivity	No Orifice Wetting	No Particle Formation	Lightfastness
Long Shelf Life	Non Crusting		Waterfastness
Flammability	Non Clogging		Solvent Resistance
Nontoxic	Thermal Properties		Smear Resistance
No Biological Growth	& Melt Point		No offsetting
Dye Solubility	(for thermal and		Crack Resistance
	solid ink jet)		Media Sensitivity
			Spreading
			Feathering
			Dry Time

Source: adapted by DRA from a Hewlett-Packard public presentation

by transferring the toned image and fixing it all in one step, sometimes termed "transfixing." The conductive toner is nonmagnetic. The paper is squeezed through the transfer station between the hardened steel dielectric cylinder and a pressure roll. The toner is forced into the paper under a pressure of around 300 pounds per square inch.

To confuse things a bit, in 1989 at a technical seminar, Delphax scientists presented a paper describing their core technology as "electron beam imaging" rather than ion printing. Their "ImageFast" series of printers were described as using EBI technology. The basic imaging process does not seem to have been changed, but cold pressure was augmented by a flash fusing station, a system said to avoid the limitations of cold pressure fusing alone. According to Dick Fotland, the name change came from marketing. "When Delphax engineers found that it was mostly electrons that

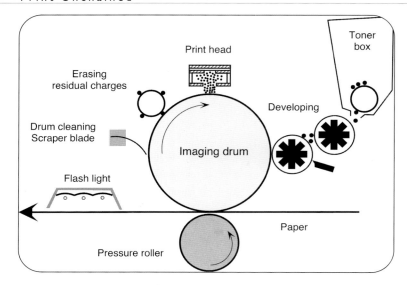

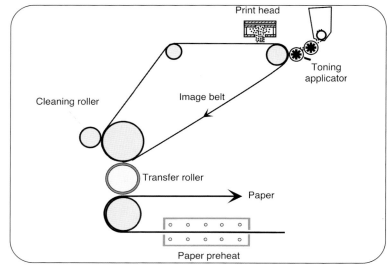

Delphax ion printing (left); Xerox/Delphax Gemini EBI system. (Courtesy of Océ Printing Systems GmbH)

provided the latent image," he said, "the marketing people thought it would be good to come out with a new name for this improved version of the basic engine."

In 1995 Avery Dennison introduced a second generation engine called the Gemini. As the name implies, the Gemini replaces the drum with a pair of belts, the first to hold the latent image and the second, which is heated, to transfer the image to preheated paper.

This technology seems to be solidly entrenched in a number of specialized, high volume applications. Moore Business Forms has been licensed and, working jointly with Delphax, has developed a version called the Midax 300. With this system resolution was improved to 300 dpi with throughput at 300 feet per minute. About 300 of the printers are used around the world in stand-alone applications for addressing, brokerage statements and billing, and as an on-press print station for variable information. Recently more advanced systems were introduced for data processing printing and short run publishing.

Xerox/Delphax believes their duplex 1800 engine, which can print 1300 letter-size images per minute at a resolution of 600

dpi, as of this writing ranks as one of the world's fastest printers. They have a number of major OEM customers worldwide. This technology, in short, may well be the most promising alternative to electrophotography for high volume applications. Color has been a hurdle. They have been working on it, but to date it has yet to be demonstrated in commercializable form.

Magnetography. Another alternative to EPG printing, promising the usual reduced complexity and better reliability, is magnetic printing. The first computer printer implementation was probably by Data Interface and Ferix in the 1974-75 timeframe. The Data Interface DI-180 printer went 180 lpm at a resolution of 180 dpi. The latent image was applied to a magnetically coated tape and toned with a magnetic powder. When a full line had been imaged on the tape, it was stopped, brought into contact with the paper, and the image transferred. Later, in 1977, Data Interface was acquired by Inforex. However, their implementation of magnetic printing never found a significant market and soon disappeared.

The major developer of magnetography has been Bull in France. Bull's magnetographic printers have succeeded in finding

some markets for high speed non-impact printing around the world. Compagnie Machines Bull, a leading European computer company that for many years built its own printers (headquartered in Belfort), seeing the future was in non-impact, refused to join the stampede toward electrophotography. They went their own way, developing proprietary magnetic printing technology in a program that culminated in the Bull 6090 printer, introduced in 1984, followed by a number of other models in various performance categories.

The original technology that led them in this direction was actually from GE, according to a former manager there. It found its way to Bull when CII Honeywell-Bull was brought into the GE-Honeywell Information Systems family in 1970. In 1982, when Honeywell reduced its interest in "C2I" to just 10%, magnetic printing development was well under way.

This technology has steps similar to EPG: a latent image formed on a transfer material. The latent image is toned and transferred to the paper, and fused. Instead of a delicate photoconductor (magnetography proponents like to use the word "delicate" along with "photoconductor"), the transfer surface is a hardened metal cylinder. Instead of a complicated light scanning system, there is a stationary imaging head running the length of the cylinder. The head consists of an array of tiny electromagnets spaced 480 per inch. Developing is with dry toner and fusing by radiant heat or high energy flash. Our understanding is that one limitation is that magnetic toner is inherently opaque and therefore does not have the potential for process color.

The magnetic component in magnetographic toner allows printing of MICR characters without changing the toner supply. Also, the technology is touted as suitable for a wider range of substrates including pressure sensitive labels, tag stock, vinyl, carbonless papers, and foils.

Three printer families are offered as of this writing from the 90 to 110 ppm 910 CE to the 1,500 ppm Varypress+. A recent enhancement is single engine or dual engine duplexing.

Bull spun off this part of their printer business as Nipson, a wholly owned subsidiary. Then, in 1999, they sold Nipson to Xeikon. Xeikon remains the most significant user of magnetographic technology and claims to now have an installed base of around 1,600 units worldwide. Nipson claimed revenues of over $200 million in 1998 and has major facilities in Belgium, France, the USA, and Germany. The only other user of this technology is believed to be Iwatsu, which markets or at least has marketed their 115 ppm, 480 dpi MG 8000 magnetrographic printer in Asia.

Seeking the inside story from a former Nipson executive, I was told, "Magnetography had a colorful past." Despite unique attributes and, according to Xeikon, great potential, the business has not grown over the past few years and has not been profitable. The future of magnetography is by no means assured.

B4. Thermal, Thermal Transfer and Dye Sublimation Printing.
These technologies have played an important role in the evolution of digital printing. They are widely used today, but do not play a significant role in general purpose page printing. Important applications remain in various narrow format listers, and in the graphic arts. Wide format direct thermal for facsimile receivers has been an important office application, a market now being eroded by ink jet and other plain paper technologies.

Direct thermal is perhaps the simplest digital printing technology of all. In this country, NCR was the main pioneer, introducing direct thermal printing devices in the early 1960s. Thermal sensitive paper is moved past an array of closely spaced resistors which when heated under computer control create dark spots on the paper. Most direct thermal printers are fixed head

Other Technologies

For each technology now in commercial use, there are alternative technologies being developed in the background. Will any of them be another xerography, destined to again revolutionize digital reprographics? It is true that a technology can make a market, but only up to a point. A new technology is no match against the deeper tides of history. Timing is everything.

Here are some of the more interesting "alternative" technologies. Some have yet to be commercialized. Others have found a place in specialized applications. Their prospects will be explored a bit more as we look to the future in Chapter 8.

Array Technologies TonerJet. This is an electrostatic technology that makes it possible to eject toner directly onto paper. The target market is currently a desktop printer with 600 dpi at 10-15 ppm, at an end-user hardware cost of under $2,000. Array Technologies, a Swedish company, has announced development agreements with Minolta and at least one other unnamed Japanese partner. In 1997 they announced a refinement called deflection control which boosted resolution from 300 to 600 dpi, and plans for a product called the Chroma Hardcopy Server. There has been commercial exploitation of the technology in a monochrome application but the envisioned color TonerJet products as of this writing have yet to appear.

Cycolor. Mead invented this process back around 1980 and its development story has many twists and turns. It is based on Mead encapsulation technology used for many years in carbonless paper. Billions of tiny microcapsules called "cyliths" are coated onto a donor sheet, the sheet is selectively exposed to colored light to create the image, and then run through pressure rollers with the media to transfer the image. Cycolor was dormant for a while in the early 1990s, but the lab and plant in Miamisburg, Ohio was acquired by CP Holdings (photo kiosks in Japan) and Vivitar in 1996. Since then, there is believed to have been some commercial application for color copying in the U.K. and as a compact photographic print processor in Japan.

printers, with the paper moving at right angles to the array. Horizontal dot resolution is determined by the spacing of the resistors. Vertical resolution is determined by the rep rate of the resistors and the speed of the paper. The paper moves continuously in contact with the resistor array.

The roots of thermal transfer printing (TTP) lie in hot stamping techniques used in the graphic arts for many years. The first digital TTP is believed to be a teleprinter developed by NCR for the U.S. Army in the mid-1950s. In the 1970s several firms in Japan developed the technology further, among them NTT, Toshiba and Oki Electric. By the mid-1980s more than a dozen TTP printers had been launched for diverse applications.

The major innovator that made this technology work has been Fuji Kagakushi Kogyo Co. Ltd. which markets under the trade name Copian. Rather than build machines, they opted for the ribbons and have succeeded in maintaining their position as the dominant supplier. For the North American market they licensed IIMAK and Armor and more recently have begun operating directly in this region. Their US factory in South Carolina is said to employ over 200 people, indicating a significant customer base using this technology.

TTP has been used for both serial and line printers. It is generally considered a non-impact technology, but it is not non-contact. The write head, an array of thick or thin-film resistors, is brought into contact with the substrate to be printed, with an inked sheet or ribbon interposed. Ink from the ink donor material transfers to the substrate when a resistor in the array is turned on. Suitable substrates have included various papers, plastic (including overhead transparencies), and fabric. The process has high tolerance for a range of papers and other substrates, but ink formulations for the donor ribbons are critical.

Closely related to TTP is dye sublimation (or diffusion) thermal printing. This variation of the original "wax" TTP tends to be

used for printing continuous tone rather than binary images. Dye diffusion is currently used for a variety of high resolution color printing applications. Low end machines use three-color, CMY ribbons, overprinting all three for black. Other printers add a fourth ribbon (or band on a multi-color ribbon) for true black. Like wax TTP, resistor density can be up to 300 per inch. However because in dye sublimation it is possible to program each dot for up to 256 gray levels, very high quality color printing is possible. Hence the major application in the graphic arts for proofing. Four-color dye diffusion is very slow, but the image quality is very high.

Wax transfer, in short, melts the ink onto the substrate. In dye sublimation, the dyes are vaporized to create virtual continuous tone images. In wax transfer the printed image more or less sits on top of the paper. In "dye sub" the image penetrates the receiver sheet. The former involves ablation, the latter, sublimation.

Still another variation was a unique resistive ribbon technology developed and fielded by IBM as the Quietwriter in 1984. Output was letter quality, well above that of other thermal transfer printers or impact matrix printers. The character format was 4 mil dots at 240 x 360 dpi. Their patented technology centered on a four layer ribbon which heats under an electrical pulse, transferring the ink to the paper. With less thermal inertia, the heating and cooling could be faster than the conventional thermal print head, offering greater potential speed. The technology was implemented as a correcting typewriter and as a printer with the full IBM 252 PC character set and graphics. Creative. IBM invested in automated manufacturing at its Lexington, KY typewriter plant. But by 1985, for word processing and PC output, ink jet was stirring and the days of the Quietwriter were numbered.

Finally, there is laser thermal transfer. In this process, a laser beam is substituted for the resistor array. A donor material is in "intimate contact" with the receiver sheet, and a precision laser beam causes the colorant on the donor material to transfer to the

Other Technologies

Elcorsy Elcography. This process is based on "electro-coagulation" of liquid polymeric ink on a drum which serves as the positive anode. The writing head has an array of negative electrodes spaced in a current prototype at 200 per inch. The degree to which the ink is coagulated is proportional to the electrical charge applied. The less coagulated ink does not cling as tightly to the drum and is scraped off by a doctor blade, then cold pressure fused to the paper. Dot sizes can be varied and the process is said to be effectively continuous tone.

The basic Elcography concept was invented in 1971 by Adrien Castegnier, working in his basement in Canada. After lying dormant for around ten years, it was revived, and is now in a development phase. Some funding is understood to have been attracted, with major participation by Japan's Toyo ink. In 1996 a 200 ppm, 200 dpi Elcography newspaper press was demonstrated

Océ Eurocolor. Océ has invested heavily in developing this non-photographic electrostatic technology. The basic technology involves a "direct imaging drum" to which the latent image is applied by a series of closely-spaced ring electrodes. Then the drum rotates past seven developer stations, each for a different color, which apply and remove toner as called for by the charge pattern on the drum. The entire image is built up on the drum in one pass and fused in one step. Plans call for using the technology for both a color printer and copier/printer. The process offers appealing advantages and has been demonstrated at several equipment shows. However, as of this writing Eurocolor has not been announced in a commercial product.

Fuji Pictography. This technology is based on a Fuji-developed photosensitive donor material which can be developed as a high quality photographic print under exposure by an array of solid state diode lasers. The CMY component colors are graded in 256 steps to provide 16.7 million color shades according to Fuji Photo Film. As of this writing, the technology has been commercialized in two versions, the Fujix Pictrogaphy 3000 and Pictrography 4000. This process for high quality color prints has already been proven and is finding a niche in color labs where it competes with Cycolor as a dry and relatively fast photographic color print system.

receiver sheet by ablation (in the case of pigmented colorants) or sublimation (in the case of dye colorants). Resistor activated thermal transfer is limited to around 300 dpi, but greyscale is possible. Laser thermal transfer can address dots at close to 3,000 per inch. Greyscale is not possible, but with this resolution is not needed. Color proofing systems based on this sub-technology for the graphic arts have been developed by Kodak, Optronics and Polaroid. In addition, Creo Products recently reported that thermal CTP (Computer to Plate) technology is beginning to look like the dominant technology in that application. The Presstek on-press CTP technology is also thermal. Others argue that proofing in the future will swing to ink jet.

C. Color, PDLs, and Software

During the 1980s, color was a somewhat elusive Holy Grail. Printer technology seemed ripe for color. Non-impact printing with superior graphics and color potential was emerging. Processors now had the power and memory to handle and store the huge quantity of information needed for a bit-mapped, color page. Color and graphics go together. The growth of printable graphics was accelerated by the availability of affordable application software for desktop computers, especially the amazingly successful Lotus 1-2-3. Various graphics standards such as GKS, VDI and VDM were becoming accepted.

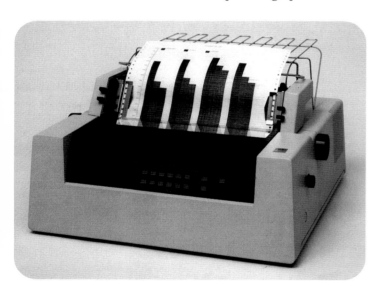

Early desktop color: IBM's Model 3287 impact serial color printer was announced in December, 1979. Utilizing a 4-color cartridge ribbon, this printer created hard copy of graphic data displayed on the IBM 3279 color monitor. (Source: IBM publicity release, 1979)

Earlier, screen color soft output was well established, but color hardcopy remained rudimentary. A main vehicle for computer generated color hardcopy was the pen plotter. Some resourceful vendors were also offering impact dot matrix and thermal transfer printers with multi-color ribbons which could print out a limited number of colors, adequate for charts and graphs, overhead transparencies, and other business graphics. The photographic quality output streaming from today's wide format and desktop ink jet printers was beyond anyone's dream. The color printers of the time, except for the pen plotters, were primarily monochrome printers adapted to print rudimentary color. An early example was the IBM 3287 Color Printer, a serial dot matrix printer introduced in late 1979.

With the advent of ink jet and dye sublimation printing, the impact color devices quickly disappeared. However, it took much longer than expected for color digital printing to drive the market explosion some analysts had been predicting for years. In fact, color has only emerged as a major new digital printer market within the past five years or so.

There are still challenging issues, especially color management. The soft, RGB image on the monitor needs to be converted to CMYK for the hardcopy device. The pigments and dyes used in toners and ink jet inks will not produce the same visual color response as the monitor. RIT's Frank Cost touches on the challenge, writing, "Every color printing process has its own peculiar way of rendering color. Nearly all of them use cyan, magenta, yellow and black colorants. However, there is no standard definition of any of these colorants. Among the hundreds of digital color printing devices currently on the market, many different colorant sets are in use."[*]

The color potential of the various technologies has been touched upon above. The ability of ink jet to deliver slow but surprisingly good color has helped drive demand for that technology.

[*] Frank Cost's *Pocket Guide to Digital Printing* contains an excellent semi-technical introduction to digital color (see Bibliography).

Color electrophotographic printers are generally faster but also much more expensive than color ink jet. Rapidly declining prices, however, have been closing the gap. Color is nevertheless expected to continue to work in favor of ink jet and accelerate the growth of its technology share over the coming decade and beyond.

Software and Fonts. Non-impact printing unchained digital print from the restrictions that characterized impact printing for three decades. Output was no longer constrained by a fixed set of characters, or by the relatively large spot size and positioning increments of impact matrix printers. The new technologies offered the potential to place ink almost anywhere on the page, with greatly increased resolution. Graphics and color were now possible.

But not with the familiar old set of ASCII characters and control codes. A major enabler gating widespread, affordable graphics and color was the development of page description languages, scalable fonts, and related software. This was as big — or perhaps a bigger challenge — than the print engine itself. The major vendors invested heavily. Xerox came up with Interpress. Hewlett-Packard with PCL — Printer Control Language. IBM with IPDS (Intelligent Printer Data Stream) for high end printers and PPDS (Personal Printer Data Stream) for small printers. Adobe around 1985 introduced PostScript. In Part Two innovators Dick Hackborn of HP and John Warnock of Adobe offer some perspectives on the development and significance of such page description languages.

Many vendor-specific page description languages are still widely used. Each vendor is faced with decisions that include how adaptable the printer will be to a variety of applications and host systems, at the time of introduction and in the future. Michael Osborn is a consultant who as Granite Systems specializes in printer software. A while back he described how he saw HP's software strategy with their PCL page description language — how they

have managed to move beyond the simple choice of "open" vs. "closed."

"HP, like everyone else, when developing a new system, is faced with the question of whether to hold it close or open it," he explained. "Now, with their PCL-6, they are effectively playing both sides. Since they control the PCL specification they can make enhancements and be the first to market with them — one of the chief benefits of a closed system. Subsequently they can make the new PCL specification available to external developers so that the new enhancements and supporting applications proliferate, gain market acceptance, and become essential customers — one of the benefits of an open system."

The nearest thing to a standard PDL over the years has been Adobe's PostScript, at least among high end printers. It helped break a logjam, serving as the interface between application software and the printer — a common point of reference. Software engineers could finally write their applications keyed to PostScript. Printer engineers could design their RIPs or other printer-resident software keyed to PostScript. Previously, application software engineers had to figure out the requirements of every printer that might be asked to run a given application. Some printer vendors at the time argued that their PDL was better than PostScript. What is really better for the industry as a whole, and the user community, is the protocol that for whatever reason becomes a standard.

With the introduction of Postscript in the mid-1980s, device-independence was achieved. The language was forward looking (John Warnock's word was "extensible") in that it brought graphic arts standards to office printing. The original and the many upgrades since accommodate various dot resolutions and color as well as monochrome applications, integrating fonts with graphics, and scalable, outline fonts. With the initial release, Adobe was able to offer a wide selection of fonts under contract with Linotype and Bitstream.

At the time, many in the industry thought PostScript was overkill, that the office had no need for graphic arts quality, that licensing Postscript was not necessary and that it placed too great a processing load on the printer. At the time, for many applications, all that was true. It still is for many printer applications. But time subsequently justified Adobe's forward-looking vision.

Now that is changing, Marco Boer of I. T. Strategies believes. "Today PostScript is overweighted," he recently noted. "It got too bulky, it was like adding too many floors to a building without shoring up the foundation. HP's PCL also got too bulky, so when they introduced PCL-6, they started over with a blank slate and made it modular, hoping it will be continually upgradable to accommodate ever-greater processing loads."

This first spec sheet for the original Centronics dot matrix printer was issued in 1970. The graphic is an airbrush rendering because the printer had not yet been built. Introduced as the Centronics 101, it delivered 165 cps for a quantity one OEM price of $2,400 (over $8,000 in today's dollars): breakthrough price/performance at the time. We have come a long way since then. (Courtesy of Robert Howard)

How Far Have We Come?

Some Perspectives on Speed, Specsmanship and Money

How far have digital printers come over the past fifty years? Asking this question, it's natural to compare them in terms of price-performance. Sounds simple. But once you get analytical, things get pretty slippery. It becomes apparent that however this problem is approached, subjectivity and over-simplification are inevitable. Instead of neat, objective sets of comparisons, the value of this exercise becomes primarily the questions raised and terms clarified, useful in their own way to both vendors and users.

From the beginning, "print speed" has been a basic printer spec, one that presumably relates to performance or value. We all love speed. So be it cars or printers, one of the first questions likely to come to mind is How fast is it? Or, What's the world's fastest digital printer? Or, over the past fifty years, how has printer speed progressed?

Representative Fast Printers, 1999		
Manufacturer and Model	Technology	Rated Speed
Scitex (formerly Diconix) VersaMark 90/500	continuous ink jet	706 ppm
Xerox (Delphax) ImageFast 900 (duplex)	electron beam	900 ppm
Océ (Siemens Nixdorf) PageStream 1060 Twin (duplex)	electrophotographic	1060 ppm
IBM InfoPrint 4000 (duplex version, Hitachi engine)	electrophotographic	708 ppm
Nipson/Bull Varypress M700	magnetography	700 ppm

In the case of today's high end digital printers, speed is usually specified in pages per minute (ppm). For some of the models which in 1999 were near the top of the speed heap, see sidebar, page 65. More recently, Scitex claimed 2,000 ppm for a new VersaMark model, which may well be the world's fastest to date.

Immediately, questions come to mind. How big is the "page?" How dense can the image be? How rich? Are we talking text only, or graphics, or graphics including color, and what quality color? And what about archivability and the kinds of media that can be handled and does it print sheets ready to bind or continuous paper that needs more post-processing? Is there automatic duplexing and collating?

In short, a broad view across all types of printers and back into history shrinks "speed" to a very limited concept, a subset of performance. Like so many terms, speed means different things to different people and presents rich opportunities for specsmanship by the vendor. Bringing up the question with Dick Fotland of Illuminare, Inc. triggered some helpful reflections (see sidebar).

Speed

One problem of comparing printers over time and among differing technologies and performance categories is that vendors have used different bases for specifying speed. The most common have been characters per second/cps (serial printers), lines per minute/lpm (impact line printers), and pages per minute/ppm (most non-impact printers). At the very high end, some digital production printers and digital presses are speed specified like the traditional presses they hope to replace, in paper speed, i.e. feet per minute times web width.

From the overall system standpoint, the system engineer might see speed as the rate at which digital data can be formatted and stored and fed to the printer. To echo Dick Fotland, How many megabits per second of printable data can the printer gobble up? This system perspective brings up another dimension of value, i.e. what is the distribution of intelligence between the printer and the host system? Do we need a standalone controller/RIP or major system upgrade to feed the printer?

Electronics aside, many of the mechanical and format variables that bear on speed are discussed in the Technology Overview section of this chapter.

For the purpose of this comparison, there needs to be a common denominator that will apply to all technologies over the entire fifty years of this history. What seems to work best is the area that can be printed over time: printable square inches per second, or, as suggested by Fotland, "SIPS."

Converting specified lpm's, cps's and ppm's to SIPS means making a number of assumptions. Not a problem, since after all, the vendors' speed specs are based on assumptions as well.

Looking at the question of speed and performance, Dick Fotland recalls positioning a non-impact page printer against impact line printers in the late 1970s and early 1980s. Among his observations:

We developed ion printing at Dennison Manufacturing for variable tag and label printing applications. Dennison did not want to get into the hardware business so we formed Delphax Systems with Canada Development as a partner. Delphax was started as an OEM manufacturer and their early customers (Datagraphics, Southern Systems, and Xerox) thought this non-impact machine might be a replacement for impact line printers. The Delphax machine, after all, was quiet, printed multipart forms without carbons so every page was an "original", and had a speed in the 60 page per minute range. A disadvantage was that it cost more, around $50,000 or about double the price of a high speed line impact printer.

The ion printer appeared faster but only when the line printer was printing a large number of characters per page on a single part form. When there were no characters on a line, a line printer would slew at high speed to the next line that did have characters. If the job called for a six-part form, that effectively increased the speed to 3600 lines per minute. And the effective speed would increase to 120 pages per minute when printing about half a page, or 30 lines of characters, of the six part form.

But with impact multipart form printing there is waste, both in labor and paper. With carbon forms, every page in a print run

To start, we need to assume the content is monochrome characters because that is all most early printers could print. Going to printable dots per square inch would exclude the early "full character" impact line and serial printers.

Density

The next assumption is character density. During the first few decades most printers were limited to characters. (Graphics were primarily the realm of plotters.) Character density was generally 10 per inch horizontally in lines spaced 6 or 8 per inch which comes to 60 or 80 characters per square inch. Then, as matrix and non-impact printing grew, ASCII character codes and monospaced characters gave way to proportional character spacing, scalable and outline fonts, bit image page description, graphics and color. With page printers, dot density has climbed from 240 or 300 dpi to 600. It goes higher with the latest ink jet printers, such as the Epson Color 900, rated at 1440 x 720 dpi in photo mode. All of this lets you pack in many times more information.

In the "Then and Now Gallery" that follows, two low-end printers – the Digital LA 36 DECwriter II and Lexmark 1100 – are compared factoring in dot density: i.e. dots per square inch per second rather than characters per square inch per second. This greatly enhances the speed gain between 1975 and 1999. Impact matrix printers typically could print perhaps 5000 dots per square inch. This compares with 600 x 600 for the Lexmark and similar ink jet printers today, for 360,000 per square inch. Multiplying these packing densities by their respective SIPS, we get an improvement of well over a 400 times.

Money

Next comes the dimension of value. Printer pricing comes out even more slippery than speed. Announced list price? Street price? Wholesale or retail? To the end user as part of a system or standing alone, or to an OEM? What quantity? Including software and maintenance? Does it need a separate controller or RIP? And increasingly important, what are the operating costs: supplies cost for media and chemistry, labor, maintenance? For the "Then and Now" gallery, again, such questions are deferred and prices given are generally the announced end user single unit list prices at introduction. And to offset inflation, all prices are adjusted to 1998 US dollars as per the change in the consumer price index.

The resulting price-performance index is a formula that looks like this in words: printable square inches per second over printer price (in 1998 US$), or, for convenience, SIPS/$, or just the "Index."

Looking again at the two printers compared in terms of printable dots per square inch (the 1975 DECwriter II and 1999 Lexmark 1100), the gain in terms of value has been even greater. How far have we come? In this case, in terms of speed over printer cost, the improvement is over 4000 times!

One somewhat puzzling revelation from would have to print the same number of parts. Years ago at Dennison I'd get six parts of a report and throw five away. The systems manager informed me that they had to print all six parts since someone in finance needed six parts of another report printed in the same run.

With a page printer, each page can be printed in the number of copies that are needed. It was this feature, together with report distribution software that really made the Delphax line printer replacement machine a great success in the mid 80s. And speed continues to be an issue, even where print volume is low. Look at transaction printing. People are waiting. High burst speed is required in applications such as ticketing and airline reservations.

With page printing and graphics, speed is generally specified in pages per minute. This is basically an index of surface area over time, but doesn't tell much about how much data can be converted to a visual image. In the new highest speed printers, it seems more appropriate to talk about the speed at which we can accept and print data. In these terms, we might say the new Delphax/Xerox 1300 duplex web printer eats data at the incredible rate of 720 megabits per second!

this exercise is not so much the speed/$ gains over time, however, but rather that low end printers now seem to deliver so many more SIPS for the dollar than the high speed printers. How can this be?

First of all, the results are skewed by the jungle of assumptions on which they are built. But beyond this, some conclusions seem safe. One is that SIPS per hardware dollar reflects in part (1) the economies of mass production and (2) the degree to which low-end printers are sold at or below cost to gain market share and high margin supplies revenues. But more importantly, this exercise demonstrates that speed per dollar of hardware cost is an interesting analysis, but from the point of view of value, far from the whole story. We have mentioned some of the larger performance considerations. A more complete value analysis would look at the life of the printer and the cost per page including supplies, service, labor, floor space and many other factors.

Dick Fotland, from his perspective at the high end of the industry, is very clear on this. "If you look at hardware cost amortized across the life of a high-end system and the millions of pages they produce and the supplies costs of the pages, you certainly have a much lower cost per page. It's cost per page that

counts," he asserts. "The limitation of the 'SIPS per hardware dollar' analysis is especially evident when you look at the low-end machines where you have to purchase a new ink jet or laser cartridge — in effect the heart of the print engine — every few thousand sheets." Cheap printer; expensive supplies. Like so much in life, a trade-off.

The importance of machine cost vs. operating costs of the life of the machine, in short, varies depending on which end of the market you are looking at. For the small volume, home user, cost per page means something, but the up front cost of the machine will be much more important. At the other end, for a high end production printer, as stressed by Fotland, it is total operating costs over the life of the system that really counts.

The Gallery

As a measure of progress over time, SIPS, though limited, does tell a useful story. The price/performance gains indicated by the Index are real. But they have been nowhere near that of computers. Although printers have become ever more electronics-intensive, there are still physical materials to move, so we are limited by inertia and other immutable laws of physics. And a great deal of value has been added in terms of non-speed features, which negatively impacts the speed-to-printer-cost ratio.

How to pick the printers to be compared? One indication of the amazing number of printers that have been released over the years is the SpecNet program of the business forms section of Printing Industries of America, Inc., the major trade association of business forms manufacturers. There are over 17,000 printers and other forms processing devices now included in this database.

The models selected out of this huge population are felt to representative of the various eras, or to have advanced price/performance, or to have enjoyed exceptional commercial success. It seemed to make sense to try to include printers from a major ven-

---

SIPS Metaphysics

The vendor's ppm spec for most page printers is based on a given printed format: area of the page printed and density of the printing. The lpm and cps spec for most impact printers is based on the printable format rather than the normally printed format. So, if SIPS is chosen as a common speed denominator, in converting from ppm to SIPS, what format do we assume? The question seems almost a Zen riddle: printable square inches vs. typically printed square inches.

To convert ppm to SIPS, a print area of 6 x 9 inches of the 8.5 x 11-inch page is assumed, with 100% coverage of that area with characters. This departs from most benchmark formats used today which assume perhaps 5% or 20% coverage (although this may be solid ink, as in a photograph, rather than characters.) Although these assumptions are arbitrary, they are applied to all technologies, so the playing field is more or less level. This makes it possible to come up with a rough speed comparison regardless of printer technology.

dor, which often led to IBM. Particularly in the early years, IBM was both an impressive innovator and the dominant vendor. It also helps that IBM has not generally marketed printers to other systems companies (OEMs) so its published price has normally been the single unit, end user price.

Looking back over the decades, one insight that emerges is how unbelievably hungry we were for printed hardcopy. The answer on the CRT screen just didn't do it. When adjusting for inflation, it comes out that users were willing to pay over $300,000 in today's dollars for a noisy 1100 lpm IBM 1403 impact line printer. And hundreds of thousands of users plunked down today's equivalent of over $5,000 for a keyboard terminal such as the DECwriter LA 30 which could produce just 30 black, coarse matrix characters per second.

THEN AND NOW GALLERY

Date: The year of introduction in most cases;
Price: (a) Quantity one, end user list price in US$ (deduced from OEM price in some cases);
(b) Inflation-adjusted to 1998 equivalent US$*;
Speed: Speed as specified by the vendor;
SIPS: Printable square inches per second, for black characters. When the vendor gives speed in ppm, 8.5 x 11 inch page size with 100% coverage of a 6 x 9 inch area is assumed;
SIPS/$1000 Computed by dividing SIPS by the quantity one list price in US$1,000's.

*Inflation adjustment:
Pre-1975, Consumer Price Index (CPI) as documented in "Historical Statistics of the United States," (US Government Printing Office, 1975) 1975 ff: "Statistical Abstracts of the United States."

A. Super Printers

1. Radiation, Inc. Series 690
1963
For raw speed, this production printer was way ahead of its time. It was configured somewhat like a printing press, printing from a roll of paper. Technology was electric discharge, meaning sparks were emitted which literally blasted off a light colored surface coating on the paper to reveal a dark substrate. The trade-off was speed for marginal print quality on an unpleasant paper. Few, if any, of these printers actually got sold.
$300,000+; 1998 equivalent $1,581,336+
Speed: 31,250 lpm (assume 8 lpi, 12 in print line @10 characters/in.)
781 SIPS
.49 SIPS/$1000 (1998 $)

1963: *Radiation 690*

1976: IBM 3800

1999: IBM InfoPrint 4000

2. IBM 3800
1976

A pioneering IBM non-impact production printer. Laser
electrophotographic technology on roll paper. As of 1985 over
2500 systems had been installed worldwide.

$310,000; 1998 equivalent $918,463

Speed: 13,360 lpm at 8 lpi (13.6-inch print width);
paper speed 32 ips less stopping each 72 inches to
accommodate 2-inch slot in drum for photo
conductor; assume net of 28 ips; 13,360 lpm at 8 lpi correlates
to 27.8 inches per second); all this said to translate to around
215 ppm.

193 SIPS

.21 SIPS/$1000 (1998 $)

3. IBM Infoprint 4000, Duplex version (Hitachi print engine)
1999

Laser electrophotographic technology; duplexing system consists of
two printers working in tandem.

$877,000.

Speed: 708 ppm (letter size) 2-up x2 printers, up to 17" print line
each printer

637 SIPS

.72 SIPS/$1000

B. Mid-range Production Printers, Work Group Printers

1. IBM 407/716 Printer
1958

IBM's first transistorized large scale computer, the 7090, announced in
1958, offered up to eight input-output channels, each of which
could serve up to ten magnetic tape units and other peripherals
including one printer. The printer was normally the 716, an
adaptation of the IBM 407 accounting machine, an impact line
printer with 120 print positions.

Est. EU price: $30,000 (no controller); 1998 equivalent $169,491

Speed: 150 lpm across 120 print positions

5 SIPS

0.03 SIPS/$1000

1958: IBM 716

2. IBM 1403 Model 2
1961

Impact line printer; type
element is a horizontally
rotating chain of type slugs;
a "back printer" in which
hammers strike the paper
from the back, pushing it
against the inked ribbon
(wide, "towel" ribbon) and
the selected character slug
as it moves by on-the-fly.

$39,600 plus one third the cost
of the controller ($63,600)
which can serve up to three
printers, or $60,800. 1998
equivalent is $332,240.

Speed: 1100 lpm across 132
print positions, or up to
1285 lpm, with reduced
character set (15, numbers
and five symbols); 10 characters per inch, 6 lines per inch.

40.3 SIPS

0.12 SIPS/$1000

1961: IBM 1403

3. IBM 3820
1985

Laser electrophotographic
page printer, based on
engine believed
sourced from Minolta.
Duplexing capability.
Networkable (LAN
and PC via
SNA/SDLC); Advanced
Function Printing.

$30,000. 1998 equivalent
$45,455

Speed: up to 20 ppm
simplex

18 SIPS

0.4 SIPS/$1000

1985: IBM 3820

4. IBM InfoPrint 20 Laser Printer
1998

Workgroup printer announced 3/98, based on Fuji-Xerox engine. Laser electrophotographic page printer. 600 x 600 dpi plus resolution doubling print mode (1200 dpi "emulation") with no impact on throughput. 70,000 page per month duty cycle. $2,259 quantity one list price at introduction.
Speed: 20 ppm on letter-size sheets, simplex
18 SIPS
8.2 SIPS/$1000

1998: IBM Infoprint 20

❖

C. Low-End Printers

1. Teletype Model 33
1962

Teletype's most widely used printer, with over 750,000 shipped from introduction until discontinued in 1981. Serial impact printer; small, non-interchangeable type cylinder scanned across the print line printing through an inked ribbon.
OEM price around $600. Assume EU price $1800. 1998 equivalent $9,730
Speed: 10 cps. (assume: 6 lpi, 10 pitch , for 60 characters per square inch)
0.166 SIPS
0.017 SIPS/$1000

1962: Teletype Model 33

2. LA36 DECwriter II
1975

One of Digital's most successful printer products, claiming 200,000 shipped by 1982. Wire matrix serial printer, single head scanned across print line, printing through inked ribbon.
$2,970 quantity one price in 1981. 1998 equivalent $5,322
Speed: 30 cps. (assume 60 characters per square inch)
0.50 SIPS
0.09 SIPS/$1000

1975: LA 36 DECwriter II

3. IBM 4201 Proprinter
1984

An IBM gambit to counter low end printers from Japan, aggressively priced and manufactured in the USA at a highly automated plant in Charlotte, NC. Named "Printer of the Year" by Printout newsletter. Serial impact matrix printer.
$549 quantity one list price. 1998 equivalent $862
Speed: 200/100/40 cps (top speed is a burst speed, 40 cps is NLQ mode; assume 100 cps and 80 characters per square inch)
1.25 SIPS
1.45 SIPS/$1000

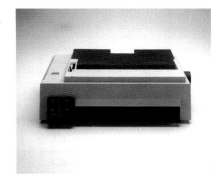

1984: IBM 4201 Proprinter

4. Lexmark 1100 Color JetPrinter
1998

Successor to the Lexmark 1000, one of the first sub-$100 printers. Serial color ink jet technology. Resolution up to 600x600 dots per square inch. Pacesetter, at the time, in affordability, for home and home office.
$89 after rebate
Speed: up to 3.5 ppm black, 1.5 ppm color (for this comparison, assume 3 ppm black character printing)
3.15 SIPS
35.4 SIPS/$1000

1998: Lexmark 1100 Color JetPrinter

COURTESY OF COMPAQ COMPUTER CORP.

Hewlett Packard LaserJet "Classic"

1984

Breakthrough low-end laser printer based on an engine derived from the Canon Personal Copier with a throwaway cartridge. The first desk-top electrophotographic page printer. A new price/performance standard. But still subject to that immutable trade-off: cheap printer, expensive running cost.

This printer was such a great leap forward in price/performance that it didn't fit into the Gallery proper. Introduced in May, 1984, the LaserJet offered 8 ppm at a list price of $3,495. In performance, it fell between the mid-range work group printers and low-end printers; it represented a new class of printer. The closest page printer in the Gallery is IBM's 3820 (vintage 1985) rated at 0.4 SIPS/$1000 in the mid-range class. The LaserJet beat that by 450% (in terms of its list price in 1998-equivalent dollars).

$3,495 list price. 1998 equivalent $5,295
9.6 SIPS
1.8 SIPS/$1,000s

Photo credits: Radiation 690, Teletype 33, and IBM 716, 1403, 4201, and 3800 are photos from contemporary news releases (DRA Archives); other printer photos are reprinted courtesy of their respective manufacturers.

The SCI Rotary Printer

One anomalous printer is included in this Gallery simply because the concept is so ingenious and price/performance so far above the norm for the time. This family of printers was introduced in 1977 and discontinued in 1989.

Three multi-wire stylus assemblies are mounted 120° apart on a plastic rotor. The paper is formed by the platen into an arc and moved over the rotating print head which prints dot matrix characters by selectively removing the paper's aluminum surface coating to uncover the dark substrate. High speed, low mechanism cost, and reliability are featured in part because there is just one motor which drives both the paper and the rotating print head in one continuous, synchronized motion. The head speed in the Model 1100 is 1,800 rpm and paper speed 10 inches per second. Were the paper wrapped completely around the print head, the characters would be configured in a continuous spiral.

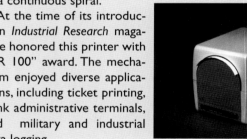

At the time of its introduction *Industrial Research* magazine honored this printer with "I-R 100" award. The mechanism enjoyed diverse applications, including ticket printing, bank administrative terminals, and military and industrial data logging.

SCI Rotary Printer, Model 1100

1977

Ingenious non-impact printer designed and patented by SCI Systems of Huntsville, AL
$300 for OEM mechanism, $995 EU list price. 1998 equivalent $2,236
Speed: 2,200 cps (70 characters per square inch)
31.4 SIPS
14 SIPS/$1000

Synergy and pragmatism: IBM in the early 1960s used an adaptation of their Selectric typewriter for the I/O terminals in the American Airlines SABRE reservation system. To win the contract for this groundbreaking commercial real-time network, IBM leveraged computer and terminal resources with their earlier involvement in the SAGE air defense system. Photo courtesy of IBM Archives.

The Industry

Industry by nature is dynamic. A fifty year overview of any industry is likely to reveal a seething complex of entities – gambling, winning, losing, bursting into view, resting on their laurels, merging, divesting, fading into obscurity. An industry is an amorphous, moving target. Any analysis is likely to call up visions of that overused metaphor of the blind men trying to describe an elephant. The printer industry is certainly no exception.

In this section of the overview, we will describe at least some aspects of the elephant by getting a feel of three parts:

First, how big is it? This question serves as a pretext to ruminate on the "numbers game" played in the information industry, as an introduction to our quantitative, 50-year industry growth overview.

Next, what has been the structural and competitive evolution? Which companies have dominated various decades?

And finally, who have been the company players in this drama? What kinds of companies have developed and commercialized products, and how has the number and geographic base of companies changed, revealing the extent of a remarkable global shift over the years? Then – depending on your viewpoint – we view the industry as either well along its way to maturity or currently opening the door to a new wave of growth by merging with or infiltrating what to date have been considered other industries.

THE NUMBERS GAME

Humans looking for answers seek out numbers for the certainty they seem to offer. In accounting there is no arguing with the bottom line. But what about something as dynamic as an industry? Here numbers have their limits. Nevertheless, industry spends billions of dollars trying to quantify supply and demand, looking back as a way to project forward, with their own in-house analysts or, more and more, outside market research groups.

The first research companies tended toward engineering, helping client firms with product development and manufacturing techniques. Leading examples are Battelle Memorial Institute and Arthur D. Little. But products mean little without markets, so these same companies soon found themselves researching markets as well. Later they were joined by a new genre of market researchers who were more management oriented, who found they could grow a business with custom consulting for the industry and publishing multi-client or broadly marketed printer market reports and newsletters.

Among the first to track the printer industry were Frost and Sullivan and Auerbach Corporation, both of whom published some printer shipment estimates beginning in the 1960s. In succeeding decades a host of other industry watchers emerged and faded, some with colorful names such as Creative Strategies, Gnostic Concepts, and Quantum Science. Perhaps a half dozen grew and diversified and continue to live on as multi-million dollar corporations. Like the industries they track, the research companies can be a complex tangle of acquisition and spinoffs and their entrepreneurs as colorful as they get (see sidebar, page 78).

Damn Lies. "There are lies, damn lies, and statistics." This is how an industry editor opened a column in which he vented his frustration with the market research industry. "Statistics can reveal almost anything, he continued. "The competition in selling market reports is a tough one. Too often, it seems, there are market reports, industry outlooks, economic forecasts or sector analyses that reveal opposing results. Which are the industry to believe and, indeed, base business decisions on?"*

Market statistics, in short, have their limits. Yet anyone reading an industry history deserves at least a bit of quantitative documentation addressing questions such as How big is it? How fast has it grown? So we offer our analysis, below. But not without first putting market numbers in their place with the following autobiographical interlude. It is a story of another research company, quite small, which served the industry for around ten years.

Datek Information Services was formed in 1978 by this author to publish *Printout* newsletter. Early on, unexpected liabilities arose. To keep publishing I decided to sell the newsletter to an investor, a valued Wall Street mentor. This helped, but after a while it seemed I was doing most of the work and he was getting most of the money.

Fortunately, I didn't give up. The newsletter, I reasoned, was a means to an end. It provided a pool of prospects who knew and respected us and to whom we could market more profitable research products. In 1978 Datek published its first report, *IBM 3800 and Competing Page Printers*. The price was $200. It sold well, went through three printings. Other reports followed, and then conferences. Things still tended to be underpriced, but our customers loved us for it. We worked hard to give value as well as quality. A good basis on which to grow a business.

But compared with some of the more aggressive competitors springing up, we didn't run as fast with the formula as we might have. In a few years Datek expanded to a friendly group of five or six and had outgrown the in-home office. I had hired Jonathan Dower early on. He had no experience as an industry researcher-writer. But he did have a quick and analytical intellect, what

---

* Duffy, Cameron, in *Packaging Today*, September 1996 issue.

seemed to be a photographic memory, and best of all, a certain fire in his eyes. Through the Datek years he was a key colleague.

It was probably around 1981 that we had the "numbers" conversation. It went something like this:

"Ted, if we're going to grow we have to get into numbers."

"Numbers?"

"Yeah. The newsletter makes some money, but Bob gets most of it. The reports and the print quality conference made some money. But the real money is in market numbers."

"You think we should compete with all those big market research groups like Frost & Sullivan and IDC? I'm not sure we have the resources. We're more qualitative than quantitative. Anyway, the numbers is a sort of hokey game. People come up with statistics. Their definitions are unclear. No one can prove they are wrong. The government with the Census of Manufactures gives market numbers away almost for nothing. And these research companies sell it for big bucks."

"Sounds good to me. Ted, we've got to do it to grow. Look, it's an *informed* best guess. That's a legitimate service. Our best guess can be as good if not better than theirs."

Hmmm, I thought. Best guess. No one can prove you are wrong. It's where the real money is.

So it was that Datek began publishing market numbers as the feature article in the *Printout Annual*. The *Annual* was free to newsletter subscribers. We were giving the numbers away. But the *Annual* was well-supported by advertisers. And our image as a provider of thoughtful quantitative as well as qualitative information to the industry grew and opened the door to some market research consulting as well. It all started from a relatively unprofitable newsletter we did not own.

So much for confessions. What about this effort to quantify fifty years of industry growth? Here's how it went.

The larger research companies that have been tracking printers over the decades were generally uninterested in opening up their archives and may not, in fact, have maintained official archives. But there were other resources. I T Strategies has developed a deep pool of industry statistics over the years and made their data available. A good source for the 80s were archived Datek *Printout Annuals*. And there was also a collection of other market estimates by various researchers, some recent, some dating back to the 50s.

Most give just a thin slice of the picture. There is the geography slice. Some researchers cover only the U.S. or North America. Others the global market.

Definitions, as always, are key. What is the scope of this industry? What is a "manufacturer?" Do dollar value totals mean manufacturer revenues or final (after channel mark-ups) sales volumes? Are revenue figures inflation- and exchange-rate adjusted? Such questions are often left unexplained in reports.

The government's Census of Manufactures has been pretty consistent on defining the "industry" as the companies that add value by actually fabricating tangible materials. In these terms, with most US vendors now outsourcing their print engines, it would appear the US printer industry has been shrinking down to almost nothing. By this definition, is developing a RIP or printer driver "manufacture?" How about attaching your company nameplate on the cabinet? Asked this question, the Census analyst with oversight for the relevant SIC responded with something like, "We're working on that."

Anyway, by putting all the slices together and making some judgments, it is possible to come up with informed estimates. For example, over the years there have been more studies of computer

23 HURT AS COLOR MARKET EXPLODES

(Newtonville, MA) A crowd of more than 100 industry pundits watched in horror yesterday as 23 forecasters were injured during an unexpected color market explosion at the annual convention of the Flat Earth Society. Noted seismologist and ex-newsletter publisher Ted Webster attributed the explosion to upper plate tectonics. "In spite of the confidence you may hear in our voices, we don't really know what's beneath the surface of such fragile forecasts," Webster noted. The breakfast-hour blast left egg on the faces of diners in the nearby Cottage Doughnuts restaurant.

The nature of the injuries was severe – particularly gruesome were injuries to those whose necks were on the line. Four received cuts requiring sutures as crystal balls shattered and sprayed the scene with shards of glass.

Most of the injured were waiting for busses and cabs near the entrance to the convention center. The number of casualties no doubt would have been much greater had not many fortunate forecasters just departed the area on a Paperless Office bandwagon.

Clipping excerpted from a mock Printout *newsletter written by the employees of Datek Information Services, Inc. in March, 1988.*

shipments than printer shipments. A ratio of printers per computer can be used to extrapolate. As can estimated ratios between the domestic and the global market.

Double counting can be a trap (see box). HP sells a huge volume of LaserJets. The basic print engine was developed and is manufactured by Canon. Who is the "manufacturer," Canon or HP or both? Sometimes there are three layers. Dataproducts bought their laser engine from Fuji Xerox, added drivers, and sold it to Apple. The last layer in this process is the OEM – the vendor who places their label on the printer and is responsible for distributing and supporting it. Each of these layers is an "industry company." But if you add up the printer-related revenues of each, rather than their *added* value, you come up with an inflated total. Today industry companies are buying and selling to each other more and more, making the dollar value of the global industry even more difficult to establish, especially if the basis is to be consistent through all five decades.

In the first two decades, research was tough because so many printers were bundled with the system. A trend now that will make it tough in the future is the "convergence" noted elsewhere.

A Hypothetical Example –

Laser print engine, price to US customer	$1,000
Complete printer with drivers, software, to OEM customer	$2,000
OEM to dealer	$3,000
Dealer to corporate end user	$5,000

If this one, three-tier example were the whole industry, what would be the $ volume? It depends in part on who is included in the "industry." Is the industry total $5,000? Or should dealer mark-up be excluded? Maybe it should be $6,000, the revenue total for each of the three manufacturer levels. Or maybe $3,000, the top level OEM's revenue.

"Digital printing" is fast becoming personal computer printing, digital photography printing, CAD and plotting, digital publishing, copying, and on and on. Should a multifunction device be reported as a printer, a scanner, or a copier? If copier industry market numbers are added in, for example, the printer market figures will make a quantum leap. (Our estimates do not include copiers.)

Most market studies today deal with the estimated market value of printer shipments, sometimes termed the "if sold" value. This avoids the double counting pitfall, and this is the basis of our industry growth estimates.

It might be be added that the slices of market data in our archives – data published over the years by various analysts – may not be really so hard in any case. A seasoned independent research subcontractor to some leading research companies has over the years learned that sometimes a client's business imperatives overrule research and analysis in any case.

"Look, just double all those numbers," she was told by one of her clients for whom she had just completed a report. "We just can't tell our client their targeted market is so small. Heads would roll." So the numbers got doubled.

We acknowledge these and other pitfalls and have done our best to deal with them. Although all market numbers are informed estimates, the trends are solid and tell a tale of rapid growth and change, such as average inflation-adjusted cost per unit plummeting from $53,000 to under $800 over the decades.

Please don't take this excursion through the numbers game as an attempt to discount quantitative research. This is a vital facet of market research and planning. The goal is rather to demystify the results and help those who rely on such reports to be more discerning research consumers.

Fifty Year Industry Growth

Products covered are general purpose printers (A4 and larger page sizes), from desktop serial printers on up to digital presses. Industry dollar volume is for annual hardware "if sold" market value (not manufacturer revenues). Other current studies size the industry as around $100 billion worldwide. This larger figure includes, in addition to hardware, media and other expendables (toner, developer, ink jet and laser cartridges, ink jet inks, etc.).

Worldwide Shipments

Grey bar: Units, Millions

Black bar: $ Volume in Billions, inflation adjusted to the value of the 1998 US$.

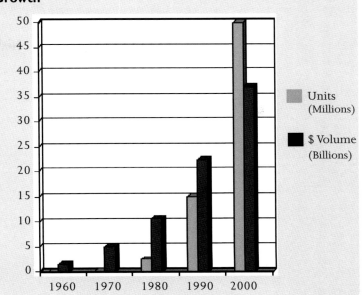

Units (Millions)

$ Volume (Billions)

Numbers: The Data

	1960	1970	1980	1990	2000
WW Shipments Millions of Units, all printers	.03	.20	2.5	15	50
WW $ Volume Billions, US$, all printers	.3	1.2	7.7	18	37
WW $ Volume Billions, US$, all printers adjusted to 1998 US$*	1.6	5.04	10.8	22.4	37
Average $/unit, all printers	10,000	6,000	3,100	1,200	740
Average $/unit, all printers adjusted to 1998 US$*	53,000	25,200	4,320	1,500	740

*Using the U.S. Consumer Price Index for urban consumers

Source: Digiprint Research Associates and IT Strategies.

Question:
How far and how fast can you go as an industry research vendor?

The answer: Very. Companies such as IDG, CAP Ventures, Dataquest, and Gartner Group over the years have thrived feeding the industry's information hunger. Here is a quick look at some information industry entrepreneurs and their companies.

International Data Group and Patrick J. McGovern. Of the market research/publishing companies covering computers, including the printer industry, IDG is certainly one of the most impressive success stories. Patrick J. McGovern, current Chairman, began his business trajectory in the late 1950s as an MIT student earning a degree in biophysics. While still at MIT he got a job with computer publishing pioneer Edmund C. Berkeley and became associate publisher of Berkeley's magazine *Computers and Automation* upon his graduation in 1959.

In the early 1960s Ed and Pat began to collect info on where computers were being installed and print listings each month in the magazine. It was perhaps McGovern more than his boss who saw how valuable such information could be if it were accumulated as a computer database. So he split with Ed and formed International Data Corporation in 1964 to continue building the database and publish the still thriving newsletter nick-named "The Gray Sheet." Three years later he collaborated on the launch of *Computerworld*, which was destined to become the industry's premier weekly newspaper and IDC's flagship publication.

IDC was the seed of today's global IDG empire which now claims 290 computer magazines in 75 countries, IDG Books with 700 titles, untold numbers of conferences, market research for 4,000 clients in 42 countries, 49 offices around the world, 1998 revenues of $2.35 billion and more than 9,000 employees. From the 1970s to the present IDC has published printer industry market statistics in various reports and studies. As a monument to his success, McGovern recently donated $350 million to his alma mater MIT to fund a brain research center, said to be the largest gift ever made to a university.

CAP International/BIS Strategic Decisions/CAP Ventures and Charles A. Pesko, Jr. Mr. Pesko has been good at growing companies, too. He has focused more specifically on digital printing and imaging so his companies are much smaller than IDG and Gartner. He founded CAP International in 1979 after working for some years in market planning and product development at Itek and Xerox. CAP International acquired both Datek Information Services

and Institute for Graphic Communications in the late 1980s, and then was acquired by NYNEX in 1988 and folded into that firm's BIS Group. The company was included twice in *Inc. Magazine's* annual list of America's fastest growing companies.

Pesko resigned from BIS after the sale of his company to NYNEX. However he chose not to retire to Florida with his

winnings. In 1991 he formed Charles A. Pesko Ventures to serve the corporate printing and imaging industry. As of this writing, eight years later, CAP Ventures claims annual revenues of around $12 million.

IT Strategies, Lyra Research. Several other successful printer industry information entrepreneurs had a period of incubation within Pesko's rapidly evolving enterprises. These include Charles E. Case and Mark Hanley, founders of I T Strategies. Charles LeCompte, founder of Lyra Research, learned the trade at Datek Information Services, then formed his own company, Lyra Research, soon after Datek became part of BIS Group. Lyra's publication *The Hard Copy Observer* is probably the industry's most widely read newsletter and has formed the core around which they have built the usual menu of printer industry information services.

Gartner Group/Dataquest/Giga Information Group and Gideon Gartner. Gartner founded the Gartner Group in 1979. The Group now calls itself "The world's leading authority in information technology." They claim to sell products and services to over 9,000 organizations worldwide, to have more than 3,000 "associates," (employees?) and revenues of $642 million in their fiscal 1998.

> IDC was the seed of today's global IDG empire which now claims 290 computer magazines in 75 countries, IDG Books with 700 titles, untold numbers of conferences, market research for 4,000 clients in 42 countries, 49 offices around the world, 1998 revenues of $2.35 billion and more than 9,000 employees.

Dataquest was founded by Dave Norman, Bill Coggshall, and Ron Miller in California in 1971 and their Electronic Printer Service (now "Printers in North America") has been a well regarded source of printer industry market data through the succeeding decades. In 1978 Dataquest was acquired by the research giant A. C. Nielsen, which was in turn acquired by Dun and Bradstreet in 1984, at the time a $2.3 billion company. In 1995 Dataquest was merged into Gartner Group, which Dun and Bradstreet had also acquired.

What happened to these founders? Presumably they did quite well financially. Norman went on to found the since defunct Businessland and is understood to now be at NetRatings. Gartner left Gartner Group in 1995 to form Giga Information Group. His new organization has already grown to claim revenues of over $30 million, a global client roster of 1,000 organizations, and offices in a dozen countries, a record they believe makes them the fastest growing company in the "IT advisory" industry.

And now the genealogy has become even more convoluted. BIS Group (which had acquired Mr. Pesko's first company, right?) was acquired by Giga a few years ago, including the still published newsletter Printout. But the Printout story doesn't end there. Giga sold the newsletter to consultant John Henry of Peripheral Insight, and he recently sold it back to CAP Ventures, taking it almost full circle.

INDUSTRY STRUCTURE

Developing technology and building printers is only the beginning of the battle to successfully commercialize product. As will be seen in some of the "how we did it" narratives in Part Two, new products required rethinking distribution. Distribution and industry structure underwent repeated transformation over this fifty year history in parallel with technology transformation. These trends are also touched upon as we visit each decade in Part Two. For now, here is an short overview of distribution and the changing competitive environment.

In the beginning, in the 1950s, the industry consisted primarily of a group of captive manufacturers who developed printers bundled and sold or leased with their systems, and some independent manufacturers who supplied the basic print engine to the systems companies. Companies such as Teletype and Anelex were important suppliers to the so-called "OEM" market. The OEM — Original Equipment Manufacturer — was the company that adapted the printer to its system, affixed its label, marketed it and stood behind it in the field. The term is a bit strange since in a sense the OEM was not really the manufacturer of the print engine itself.

The distribution path was relatively straightforward: printer manufacturer to OEM to end user, normally bundled with the system.

A breakthrough that revolutionized distribution was a decision by IBM to unbundle peripherals, a decision that was soon copied by most of the other systems vendors. This opened the door to the "plug compatible" segment of the industry. These were the "independents" such as Dataproducts which in the 1970s enjoyed rapid growth developing printers which they could market directly to computer end-users. During this second phase, a dominant game was shifting protocols by the systems companies to make it difficult for the competing plug compatible peripheral suppliers.

Over the most recent two decades, much more complex and volatile distribution patterns have emerged. At the low end, open standards and the domination of just several operating systems has unleashed a tumultuous retail channel. When the consumer market was young, it was often the retailers and distributors pleading, "Please, may we carry your products?" In a phase that may have peaked in the second half of the 90s, more often it has been the printer manufacturer pleading, "Please, please, please give us just a bit of shelf space for our wonderful printers."

The power relationship had reversed. Now, with the emergence of e-commerce, it may be shifting again with the stores worrying about on-line competition. We explore this impact of the Web a bit more in Chapter 8.

At the higher end, some direct sales persist. For example, Xerox still has a direct sales force. But more and more volume is flowing through the variety of multi-layered distribution channels shown in the Distribution Overview Chart. In terms of units, by far the most now move through computer stores, discounters, and mass merchandisers. There continues to be a direct market at the high end, and in the middle are various VARs and other distributors.

Derived industries warrant attention. Inks and media are part of the system. Papers engineered to get the best image quality from ink jet printers had a lot to do with the growth of that technology. Inks and toners are an integral part of the development of non-impact printers. Both media and chemistry companies have the option of working as partners or competitors to the printer vendors. They may be partners in the development cycle. They may offer their products to the printer vendor for private labeling. Or they may link with the multitude of independent aftermarket companies underselling the printer vendor's own factory-approved supplies. One of the more occult considerations in printer design is how to formulate and package supplies in a way that locks out competing aftermarketers. That most such efforts are only marginally successful is testimony to the power of human ingenuity.

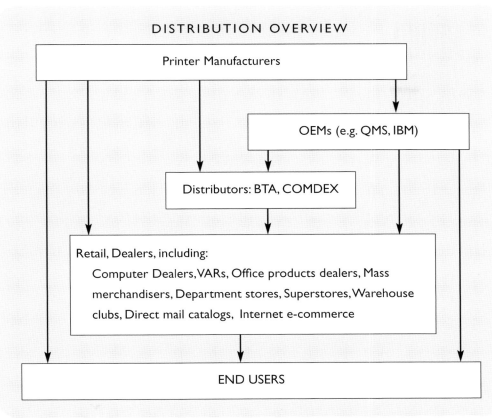

DISTRIBUTION OVERVIEW

Printer Manufacturers

OEMs (e.g. QMS, IBM)

Distributors: BTA, COMDEX

Retail, Dealers, including:

Computer Dealers, VARs, Office products dealers, Mass merchandisers, Department stores, Superstores, Warehouse clubs, Direct mail catalogs, Internet e-commerce

END USERS

Source: Digiprint Research Associates

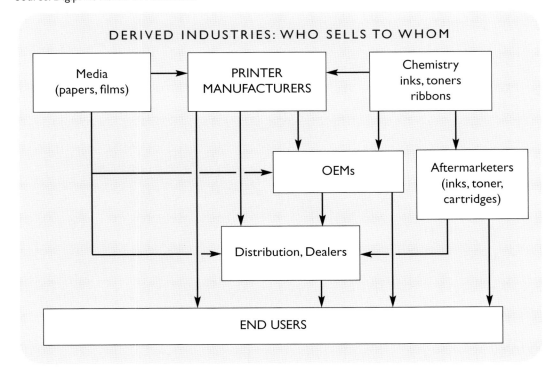

DERIVED INDUSTRIES: WHO SELLS TO WHOM

Media (papers, films)

PRINTER MANUFACTURERS

Chemistry inks, toners ribbons

OEMs

Aftermarketers (inks, toner, cartridges)

Distribution, Dealers

END USERS

Industry Concentration

The vendor mix charts for 1980 and 2000 reveal a remarkable shift. The charts give the estimated market share for the leading companies in units and dollar value of shipments to the U.S. market. Only companies that develop or manufacture and market the basic printer marking engine are included.

What a difference twenty years can make! Almost every company which held significant market share in 1980 had disappeared by 2000. The revenue threshold for including companies in the 1980 chart was $100 million. The only company appearing on both charts is Xerox.

Where was mighty HP in 1980? Where is mighty IBM now?

Back then, HP had only recently launched its marginally successful 2680 laser printer. Their medium and high speed line printers were OEMed from Dataproducts and Data Printer. They had a number of impact and thermal serial printers and printer-plotters, but they were developed specifically as output devices for users of HP computers. At that point HP was already bringing in over $3 billion in revenues. But their U.S. market for in-house manufactured low-end printers was probably not quite to our $100 million threshold.

IBM was the HP of 1980. IBM still held by far the largest share of the market in terms of value of shipments and units. IBM was the leader in all three major segments, impact serial, line, and non-impact page printers. They owned a good share of the mainframe computer market and along with that, despite competition by the plug compatible vendors, they supplied printers and other peripherals to most of their customer base. At the low end they offered at least thirteen serial matrix printers and a teleprinter based on their Selectric mechanism. The Selectric was also sold to other OEMs who adapted the mechanism for console printers, remote terminals, and the booming word processing market of that era. At the high end, by 1980 IBM had been shipping its $300,000 3800 production printer for four years and volume was still believed to be building.

But already the field was shifting. The installed base of old line companies such as IBM, Teletype, and GE in 1980 was significantly higher than their share of current shipments. Newcomers were eating into the pie. In serial printers, Centronics was shipping almost as many as IBM. Digital Equipment Corporation had become a teleprinter leader. By 1980, according to one report, they had shipped more than 200,000 of their LA36 II DECwriters. And newcomers with new technologies were beginning to make their presence felt.

The hefty 24% "all other" slice of the 1980 pie includes many of these newcomers. Who were they? You can find them in the two-page, decade-by-decade listing of active companies (see pages 88-89). This was when the number of contenders had peaked. Consolidation was already in the works.

Cut to the present. HP has taken the place of IBM as the leader by far, at least in terms of unit shipments. IBM, still a major printer vendor, but now an OEM, having given up the development of print engines years ago. This new posture by the company which pioneered so much of printer technology is a telling sign of one of the major themes running through this history — the "componentization" of the marking engine.

MARKET SHARE EVOLUTION, 1980 TO 2000
U.S. MARKET, Printer Hardware Only

1980

2000

UNITS

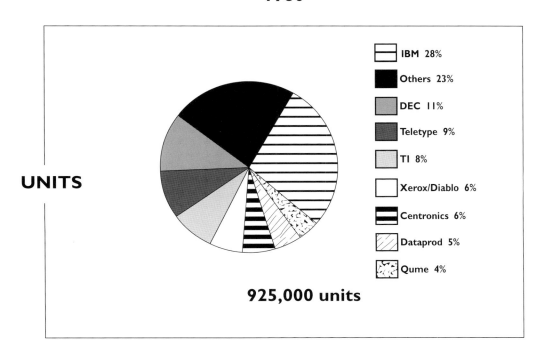

IBM 28%
Others 23%
DEC 11%
Teletype 9%
TI 8%
Xerox/Diablo 6%
Centronics 6%
Dataprod 5%
Qume 4%

925,000 units

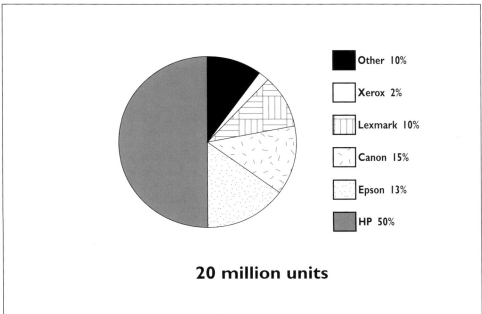

Other 10%
Xerox 2%
Lexmark 10%
Canon 15%
Epson 13%
HP 50%

20 million units

VALUE OF SHIPMENTS $US

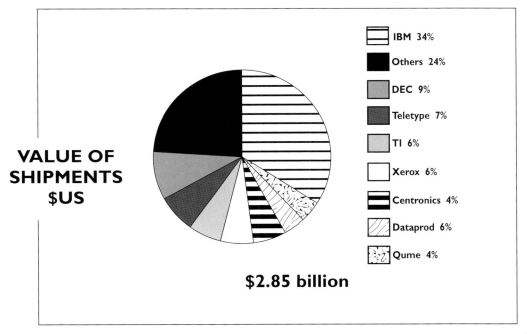

IBM 34%
Others 24%
DEC 9%
Teletype 7%
TI 6%
Xerox 6%
Centronics 4%
Dataprod 6%
Qume 4%

$2.85 billion

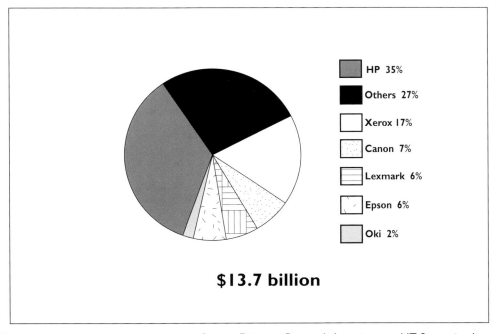

HP 35%
Others 27%
Xerox 17%
Canon 7%
Lexmark 6%
Epson 6%
Oki 2%

$13.7 billion

Source: Digiprint Research Associates and IT Strategies, Inc.

INDUSTRY MATURITY, GLOBAL SHIFT

It is the sum of details that most reliably reveals sea changes. One of the more interesting historical exercises was to try to tabulate all the companies that developed and successfully commercialized printers during each of the five decades (chart, pages 88-89).

This listing of the cast of company characters from each decade seems to confirm that the digital printer industry is not immune to the inevitable flow that characterizes most industries. The total number of companies swelled through four decades from just 12 to 93 by the end of the 1970s. During the following decade growth slowed considerably, and then in the 1990s the number of active manufacturers of print engines plummeted by almost 50%.

Clearly the core industry, at least as defined in this history, is well on its way to maturity, a trend which does not often reverse. The explosion of inventive companies gambling on varying levels of technology innovation (or just jumping on to what they saw as a bandwagon) has now collapsed. Today's printer industry is characterized by fewer vendors grabbing larger and larger shares of a market which is slowing in terms of unit shipments and revenues. It is an industry increasingly dominated by an ever smaller club of giants.

Exactly where the industry is perceived to be on its trajectory through maturity to eventual decline depends in part on definitions. How do we define "the industry?" Probably the best industry directory is *The Hard Copy Observer Guide to the Printer Industry* from Lyra Research (which unfortunately hasn't been published since 1996). That last edition of the *Guide* listed over 2000 companies! Lyra defines the "industry" in its broadest sense, to include vendors engaged in the many derived markets such as software, materials, paper handling gear, accessories, books and services.

Guidelines for our listing include,

- Manufacturers of marking engines for general purpose printers and other page-size documents. This means excluding companies that build only wide format printers, small industrial and POS listers and label printers, graphic arts proofers, militarized printers and other specialized printers.
- Primarily companies who design and perform at least the final assembly of the marking engine, and who brand and market their products.
- Primarily vendors active in or at least visible in the U.S. market during at least some of the decade.
- Geography assignment is governed by the location of the home office of the parent company.

These guidelines may have been bent just a bit in a few cases, mostly because it is unknown how much actual "manufacture" is done. If it is a design contracted to an offshore fabricator, does that eliminate the company which designed and is marketing the product? To be faithful to this definition, even IBM, in the wake of its spinoff of Lexmark, should probably be stricken from the listing. But that somehow seems overly compulsive. Many of the companies that pioneered the industry in the 1950s still exist, but only IBM has been a major printer player through all five decades. The brand is so strong, the value they add so significant, and their printer heritage so impressive that they are in a class by themselves.

In the listings we have also shown, where known, mergers and acquisitions. One of the more challenging pieces of the research was ferreting out what happened to all those companies which have disappeared from the roster. Many still exist, thriving

in other product pastures. Others have disappeared. A typical pattern is decline and acquisition, with the new parent company not investing further in the products or in marketing, but rather just milking the current product line and customer base, selling supplies, service, and replacement parts for as long as possible. Companies are born with a flourish, but die with barely a whimper.

Trends. It could be said that just listing dozens of mostly small companies doesn't accurately demonstrate the extent of the global shift or path to industry maturity. True, the bulk of the companies listed represent only a few percent of the market. (We have intentionally left out some that introduced perhaps only one or two products and faded quickly from the scene. And it can be assumed companies have been overlooked, some of whom may have made significant contributions. For this we apologize.) As a gesture toward weighting the companies listed, we have boldfaced those believed to have held significant market share for at least some of the decade.

Listing all the companies, regardless of their volume, is a significant indicator of the maturity of the industry. Youth is characterized by vendor proliferation. In addition, the kinds of companies are indicative of other industry trends in given decades.

As we have seen, the 1950s and early 1960s were times of domination by the computer vendors who in general developed and built their own peripherals, including printers. The 1970s saw an explosion in the number of companies developing and marketing plug compatible printers, a pattern which continues to this day.

The listing reflects the trend throughout industry toward diversification which perhaps peaked in the 1970s, although it has certainly not completely subsided. It is interesting to recall past diversification attempts by giant manufacturing conglomerates, most of which were doomed. Companies such as Exxon (oil),

Electrolux (vacuum cleaners), Mead (paper), and Ericsson (electrical equipment) plunged in. Some quickly exited, but others, such as Exxon, persisted quite a while. Exxon Office Systems included Qwip, Qyx and the word processing company Vydec. Exxon's investment in this effort before giving up and selling the remains to Lanier in 1985 is believed to have been around $2 billion. Optical and camera companies (at least their focus is imaging) have rightly moved toward digital printing with varying degrees of success, including Kodak, Canon, Casio, Agfa, Polaroid, Minolta, Seiko, Fuji and others.

Digital Printer Fever

At the height of the explosion of innovation and applications in digital printing, with word processing and minicomputers fueling demand, some of the world's largest corporations sought a piece of the action, a trend that peaked in the 1970s. For some, such as RCA and Honeywell, printers grew naturally out of their earlier diversification into computers. But for others, it was an odd coupling. Their most common route was acquisition, although a few attempted in-house development. Here are some of the giants, other than the computer companies, who caught the fever.

Exxon	ITT	Philips
Citizen Watch	Western Electric	Kodak
Electrolux	GE	Mead
3M Company	Litton Industries	General Dynamics

Global Shift. As usual, Japan breaks all the rules. Conglomeration, with many diversified and more or less loosely organized manufacturing "families," is an ancient pattern which generally seems to work for the Japanese. In printers, Canon and Epson (Seiko) are two of the most impressive success stories. What is highlighted most vividly in this overview is the global shift of the industry toward Japan.

This Global Distribution chart shows the number of active companies by region. Here again in a few cases there needed to be arbitrary classifications. Where do we put Fuji Xerox (50-50 ownership shares), for example? Should they even be listed as a separate company? In general, companies are assigned countries according to the location of their home office or that of their parent firm. For example, this makes Scitex Digital Printing in Dayton, OH USA a Dutch company. Their parent firm is Israeli, but Scitex lists its home base as The Netherlands.

Japan now represents well over half of the world's printer companies. This has been perhaps the major sea change of the last decade. The shift was reported in almost pensive terms by one observer of the 1998 CeBIT industry exposition in Hannover, Germany: "Proud firms such as Bull Compuprint, Facit, Olivetti, Olympia and Tally that once actively developed and marketed their own print engines and controllers have now mostly retreated into shrinking niches such as impact printing, or have abandoned printer development altogether."[*] Were this trend to be weighted with the volume of printers produced, Japan's percentage would certainly be well over 75% in terms of marking engine volume.

On the other hand, if the industry is defined more broadly, the global shift becomes much less pronounced. Software, controllers, supplies, and the myriad other derived, printer-driven sub-industries remain in North America.

These megatrends, in short, include industry maturity, concentration, and a strong global shift. However, if we see the industry as including myriad derived industries, and expanding into new applications, these sea changes are less clear.

IT Strategies consultant Marco Boer sees this sea change as expansion, not contraction. "If you have narrow sights, looking backward, it can look like contraction," he said during a briefing at their Hanover, MA offices. "This trend is actually wonderful. Instead of the maybe $100 billion office automation printer market, the printer industry is becoming part of maybe a trillion dollar set of markets. Commercial graphic arts printing will become part of digital printing. Industrial printing will more and more be digital printing. It all depends on how you look at it."

Clearly, the symphony is far from over. In the opinion of some observers, the industry could experience a new wave of growth. There could be new players once again based in the USA or Europe. Explosive energy and opportunity still awaits the adventuresome in these frontiers. We will follow this thread further, exploring the future in Part Three.

The Shift (pages 88–89): *Companies understood to have developed and marketed marking engines for general purpose printers active in the US market during each decade. Companies in boldface are those which held significant market share at some point in the decade. Companies which introduced products that did not make it commercially in general are not included.*

---

[*] *The Hard Copy Observer, 4/98 issue*

GLOBAL DISTRIBUTION OF PRINTER
MANUFACTURING COMPANIES

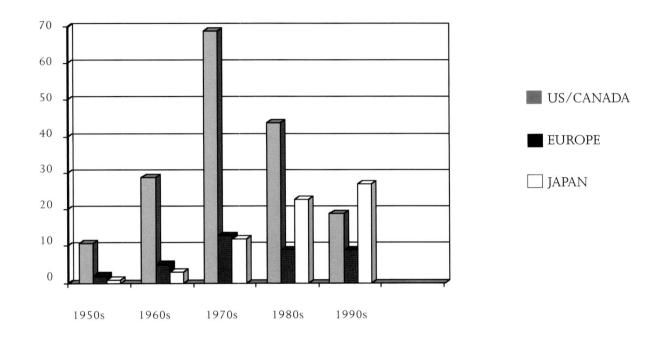

Number of companies manufacturing general purpose print engines by geographic region of their home office during each decade. Europe includes Israel. Japan includes all the Far East. USA includes Canada. Includes primarily companies active, or at least visible, in the U.S. market during any part of the decade. If tabulated at the end of the 1990s, the roster of US-based companies would be still smaller.

Source: Digiprint Research Associates.

Manufacturers of General Purpose Print Engines

(Primarily those active in the U.S. Market) 1950-2000

Source: Digiprint Research Associates

50s	60s	70s	80s	90s

70s column:

A. B. Dick Co.
AM Multigraphics
Anadex
Anderson Jacobson, Inc.
AT&T Teletype Corp.
Brother Industries
Bull Groupe/Bull GE
Burroughs
Canon
Casio Computer Inc.
Centronics Data Computer
Citizen Watch Co. Ltd.
Computer Devices, Inc.
Computer Peripherals, Inc. Joint venture among CDC, NCR, ICL; CDC buys controlling interest, 1978)
Control Data Computer Peripherals Division
CTSI
Bright Industries
Data-100 (acquired ODEC)
Data General
Data Interface, Inc. (acquired By Inforex, 1977)
Data Printer (acquired by Printronix in 1983)
Datapoint
Dataproducts
Dataroyal, Inc. (merged with Facit in 1982)
Datasouth
Dennison Control Print /
Delphax Systems
Diablo Systems (acquired by Xerox, 1972)
Di/An Controls
Digital Equipment Corp. (DEC)
DIP, Inc.
Documation (founded 1970, acquired by Storage Technology Corp. in 1980)
Epson America (Shinshu Seiki Co. Ltd./Seiko)
Extel
Facit-Odhner
Florida Data
Friden Divison, Singer Business Machines
Fujitsu
GE
Harris/Radiation
Hermes
Hewlett-Packard
Hitachi Ltd. (Nissei Sangyo America Ltd.)

80s column:

Anderson Jacobson, Inc.
AT&T Teletype Corp.
Brother Industries
Bull Peripherals/
Groupe Bull
Cii-Honeywell Bull
Burroughs (merged with Univac to become Unisys in 1986)
Canon
Casio Computer, Inc.
Centronics Data Computer (Merged with Computer Peripherals, Inc. 1982; Printer operations acquired by Genicom in 1986)
Citizen America
C. Itoh Electronics, CIE Terminals (Tokyo Electric)
Computer Devices, Inc.
Computer Peripherals Inc.
Control Data Computer Peripherals Company
CTSI
Data General
Datapoint
Dataproducts Corp.
Datasouth Computer Corp.
Decision Data
Delphax Systems
Digital Equipment Corp.
Durango Systems
Epson America
Exxon Office Systems

90s column:

Advanced Matrix Tecknology
Ahearn & Soper (acquires Facit Printers, 1993)
Alps Electric
Asahi Pentax
Brother Industries
Groupe Bull (includes Bull HN, Nipson Printing Systems; Bull Compu-print (Xeikon acquires Nipson, 1999)
Canon, Inc.
Casio
Citizen Watch
Dataproducts (acquired by Hitachi 1990)
Datasouth Computer Corp. (acquired by Bull Run Corp., 1993)
Delphax Systems (acquired by Xerox, 1998)
Digital Equipment Corp. (sells printer business to Genicom in 1997; acquired by Compaq, 1999;)
Epson America

ANelex Corporation
AT&T Teletype Corp.
Burroughs/ Control Instruments
Compagnie Bull GE
Friden (Flexowriter)
IBM Corporation
ICL
NCR
Potter Instrument
Rank-Xerox (Xeronic, 1958)
Sperry Rand (UNIVAC)
Shepard Laboratories

A. B. Dick Co. (Videojet)
Anadex Inc
Anderson Jacobson, Inc.
ANelex Corporation (acquired by MDS, late 60s)
AT&T Teletype Corp.
Bull Groupe/Bull GE
Burroughs
Computer Peripherals, Inc (joint venture between Holley Carburetor Co. and Control Data Corp.)
Computer Devices, Inc.
Computer Transceiver Systems, Inc. (CTSI)
Control Data Computer Peripherals Division
Dataproducts
Digital Equipment Corp.
Diablo Systems (est. 1969)
Friden
GE/Waynesboro VA
Hitachi Ltd.
Drum line printer (ANelex Design, 1964)
Holley Computer Products (formed 1962 as joint venture between CDC and Holley Carburetor Co.)
Honeywell
IBM Corporation
ICL
Mohawk Data Sciences (acquired ANelex, late 60s)
NCR
Nippon Electric
Olivetti, Olivetti GE
Philips (Mosaic Printer, 1964)
Potter Instrument Co.
Radiation, Inc.
RCA (acquired by GE, 1986)
SCM Corp.
Shepard/Vogue
Shinshu Seiki Co.
Siemens AG
Sperry Univac
Stromberg Carlson
Texas Instruments
Versatec, Inc (acquired by Xerox in 1975)

Holley Computer Products (Control Data buys out Holley Carburetor's 50% share, 1964)
Honeywell Information Systems
Hydra (acquired by Lear Siegler, 1977)
Inforex
Infoscribe (England)
Integral Data Systems (acquired by Dataproducts 1982)
Interdata
IBM Corporation
Konishiroku (U-Bix)
Lear Siegler
Litton Industries Datalog Div.
LogAbax S.A.
Mannesmann
Mead Digital Systems
Mohawk Data Sciences
NCR
NEC Information Systems (Nippon Electric)
Nixdorf Computer
Nortec
ODEC Computer Systems (acquired by Data-100)
Okidata Corp, Oki Electric
Olivetti
Olympia
Perkin Elmer
Pertec
Philips Peripherals
Plessey (US daisywheel operation acquired by Dataproducts)
Potter Instrument
Practical Automation
Primages, Inc.
Printer Technology
Printronix, Inc.
Qume Corp (founded 1973, purchased by ITT, 1978)
Ricoh
Sanders Technology Systems
SCI Systems
SCM Kleinschmidt
Sharp Electronics
Siemens AG
Silonics, Inc.
Sperry Rand/Univac
Star Micronics, subsidiary of Star Manufacturing Co., Ltd.
Sycor
Tally Corp. (acquired by Mannesmann, 1979)
Telex
Texas Instruments (forms Industrial Products Div. 1971)
Triumph-Adler
Uppster Corporation
Varian Associates
Wang Labs
Xerox Corporation (acquired Diablo Systems, 1972)

Facit, Inc. and Dataroyal (both companies subsidiaries of A.B. Electrolux)
Florida Data Corp.
Fuji Xerox
Genicom (Founded 1983 as buy-out from GE Data Comm'n Products Dept.)
Hewlett-Packard
Hi-G Company, Inc.
Hitachi Ltd., Hitachi Koki, and Nissei Sangyo America Inc.
Honeywell Information Systems and HIS Italia
Howtek, Inc.
IBM Corporation
Indigo
Kentek Information Systems
Konica Business Machines
Konishiroku
Kyocera
Mannesmann Tally
Mead Digital Systems (acquired by Kodak, 1983)
Miltope, Inc.
Minolta Corporation
Mohawk Data Sciences
NEC Information Systems
Nixdorf Computer (merged with Siemens, 1990)
Okidata Corporation
Olivetti & Co. S.p.A.
Olympus
Perkin Elmer
Pertec Computer Co.
Practical Automation
Primages, Inc.
Printek, Inc.
Printronix, Inc. (acquired Data Printer, 1983, Anadex, 1984)
Panasonic/Matsushita
QMS (acquired by Minolta, 1999 – "strategic partnership")
Qume Corp./ITT
Ricoh
Sanders Technology Systems
SCI Systems
Shinko
Siemens AG, Siemens-Nixdorf
Silver Reed/Silver Seiko
Sharp Corporation
Sperry Univac (Unisys)
Star Micronics, Inc.
Storage Technology Corp. (acquired Documation in 1980)
TEC
Tektronix, Inc.
Telex Computer Products
Texas Instruments
Triumph Adler
Toshiba America
Uppster Corporation
Wang Labs
Xerox Corporation

Facit (Div. Ahearn & Soper)
Fuji Xerox
Fujitsu, Ltd.
Genicom Corporation (buys Texas Instruments printer business in 1996 and DEC printer line in 1997)
Hewlett-Packard
Hitachi Koki Co. (subsidiary of Hitachi Ltd., parent firm of Dataproducts/Hitachi Koki Imaging Solutions)
IBM Corporation
Indigo, NV
Kentek Information Systems
Kodak (Mead, Diconix; acquired by Scitex 1993)
Konica Business Machines
Konishiroku
Kyocera
Lexmark International (buy-out from IBM 1991)
Mannesmann Tally (Tally Corporation effective 7/96)
Minolta (buys controlling interest in QMS 7/99)
Mitsubishi Electronics
Memorex Telex N.V.
NEC Corp.
Océ (acquires Siemens-Nixdorf, printer opns, 1996)
Oki Denki Kogyo, Okidata
Olivetti & Co., S.p.A.
Olympus Image Systems
Panasonic/Matsushita
Pentax
Printek
Printronix
Ricoh
Samsung
Scitex Digital Printing (acquired Kodak Digital Printing, 1993)
Seiko Epson Corp.
Sharp Corporation
Shinko
Siemens-Nixdorf (printer operations acquired by Océ, 1996)
Star Micronics
Storage Technology (acquired by Siemens-Nixdorf, 1993)
Tally Corp. (divorced from Mannesmann, 1999)
TEC Corporation (acquired by Toshiba, 1999, to form TTEC)
Tektronix (printer business acquired by Xerox, 1999)
Texas Instruments (printer business bought by Genicom, 1996)
Xeikon N.V. (acquired Nipson, 1999)
Xerox Corp. (acquires Tektronix printer business, 1999)

ℐℕ𝒯𝐸ℛ𝓜ℐ𝒮𝒮ℐℴℕ

RECORDS

Things move fast. Definitions are everything. So, in deference to those who would contest these "world records," they are presented as nominations and couched as questions.

FASTEST?

Scitex VersaMark™ MSP 22. (photo courtesy of Scitex Digital Printing)

Printer:	Scitex Digital Printing VersaMark
Speed:	2000 8 x 11-inch pages per minute at 600 dpi
Technology:	Continuous Ink Jet
Date Announced:	February, 1999
Price:	Starts at $715,000 for a low-end monochrome system; up to $2.9 million for the loaded Twin-66 (page wide spot color, maximum in-line post processing, etc.)

VersaMark applications include direct mail, business forms, and a variety of on-demand printing and publishing jobs. A book printing version was demonstrated at BookTech'99 in New York which could produce a completely-bound, 360-page book in around eight seconds!

This printer may qualify for two other world records. At almost $3 million for the maxi version, it may be the most expensive. But also, Scitex points out it may be the cheapest – in terms of operating cost. There is no click charge and your supplies, they promise, will be under a tenth of a cent per page. So the VersaMark may be the world's cheapest printer as well a the most expensive.

SMALLEST?

Candidate 1:	Brother MP-21C Mobile Inkjet Printer
Weight:	2.2 pounds
Dimensions:	11.81 in. wide, 2 in. deep, 4.17 in. high
Performance:	plain paper, color, 720 x 721 dpi; speed not specified
Technology:	piezoelectric ink jet
Price and Intro Date:	DX model $319, C model $269; 2000

Get color anywhere, Brother says, advertising its MP-21C. (Photo courtesy of Brother International)

Candidate 2:	Pentax PocketJet 200
Weight:	1.12 pounds with battery
Dimensions:	1.18 x 10.04 x 2.17 inches
Performance:	thermal paper, 200 x 200 dpi, 3 pages/min.
Technology:	thermal
Price, Intro Date:	$80; September, 1999

The Pentax PocketJet is small enough to lose if you're not careful (Photo courtesy of Pentax Technology)

The Brother may be the smallest full page size, color, plain paper printer, and the Pentax may be the smallest page-size printer regardless of technology. The Pentax people urge users not to be afraid of thermal technology: the paper for this printer is nothing like the "flimsy, curley" paper of the old thermal fax machines. Applications for the Pentax are selected vertical markets such as law enforcement, utilities, and medicine. The Brother machine is a general purpose printer for people on the go which plugs into a host laptop for both data and power.

CHEAPEST?

Printers:	Apollo P-1200 and Lexmark Z11
Price:	$49
Date:	August, 1999

As of this writing, these two printers may be the world's cheapest page-size, color, plain-paper printers. In early 1999 the Apollo, listed at $79, was the cheapest at launch, followed closely by the 1,200 x 1,200 dpi Lexmark at just $89. By mid-summer, a mini price war erupted, and with mail-in rebates both printers became effectively $49.

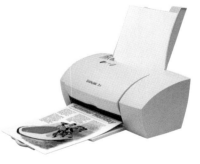

Apollo P-1200 (top) and Lexmark Z11. (Photos courtesy of Apollo Consumer Products and Lexmark International.)

How cheap can you get? How about free?

Yes, free digital printers have been also pretty easy to find. Some computer retail chains have packaged free printers with the purchase of a computer. In 1999 Circuit City was offering a free Canon BJC1000 printer with the purchase of any desktop computer and monitor.

How about a free workgroup color printer? Tektronix (now Xerox) in 1999 was promoting an offer for free Phaser® 840 solid ink jet printers. The 10-ppm machine when introduced in 1998 listed at around $2,810. What's the catch? You need to use a lot of Tektronix ink, and to qualify, you need to indicate your usage will be high enough.

Over 11 billion color pixels adorn a wall of the Cologne Kunstverein. (Photo courtesy of ENCAD GmbH, Ottobrunn, Germany)

LONGEST (ALMOST) CONTINUOUS PRINTOUT?

Printer:	ENCAD NovaJet PROe
Date:	July 17, 1998
Place:	Cologne, Germany
The Print:	180 feet long x 59.1 inches wide

The ENCAD news release was somewhat modest, claiming only that this was the longest ENCAD print produced to date. The output was a mega-poster by German artist Peter Zimmermann displayed at the Cologne Kunstverein, part of an exhibition entitled, "Eigentlich koennte alles ganz anders sein" ("Things could be completely different after all"). The final piece consisted of two strips, one 98.4 ft. long and one 82 feet long, printed unattended, except to change paper rolls, in a time of 16:59:35 hours. The print contained, according to the ENCAD release, 649,607 lines and 11,508,437,612 pixels.

WORLD'S LARGEST SIGN?

Size:
36,325 square meters
Location:
Shanghai, China
Substrate:
UltraMesh 500 by
Ultraflex, Inc.
Advertiser:
Coca Cola

This candidate for the world's largest sign has been submitted to the Guinness Book of World

Records but as of late 1999, verification was still pending. The huge Coke billboard covered four sides of one of the many empty office buildings in Shanghai. Each panel measured 62.2 x 146 meters. The sign was printed digitally on UltraMesh 500, an advanced signage fabric developed by Ultraflex (Rockaway, NJ). If Guinness confirmed the Shanghai sign record, it did not last long since Ultraflex reported that in Brazil a sign twice that size had been committed to construction in 2000.

This sign in 1999 was probably the world's largest. The sign is made up of digitally-printed strips of UltraMesh around 18 feet wide with butt seams heat-sealed or RF welded in place. (photo courtesy of Ultraflex Systems, Rockaway NJ USA)

OLDEST STILL OPERATING?

Printer:	Printronix P300
Current User:	Cal-Air, Inc. of Whittier, California
Model Year:	1974

During 1999, to celebrate their 25th Anniversary, Printronix launched a contest to find the oldest operating Printronix printer and found some impressive candidates. The Cal-Air printer was the oldest found in North America, said to be still at work and to be the 153rd printer built by Printronix. In Europe/Middle East/Africa, the oldest they found was being used daily by Braun AG in Gossau, Switzerland. The Far East winner was Novelty Philippines where the Printronix impact matrix line printer was still being used for bar code and report printing.

Around the world there may be older printers still at work, but no one but Printronix has tried to track them down. IBM in Germany is understood to be restoring a 407 Accounting Machine to working order, but this puts it into the category of an artifact rather than a production tool.

Still at work, after all these years: the Printronix P300 printer in the foreground has been going strong since 1974. (photo courtesy of Printronix, Inc.)

PART TWO

The Decades

This is the 9400 large scale computer developed by Sylvania in the late 1950s and early 1960s. Periperhals include multiple magnetic tape transports, a Shepard line printer, an IBM card punch, and in the foreground, Friden Flexowriter. (DRA Archives)

CHAPTER 3: THE FIFTIES

THE SETTING   In terms of culture in the U.S., the decade of the 50s conjures up visions of a self-satisfied populace basking in the comfort of the conservative, old fashioned values of God, Mother and Country. After years of Democratic administrations dragging the country out of the Great Depression and through World War II, it was a primarily Republican decade, dominated by father-figure President Dwight D. Eisenhower. It was a decade of crew cuts, peg-leg trousers, and the heavy beat of early Rock and Roll from the Beach Boys, Elvis, and Bill Haley's Comets. Women were still "girls," in heels, stockings and permed hair. Clean-scrubbed families in huge automobiles plied drive-in restaurants and drive-in movies

But in contrast, there were also wars, social unrest, and the beginnings of the rebellions of the Sixties. Even as the world was healing from the ravages of The War, the 50s saw the Korean War, smaller conflicts such as the Suez, and the end of the French phase of the Vietnam War. The "Movement" began in the U.S. with the 1955 bus boycott in Montgomery, Alabama, led by Dr. Martin Luther King, Jr.

There were leaps in the technology of both destruction and creation. The U.S. tested the hydrogen bomb in 1952 and launched its first atomic submarine in 1954. The Soviets beat the Americans into space with the Sputnik in 1957. The transistor began the transition from vacuum tubes to solid state electronics in everything from radios to computers. Color TV was introduced into the U.S. in 1951. The decade also saw the arrival of the laser, phototypesetting, the commercial electronic scanner, early OCR and MICR, and the Xerox 914.

The computer industry was born, led first by UNIVAC and soon by IBM, evolving from experimental, scientific number-crunchers to indispensable business machines. Electromechanical peripherals were no match for the exponential expansion of processing power, revealing the "input-output gap" which drove the rapid evolution of the printer industry. The term "printout" emerged and "printer" came to mean not just a person operating a printing press, but also a digital hardcopy output device.

A major shift between the 1940s and 1950s was the emergence of digital printers as separate, free-standing peripherals. In the accounting machines which preceded true computers, the print mechanisms were an integral part of the box. Then, with the early, pre-1950 experimental computers, printers were pretty much invisible. As noted earlier, output from the ENIAC was punched into IBM cards. The result could be read visually from the cards, or they could be carried to an IBM accounting machine to be printed out off-line. In short, before 1950, as observed by Tomash and Wieselman, printers were pretty much an "afterthought."[*]

All that changed in the 50s. Computers became commercial. This meant they became input-output-intensive rather than computing-intensive. IBM and the Remington Rand were the two early contenders. The approach of each to the problem of volume printing was colored by their earlier history.

RemRand vs. IBM. Remington Rand got started early in the decade and ran very fast. In 1950 they acquired the Eckert-Mouchly Corporation and a year later delivered the first large-scale commercial computer system, christened the UNIVAC (for UNIVersal Automatic Computer), to the U.S. Census Bureau. Shortly thereafter they bought Engineering Research Associates which gave birth to the 1103 UNIVAC scientific computer,

described as the first commercial computer with random access memory. In 1955 they merged with Sperry Corporation to form Sperry Rand. The same year General Electric processed payroll on their new UNIVAC, which was said to be the first computerized business application by a major corporation.

The only console printing terminal available with the early UNIVACs was the Uniprinter I, a paper tape-oriented, 10 cps adaptation of a Remington Rand electric typewriter. From the beginning, the Remington Rand computers were magnetic tape-oriented and all volume printing was off-line. They had a customer base using punch cards, but they rightly believed cards would soon be obsolete.

IBM entered computers committed to the punched card. They pioneered electric typewriter development, but EAM (Electronic Accounting Machine) "tab" equipment, which had evolved from the time of Hollerith, was their product centerpiece. As of 1943 the company boasted around 10,000 tab machines on rental, mostly Type 405 alphanumeric accounting machines.

The 405 was replaced in 1949 by the popular 407 Accounting Machine. They were a few years behind Remington Rand in computers, delivering their first Model 701 to the government in 1953. This was followed by a rapid series of upgrades culminating in the IBM 709 five years later, their last vacuum tube computer. A second series of smaller computers, the IBM 650, was fielded which enjoyed rapid popularity (for that era) with almost 2,000 delivered over its product life. In 1956 they introduced the 305 RAMAC which featured magnetic disk memory and abandoned the punch cards. By the second half of the decade they had decisively seized leadership from Remington Rand.

But soon IBM and RemRand were no longer alone in computers. They were joined by a number of competitors including Honeywell, GE, NCR, RCA and Burroughs (eventually acquired by

This diagram from an early Sylvania operations manual shows the main components of a typical large scale computer and the data flow. The two printers are almost lost among the ranks of freestanding processors and memory subsystems. The speed and capacity of such a system was much less than that of a typical desk-top computer today. (DRA Archives)

---

[*] "Marks on Paper: Part 1", Annals of the History of Computing, Vol. 13, Nr. 1, 1991

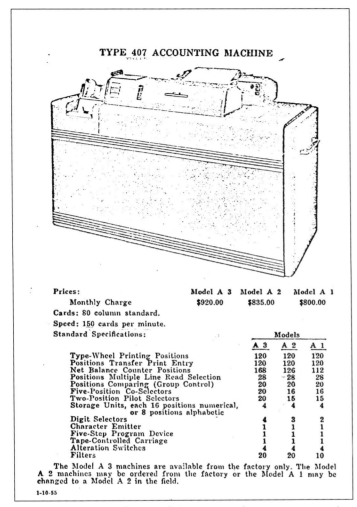

Price, performance, and features of the IBM 407 from an IBM manual
dated January 10, 1955. (Courtesy of IBM Archives)

Sperry Univac to form Unisys). Overseas there was activity as well. The U.K. developed some pioneering digital systems during the late 1940s and early 1950s. These included the EDSAC at Cambridge University, a commercial adaptation named LEO (Lyons Electronic Office), and the Ferranti Mark I and Mark I Star. Later several British electronics companies including English Electric

and Ferranti joined forces to found ICL (International Computers Ltd.). On the Continent, Bull, Siemens, Zuse, Olivetti and others began developing and fielding commercial computers during the later 50s and early 60s.

The computer systems of the day consisted of a room-full of free-standing units: the central processor and peripherals which might include punch card and/or magnetic tape subsystems, auxiliary mass disk storage, and printers. These components were usually bundled – sold and priced as a single package. The tangle of heavy-duty data and power cables that connected the units normally would be hidden below a raised floor comprised of square panels that could be pulled up with a suction device for maintenance.

In the mid-fifties, responding to growing competition, IBM assembled a team of their scientists from the U.S., Germany and France to conceptualize an affordable successor to the RAMAC and the beloved but obsolete punched card accounting machines. This "Worldwide Accounting Machine (WWAM)" program was given top priority and culminated in the IBM 1401, announced in 1959. This system quickly became one of IBM's most important products ever. According to one account, "it was undoubtedly the product that gave IBM its first realistic glimpse of the size and importance of the computer market that was unfolding."[*]

High Speed Printing. IBM had surged ahead of Remington Rand in computers, but they were well behind in the printer race.

Throughout the decade IBM made do primarily with the 407, which was modified as an off-line and on-line printer (designated Model 716 with the IBM 709 and 7090 computer systems). RemRand began developing their high speed, on-the-fly drum printer early in the decade. They set up a team of engineers working under project leader Earl Masterson, who later described

[*] Bashe et al, *IBM's Early Computers*. This book was an important (but not the only) source of the IBM printer lore in this chapter.

On-the-Fly Impact Printing:
How it was in the Field

The noise and sweat of the computer "machine room" of the 1950s and 1960s have largely been forgotten. Computer operators used hand trucks to carry cartons of continuous forms to the clattering impact line printers. As a new technology pushing the state of electromechanics, downtime was common.

One pitfall that was difficult for early design engineers to anticipate and accommodate was the possibility of transient electrical or mechanical load surges. A line printer did not really fire all hammers simultaneously in normal operation. Care was taken to minimize this possibility. Characters were often spiraled around the type drum so that even if a whole line of dashes was called for, all the hammers would not have to fire at once. But despite precautions, such surges on occasion happened. The resulting vibration could make a heavy printer walk, or blow fuses, or both.

Searching <Deja.com> on the Web pulled up a few postings by those who remember and seemed to enjoy swapping printer war stories. Here is part of one such virtual conversation.

A computer programmer remembered what happened when he wrote a short Fortran program to create an early version of computer art made up of characters. This called for combinations of characters not anticipated by the designers of the line printer.

"Well, all 120 hammers fell at the same time, GNUNK! Then they did it again, and again. By line 20 the printer was visibly rocking, and around line 25 – Pop! Tinkle! The little exciter lamp that worked the optical encoder came out of its socket and shattered."

Another veteran sympathized –

computer peripherals as "the largest, noisiest, most troublesome, and expensive" part of the computer system. The group's first printer product, announced in 1952, was the 600 lpm UNIVAC Model 16 High Speed Off-Line Printer System.

On-the-fly printing was a logical evolution beyond the type bars of early calculators and accounting machines and the wheels of the IBM 407. As long as printers used metal type elements, weight and inertia were implacable challenges. Through history all sorts of machines progressed from reciprocal, start-stop motion to rotary motion to continuous rotary motion. In short, wheels beat walking. To speed things up, you can reduce the mass (hence part of the appeal of wire printers), or eliminate inertia (hence the appeal of continuous motion). In typewriting,

spherical type elements replaced individual slugs on levers. In adding machines and the later IBM 407 Accounting Machine, type wheels replaced reciprocating type bars for each print position.

There was apparently nothing patentable with the general concept of a back printer in which the form is struck against a continuously rotating drum. This innovation stemmed from two basic insights. First, it doesn't make much difference whether you strike the paper with the typeface (as in a typewriter) or strike the typeface with the paper. If you want to minimize inertia, paper has the least mass of any part of the system. Second, if your hammer is fast enough and timed well enough, it can in effect "stop" a continuously moving type element.

So the initial impulse of the computer vendors may have been to develop their own printers. Most (wisely) did not – at least not at first – but rather looked to a small, emerging group of OEM suppliers: ANelex, Potter Instrument Company, and Shepard Laboratories.

ANelex grew from the Boston engineering firm Anderson-Nichols (hence the capital "AN"). Founded in 1952, early customers included RCA, GE, Burroughs/Electrodata, and Control Data. Later, in the early 1960s, they supplied the booming European computer market, listing as customers ICT and Elliott Automation in the UK, and Bull in France. Potter Instrument Company introduced a drum printer dubbed the "Flying Typewriter" in 1951. Their early machine was the first of a series which was especially successful meeting high reliability requirements for military systems.

Shepard Laboratories was a different sort of company. Inventor F. H. Shepard may have been first, introducing his "high speed electronic typer" in 1950. There was both manufacturing and licensing. According to one report Shepard sold full design specs for just $5,000. Among the takers were RCA, NCR and Olivetti. According to Wieselman, the Bureau of Census picked a Shepard

off-line print station for their UNI-VAC I which marked the beginning of the plug-compatible peripherals market.

A Shepard printer was installed by Sylvania in its Needham, Massachusetts Data Systems Operation that was developing a commercial computer around 1960. Employed there in Computer Operations at the time, I remember at least one occasion when Mr. Shepard himself came in with his screw driver to tweak hammer timing to help smooth out the print line that was getting a bit too wavy. We've come a long way from that personal touch! The industry at that time was incredibly small, almost intimate, from today's perspective.

IBM's Catch-Up Race. IBM's search for higher speed printing in a way was both hurt and helped by its development of the 407. Credit for the 407 and many attributes of IBM's later printers goes to one of their printer engineering gurus, H. S. Beattie, who joined IBM in 1933 and was assigned to their East Orange, NJ laboratory. Right away he began thinking about how to speed up the 100 lines per minute of their tab machine. He is credited with the concept of replacing the type bars of the 405 with individual type wheels and developing a resourceful pivot design that achieved the effect (if not the reality) of stationary wheel positioning at the moment of impact. In this way, the print quality of the type bar accounting machines was maintained.

MARCH, 1964

It was not quite this dramatic, but careless button-pushing on a tab printer such as the IBM 407 or early line printers could result in uncontrolled, high speed form skipping. This cartoon appeared in an early 60s article published in Business Forms Reporter, a monthly magazine for the business forms industry. (Reprinted courtesy of North American Publishing Company, Philadelphia, PA.)]

But it is this print quality that may have slowed IBM's development of higher speed printers in the 1950s. They – and also their customers, they believed – were accustomed to relatively crisp characters in exact horizontal alignment. As described in Chapter 2, the early drum

"This sort of problem was very common on the RCA 301 . . . which came with one of the first drum printers, built by Anelex. Each print position was hardware-associated with an address at the top of memory which was supposed to be pre-loaded with the character table. If an errant program clobbered the table then the printer might try to print 135 of some given character all at the same time, firing all 135 solenoid hammers simultaneously and attempting to fire them 63 times in one rev of the drum (as I recall, there were 63 characters round the drum). The net effect was to blow 135 fuses simultaneously, each of which had to be painstakingly replaced. [The add and subtract locations in memory] were write-protected and would cause an errant program to stop when and if it tried to write below address 0199. So I never figured out why they didn't protect the print tables at the other end of memory. Sure would have saved lots of fuses!"

To this a third veteran responded: "Never had this happen, and not only was I an official RCA CSR, I may have played one on TV. The fuses were fairly simple to replace."

Back to vet number two: "You must have been lucky. I ended up doing it on average once a shift. After analyzing it for some time we discovered that the COBOL compiler was generating a bad code for a common situation and triggering a massive overlapping move in memory which clobbered the delimiter which was supposed to terminate the move. As a result, almost any new COBOL program being run (debugged) for the first time was cause for me to pull up my chair and 'assume the position' behind the Anelex. The fuses were indeed simple to replace, but at 135 a crack, it took a little while."

But time can convert pain to nostalgia. Vet number two concludes, "I still have a segment of an Anelex print drum on my desk as I type this."

printers simply could not deliver this level of print quality. Hammering paper against the swiftly-moving type drum behind the paper was bound to create vertical blur, wavy print lines, and shadow images from adjacent character positions.

So IBM invested countless millions looking for a better answer through most of the 1950s. They wanted both speed and print quality.

Even as his 407 was undergoing testing in the late 1940s, Beattie was dabbling with other concepts. One was a single-element serial printer. The type element was a "mushroom-shaped" hemisphere, bearing 52 raised characters. It was rotated and tilted to position the selected character, then driven into the ribbon and paper. Demonstrated in 1946 but never commercialized, it presaged the IBM Selectric typewriter of the 1960s.

Next Beattie came up with a stick printer using one or more 8-sided type sticks which had a lower moment of inertia than the earlier mushroom. A single stick, 30 lpm version of this technology was offered for a while with the 305 RAMAC. A bit later, in connection with the WWAM program, a multiple stick printer achieved 300 lpm, but this too remained primarily in the laboratory.

What about wires?! Less inertia. Cheap. Dot printing: appropriate to digital input.

IBM was already there with the Type 26 interpreting keypunch machine that came out in 1949. As the keyed character codes were punched into the card, the character was also printed above that column, just below the top edge of the card. This single-line wire matrix printer had a stationary head with 35 wires. Selected wires were moved forward by a code plate, then the entire matrix of wires driven against the card through a ribbon.

The IBM Wire Printer. It was recognized at IBM as early as 1950 that print speeds in the order of 1,000 lpm would be appropriate for computers. Engineers argued the merits of faster versions of Beattie's stick printer versus the wire printer. The wire printer won. Development was begun at IBM's Endicott Laboratory, with the somewhat modest goal of building a 400 lpm wire matrix line printer for an IBM accounting machine. This initial goal was soon upped to 1,000 lpm.

Almost right away poor IBM was trumped again. They had barely started on their line matrix printer when, in 1954, the Burroughs subsidiary Control Instruments announced their own 1,000 lpm wire printer. The IBM engineers took this to be affirmation of their direction. More importantly, it reinforced the shift in their attention from print quality to speed. The drum printers of their competitors were four times faster than what IBM had to offer, and the Control Instruments printer more than six times faster. It looked like the market was opting for speed over print quality. IBM management was persuaded to upgrade the wire printer to a full-fledged product program.

But just two years later Burroughs withdrew its wire printer, finding it too complex and unreliable. IBM by that time had gone so far they couldn't go back, introducing a pair of wire printers in 1955. Both used print heads with relatively long and flexible wires, enabling them to be shuttled horizontally to share two column positions. Both printers were rated at 1,000 lpm. Model 719 had 30 heads and could print a 60-character line. Model 730 had 60 heads for a full 120-character print line. The controller permitted the printers to work either on-line with the computer or off-line in a magnetic tape subsystem.

Unfortunately these two printers turned out to be conceptually flawed. Two months after they were announced, it was decided that maintenance load, cost, and noise level were all unacceptable. They were re-engineered and performance lowered to 500 lpm. After several of the 1,000 lpm machines had been shipped and around 200 of the slower printers, IBM gave up. It

was deemed an expensive failure but a useful educational experience. One spin-off benefit was the development of hydraulic actuators for more accurate paper movement (later used in the 1403), and methods of lowering the acoustic noise level.

It was early in 1957 that the IBM planners at Endicott accepted the failure of their wire printers. Competitors had been shipping 600 lpm line printers for five years. IBM customers deserved more than 150 lpm.

Non-Impact Explorations. They began to look at non-impact printing, specifically what they termed "ferromagnetography." An exploratory program was launched which continued for five years without any product that looked viable. Relatively good image quality was achieved, but fusing was not solved. The image tended to fall off the paper. Also, at that point a magnetic toner that worked did not seem to exist.

In parallel with the wire printer program and later magnetrography experiments, others at IBM began working on a microfilm printer. Such a printer would expose film at very high speed, which could then be viewed and/or printed out off-line. It seems hard to believe this was envisioned as a viable offering for general-purpose business computer users, but it was at least viewed as a possible partial solution.

IBM's microfilm printer used the RCA "Compositron" tube which shot a large diameter electron beam through a stencil that held the printable characters. Stromberg-Carlson in the meantime was developing a microfilm printer based on their proprietary "Charactron" tube. The main difference was that in IBM's system the stencil was external to the tube and Stromberg Carlson's was internal. This gave IBM the ability to offer interchangeable character fonts. Nevertheless, the IBM product never made it. They decided to OEM the Stromberg-Carlson system instead. However, they were a bit early. Although eventually Stromberg-Carlson

enjoyed modest success with the product, they were never able to deliver under the IBM contract.

Both these non-impact printer programs seem to have been born of a certain sense of desperation.

IBM's engineers and planners, with all those revenues pouring in from their gathering momentum in computers, were nevertheless stumped by printing. They seemed fixated on going their own way, and on doing it themselves. It was not until many years later that the IBM make-or-buy pendulum began to swing in the other direction. It was maybe a kind of arrogance, a certainty that no one else could do it as well as they could.

If at first you don't succeed, lower your expectations, suggests one piece of bumper-sticker wisdom.

In this vein, in parallel with the magnetic exploration, a less ambitious program had been initiated based on some initial work by Fred M. Demer at IBM's Endicott Laboratory. A planning committee was formed to recommend a direction for an intermediate speed printer and within two weeks they came back proposing development of a chain printer.

Back in the 1930s there were designs for printers using type on a horizontally reciprocating stick or rotating chain. During a given cycle every character was presented at each print position. The main problem was how to let each hammer in the bank know when to fire. Both a wheel printer and a reciprocating stick printer were considered by IBM as the upgrade for the 402 tabulator. The lore is that a prototype of each was tested by IBM for 24 hours and by the end of that period the stick printer had only two columns in working order. Selecting the wheel printer was not a difficult decision at that time. In the stick printer, the problem was apparently the electronics, not the mechanics. But now it was the 1950s and it looked like the character selection circuitry could be both reliable and cost-effective.

The Chain Gamble. So it was that in July, 1957, IBM decided the chain offered their best hope. They could have OEMed a drum printer, or designed their own. But the advantages they foresaw for the chain printer proved irresistible —

- perceived print quality would probably be better since the human eye (or mind) is more sensitive to vertical character misalignment than horizontal misalignment;
- with a chain printer, type slugs could be spaced further apart, reducing or eliminating shadow images in neighboring print positions;
- the above two considerations could also justify less hammer flight time precision, lowering cost and/or reducing maintenance;
- the type drums were heavy, expensive and bulky since they needed to have a full set of the printable characters available to every print position;
- since fewer characters needed to be on the chain, its cost would be significantly below that of a drum;
- it offered greater potential for operator interchangeability, for special character sets or reduced character sets for faster print speeds;
- and, finally, there was its very uniqueness, seen as a marketing consideration.

Once that decision was made, things moved fast. They had made their comitment to the chain printer in July, 1957 and the product was announced as the IBM 1403 printer in October, 1959. Besides designing and gearing up for the manufacture of the printer, in 1958 the decision was made to integrate the printer into the 1401 computer program.

As they say, the rest is history. Observers agree IBM's leap forward in high speed printing added immensely to the success of the 1401 computer. It has been called one of IBM's most success-ful products ever. Within a few years the initial 1403 was followed by a train printer variation which offered up to 1,285 lpm in numeric mode and operator-changeable train cartridges. At some point in the latter 1960s, one observer estimated that half of the world's continuous forms volume was being printed on the IBM 1403 printer. The series was manufactured until 1985.

Late in the decade IBM began another line printer program to develop a lower speed printer for the 1440 System. The heart of this printer was a horizontally-reciprocating type bar. The comb-like bar carried a line of flexible springs, each with a raised character mirror image. Unlike the 1403, the IBM bar printer was a front printer: there was a bank of hammers that drove selected characters into the ribbon and paper, as in a typewriter. In early prototypes the bar was halted each time a hammer was fired. Soon it was found the dwell time of the hammer was so short that they could print on the fly, with the bar reciprocating back and forth at a constant speed without the hammer snagging the next finger of the comb.

The bar printer was introduced as the 1443 printer along with the 1440 computer in 1962. Two models were offered performing at 240 lpm and 150 lpm respectively, with the normal 52 character set. The print quality was said to be good enough so that used with a special film ribbon they could print MICR checks. These machines, like the 1403, proved to be a commercial success, although not on the scale of the chain printer.

Yes, it took them a while. But they persisted, marched to the beat of their own drummer, and in the end got it right. A lesser company might have given up. Or more likely, run out of money. It was the beginning of a roll. Through the 1960s and 1970s IBM followed this success with a string of winning hardcopy devices. The Selectric became not only a leading typewriter, but also a mechanism of choice for an assortment of I/O terminals. Then came the equally successful 1403 Model 3 train printer. The IBM 3211 band printer introduced in 1971 at 2,000 lpm, was an in-

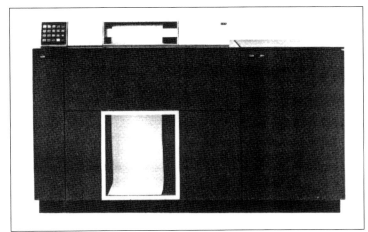

Left to right. Remington Rand won the high speed printer race with pioneering 600 lpm drum printers in the 1950s. They stayed with this technology quite a while. This is their 1971 Model 0768 Line Printer. 2. IBM's chain and train printer family, shown here with a low-end IBM system, was manufactured for 25 years. 3. The IBM 1443 stick printer is shown here with a low-end System/360. 4. IBM's high speed impact printer development evolved to a well-received series of band printers such as the 3211 printer with 3811 control unit, introduced in June, 1970. (Source: UNIVAC drum printer, IBM chain printer, and IBM 1443 photos are publicity photos from the respective manufacturers; DRA Archives; IBM 3211 photo is courtesy of IBM Archives.)

dustry pace-setter. Then they pioneered non-impact printing with the high speed 3800 introduced in 1975, and shortly after that, the sheetfed 6670 laser printer.

IBM had demonstrated that horizontally moving type was better than vertical. The rest of the world followed. During the 1960s and 70s, competitors pretty much abandoned on-the-fly drum printing and switched to an incredible variety of back and front printers with horizontally rotating or reciprocating bars, trains, belts and bands. In printers as well as computers IBM led and the lesser competition followed.

For the rest of the IBM printer story to the present, visit the 1970s chapter which covers the high-end production printer race featuring IBM, Xerox, and Siemens.

The Pioneers. What happened to the early drum printer pioneers? Apparently they were not able to make the investment needed to transition to horizontal moving-font technologies. The printer assets of ANelex were acquired by Mohawk Data Sciences sometime in the late 1960s. Shepard Labs is believed to have become part of

*Input/Output Typewriters:
Throughout the 50s and into the 60s the IBM Model B typewriter and the Flexowriter — built around the IBM Model B mechanics — were by far the most popular machines for console printers and input/output terminals, followed perhaps by the Teletype Model 28. This is the adaptation of the IBM Model B typewriter by Soroban Engineering, Inc., an OEM supplier active in the 1950s and 1960s.
Circa 1960 one of the many users of the Friden Flexowriter was Digital Equipment Corporation, as shown in this photo of an early DEC PDP (Programmed Data Processor) minicomputer. (Photos courtesy of Compaq Computer Corporation, Digital Photo Library.) Friden Flexowriter controls, from an operator's manual for the Sylvania 9400 computer system (1961). (Source: DRA Archives)*

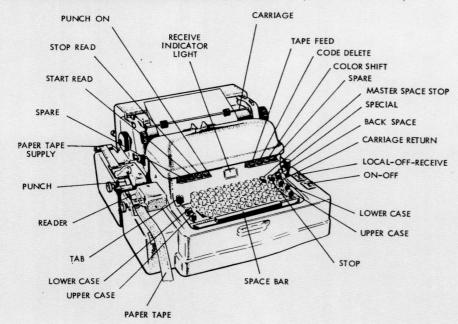

Vogue Instruments around the same time. Remington Rand UNIVAC merged with Burroughs in 1986 and is of course still with us as Unisys. They built well-regarded and cost effective high speed drum printers through the 60s, followed by a series of band printers in the 1970s. By the 1980s they were also sourcing print engines from various OEM suppliers including Dataproducts, Mannesmann-Tally and Okidata.

NON-IMPACT PRINTING

During the 50s, besides IBM, a number of other companies were working on various types of non-impact printers. As mentioned, CRT-to-microfilm printing was being developed by Stromberg-Carlson. The long defunct firm Data Display fielded the DD-80 microfilm printer around 1963. It is believed only a few of these systems were shipped.

In this decade, non-impact printing was generally seen as one of those concepts that was destined to prevail, but that was ahead of its time. Wieselman and Tomash, in their 1991 retrospective, covered non-impact first in a section headed "Nonimpact Printing." Later, covering the 60s, they wryly labeled the topic "Non-impact Printers that Worked." [*]

In the 50s technologies employing coated papers were well established, but primarily for copying (3M Thermofax, Diazo, etc.). These processes had significant drawbacks including poor archival properties, often marginal print quality, odor, and high supplies costs.

The boldest non-impact leap was made by Rank-Xerox in England which introduced an electrostatic printer named the Xeronic page printer in 1958. Rank-Xerox was created with the merger of Rank Precision Industries Ltd. with Haloid-Xerox. The Xeronic printer used a CRT to generate characters and was speci-

---

[*] Tomash, Erwin, and Wieselman, Irving, op. cit.

The Xeronic Printer, a bold but unsuccessful leap into non-impact page printing in 1958. (Source: Charles Babbage Institute Archives)

The Burroughs "Whippet" electrostatic printer, built for the U.S Army Signal Corps in the 1950s, was perhaps the first non-impact printer actually committed to manufacture. (Photo from Annals of the History of Computing, Vol. 13, Number 1, 1991: "Marks on Paper: Part I" by Irving Wieselman and Erwin Tomash © 1991, IEEE)

fied to run at 2,880 lpm or print two-across pages at 88 per minute. It had a form overlay feature and print width of 23 inches. At least one system was actually delivered to a British government agency, but that might have been about it. It was a notable feat, however, since it was another 16 years before another electrophotographic page printer of this sort was fielded.

There may have been only one non-impact printer during the decade that might be described as at least semi-successful. This was the pioneering Burroughs "Whippet," formerly named the Burroughs High Speed Electrostatic Teleprinter. Developed under

a government contract, the Whippet was an electrostatic message printer for the Signal Corps which could print at around 3,000 words per minute (or 6.5 pages per minute). Coated dielectric paper was used with liquid toner. It was announced in 1956 and around 100 of the machines are believed to have been delivered. But it did not appear to have seeded any successor products, at least not for Burroughs.

Non-impact printing continued to be mostly a dream in the 1950s. In normal computer and office applications, graphics and color were not yet even a dream. The cast of company characters

that made up the emerging industry was very small. The computer industry was young and its explosion into almost every facet of commerce, industry, and national defense had barely begun. At the decade's outset, the very need for high speed printing was realized only by a few. Yet printers evolved from 100 lpm accounting machines through 600 lpm drum printers to the remarkably successful IBM 1403. A small group of pioneering printer firms served all the computer vendors except for IBM and Remington Rand. By the end of the decade most of the other computer companies were working to develop their own printers. Industry structure was about to radically change.

The stage was set for Decade Two which would see accelerating innovation, the number of printer vendors more than triple, and the emergence of the OEM printer market — in short, the birth of an industry.

Heavy metal: print drums await final assembly on the Dataproducts plant floor, 1968. (Courtesy of Hitachi Koki Imaging Solutions)

CHAPTER 4: THE SIXTIES

THE SETTING What a difference a decade makes! It is probably not quite so clear-cut, but if the USA of the 50s can be broad-brushed as smug and integrated, the 60s can be caricatured as a decade of doubt, dis-integration, and also hope. The decade witnessed Vietnam War protest, cultural liberation, assassinations, and race riots, shattering what was left of the idealized "American Dream." For some, it really was a time of sex and drugs and rock and roll. More importantly, there was also idealism that bore fruit in terms of real gains. The walls of racial segregation began to be torn down. Betty Friedan's "The Feminine Mystique" appeared in 1963, marking an emerging women's rights movement. A "war on poverty" was launched.

Through all this, capitalism was not touched and mainstream America continued to barge ahead with business as usual. Adventuresome entrepreneurs rose and fell, who along with the Cold War-driven military, served as dynamos powering spectacular advances in science and technology. Computers and communications advanced in leaps and bounds. IBM's 1401 and the subsequent System 360 introduced in 1964 far outsold all other computers combined. The later 60s saw early word processing and the Carterphone Decision that allowed independents to hook "foreign" equipment into the telephone networks. In 1969 INTELSAT was operational and enabled live global TV coverage of the Apollo lunar landing to reach 500 million people around the world.

The first seeds of today's Internet were sown. In 1969, under a defense contract, UCLA developed a "switching computer" to control dataflow and direct messages within ARPAnet, the Defense Department's Advanced Research Projects Agency messaging network. Computers continued to be expensive. Most companies could not afford their own computer, but rather had their needs met by time-sharing companies. IBM's decision to unbundle – i.e. price hardware and software separately – in 1969 was a major development accelerating the growth of both the independent software industry and plug-compatible peripherals, including printers.

Tomash and Wieselman had a nice way of distilling the computer industry decades, writing, "Just as the 1940s were the research years and the 1950s the engineering years, the 1960s were the systems years."[*] In the early years of the decade the vacuum tube machines were replaced with smaller and faster transistorized second generation systems that came bundled with peripherals, utility programs, and compiler software so the user's programmers could write their own applications. When IBM introduced its ground-breaking 1401 computer, the total population of computers in the USA was believed to number only about 6,000. IBM shipped well over 15,000 of its 1401 computers during that system's full lifetime. IBM's later System 360 family accelerated their domination of the industry.

Contemporary observers dubbed the U.S. computer industry of the 60s "Snow White and the Seven Dwarfs:" It was IBM and the dwarfs, namely CDC, NCR, Burroughs, GE, Univac, RCA, and Honeywell. Philco was there, too, as a lesser player. Overseas, ICL was formed in the U.K., the French government funded a major computer initiative, and others such as Olivetti and Siemens brought out new systems. GE and Honeywell in the 60s and 70s, seeking to escape dwarfdom, indulged in a long series of relationships with one another and with Olivetti and Bull in Europe, a complex a tangle beyond my ability to summarize.

The typical computer of the 60s was not as huge as the pioneering computers of the 50s. But it was still big, a multi-component "mainframe" normally filling a specially designed room and tended by trained operators who interfaced with a galaxy of software engineers and programmers who developed the actual applications. With their high-end systems, IBM often stationed a maintenance engineer on-site full time. The system was so expensive that usage would be timed to the second and billed to the departments that requisitioned the application.

But in parallel with this computer mainstream flowed a stream of minicomputer development. In a sense, this was as important a leap forward as the PC revolution that was to sweep the world in the 80s. The minicomputer, in contrast to the behemoths, was still not cheap, but was more of a hands-on tool, first for engineers and scientists, and later for small data processing departments. Among the pioneers were Digital Equipment Corporation (DEC), Scientific Data Systems, and Computer Control Corporation.

Of these companies, DEC's story is the most spectacular. It was founded in 1957 by three men with $70,000 in capital who rented 8,680 square feet of space in an old woolen mill building in Maynard, Massachusetts. By the end of the 60s, the corporate profile looked like this:

5,800 employees

$135 million in revenues

68 locations around the world

8,000 computers installed

In 1974 they bought the whole 19-building mill complex for $2.25 million in stock.

The minicomputer industry was obviously booming and it served as a major driver that set the stage for an explosion in the printer market and number of firms addressing it. Other forces were the ever-expanding input-output bottleneck and IBM's unbundling decision of 1969.

High Speed Printers. Early in the decade, among the computer companies, it was only IBM and RemRand that developed their own high speed line printers. Their smaller competitors continued outsourcing from the first generation of independent suppliers of which the largest was ANelex. But as the decade wore on, several computer manufacturers began building their own. A major trigger was the IBM 1403. The drum printers developed by the first generation independents were not competitive performance-

---

[*] Tomash, Erwin, and Wieselman, I.L., op. cit.

wise. Honeywell, Burroughs, RCA, and Control Data thought they could do better and soon launched their own printer development programs.

In parallel, a second generation of independent printer vendors appeared, led by Data Products Corporation, founded in 1962. As probably the most successful and longest-lived of the independents, Data Products (later, Dataproducts) was a logical choice for our defining company of the decade. As described below by the founder, they attributed their success in part to a pro-

THE BOTTLENECK

While print speeds barely doubled, computation speeds increased almost 3000 times!

For example –

Year First Delivered	Computer	Speed (additions per second)	Printer Speed (at least 46 char.)
1951	UNIVAC I	3,500	600 lpm
1958	IBM 709	41,600	150 lpm
1961	IBM 7090	227,000	600 lpm
1967	CDC 6800	10,000,000	1,000 lpm

Source: *Business Forms Reporter*, January, 1966 issue

prietary hammer invention which let them come out with better print quality at lower hardware cost. Early customers included the minicomputer companies DEC, SDS and Hewlett-Packard. In time some of the larger mainframe manufacturers became customers as well.

Dataproducts' entry into the market, according to Tomash, created price and performance pressures which were devastating to the first generation printer companies. He relates, not without a bit of satisfaction, how ANelex, trying to keep the business, had to cut their prices and in time were forced into bankruptcy. The

assets of ANelex ended up with newcomer Mohawk Data Sciences by the end of the decade. Shepard Laboratories seems to have had the same experience, in time acquired by Vogue Instrument Company. Dataproducts was truly a company that changed the industry. This chapter concludes with an in-depth profile of Dataproducts and an interview with Erwin Tomash, the man who made it happen.

Although Dataproducts was the leading independent by far during the 1960s, in terms of shipment volume, most of the line printer action remained with the captive printer operations.

Computer Peripherals, Inc. (CPI) was an interesting enterprise set up to spread the costs. It began in 1964 when Control Data formed a partnership with Holley Carburetor Company, the auto industry supplier which somehow had developed a cost-effective, mid-speed drum printer. Later NCR and ICL bought into the company allowing the cost of research, development and tooling to be shared. CPI remained captive, supplying printers mainly to its three parent computer companies until 1982 when it was merged into Centronics.

Except for IBM, as of the mid-60s, all high speed printers were flying drum machines. The laws of inertia could not be repealed, so price/performance gains were modest. Print speed was a function of hammer speed, paper advance speed, and more sophisticated electronics that permitted asynchronous timing.

Hammer action could not be significantly accelerated. Paper advance line-to-line and especially high-speed skipping over vertical white space was accelerated. While ANelex moved paper at just around 27.5 inches/second, the IBM 1403 could perform high speed skipping at up to 75 inches/second thanks to its "dual

speed carriage." IBM's interchangeable type elements allowed higher speeds with reduced character sets. Some drum printers also came out with interchangeable drums, permitting accelerated performance (2,000 lpm with an all-numeric drum or train, for instance), or additional applications such as OCR printing. Asynchronous timing allowed paper advance and a new line printing cycle whenever all the characters in the input buffer had been printed, rather than waiting for the drum or train to advance to a fixed position.

Formatting and feeding data to the printer was not a big problem. It only needed one line's worth of characters at a time. The mechanics were the bottleneck. Today, with color and otherwise more complex images, the opposite can be the case: formatting the page to be printed can hold up a page printer. RIPing the images is often off-line from the printer.

Speed gains were modest, but quite a bit of progress was made in tailoring performance and printer cost to the application. Over the decade pricing of high speed printers remained more or less constant at between $25,000 and $35,000 (without control unit). By mid-decade a number of manufacturers offered lower speed and de-functioned printers. These printers, running in the 150 to 540 lpm range, were priced at from $10,000 to $22,000. Controls and paper handling mechanics could be stripped down because of the lower speeds. There might be just one set of tractors pulling the form up through the print mechanism with vertical tension provided by an adjustable "drag control." Instead of a swing-out gate yoke, the drum might be held by an assembly hinged on both ends. The yoke could be dropped down a few inches for threading through a new form. Some printers cut the number of print hammers in half, spaced five per inch rather than ten per inch and shuttling the paper back and forth to cover all the print positions.

In the mid-60s timeframe, the world of high speed line print-

ers remained amazingly small. One source placed the U.S. population still below 40,000 including IBM's old punch card tabulators (see side bar). Also, a look at representative printers affirms the decade's limited price/performance gains. It was not until the early 70s that printer speed made a significant jump.

Low-end Printers. During the 60s two ground-breaking low-end printers chewed into the domain of the old type-bar machines. The IBM Selectric appeared in 1961 and over the decade formed the mechanical heart of a variety of console printers and remote terminals. A year later Teletype replaced its Model 28 with the Model 33. The Selectric had better print quality and more speed, but the Model 33 cost less. It became one of the most widely used printers of all time with (depending on your source) 500,000 or 750,000 delivered before being discontinued around 1981.

Toward the end of the decade a new serial impact technology appeared, the daisywheel printer from Qume and Diablo. Their print quality was close to that of the Selectric at double the speed. They never displaced the Teletypes, Selectrics and Flexowriters for computer console or telecommunications applications. Instead, during the ensuing decade, they served as significant enablers of that decade's word processing industry.

In 1964 Epson in Japan demonstrated serial matrix printers in conjunction with the Tokyo Olympics, an almost unnoticed portent of revolutionary changes to come.

GENERAL ELECTRIC BREAKS NEW GROUND

Late in the 60s, rapidly expanding telecommunications generated demand for faster teleprinters. GE responded, pioneering another stream of low-end printer development. The story touches on a number of themes. An era of diversification by major corporations

Estimated Population of Line Printers in the USA as of Mid-1965

ANelex	1,775	IBM 1443 (all models)	1,575
Burroughs	450	Potter LP 600 and LP 1200	675
Dataproducts	200	RemRand Model 3 Tabulator	1,435
Honeywell	650	Shepard Drum Printers (RCA, NCR, etc.)	620
IBM Type Bar Tab Printers	7,050	UNIVAC III and 1004	2,265
IBM Wheel Tab Printers	9,925	UNIVAC High Speed Printers	520
IBM 1403 (all models)	10,525		

Source: *Business Forms Reporter* Magazine, August, 1965 issue

The cozy world of line printing in the 1960s: Prices climbed a bit for given speeds and lower speed, lower priced printers were fielded for the minicomputer market. Estimated printer populations, deduced from announced computer placements, remained miniscule from today's vantage point.

Representative Line Printers, 1960-1971

Printer	Year of Intro	Price*	Speed	Print Positions	Array**
IBM 1403 Model 1	1960	$35,000	600 lpm	132	48
IBM RCA 301***	1962	$30,000	1,000 lpm	120	64
Potter 3502	1964	$12,000	400 lpm	132	48
IBM 1403 Model 3	1965	$50,000	1,100 lpm	132	48
NCR 64D-2DD	1968	$45,000	3,000 lpm	132	13
IBM 3211	1971	$100,000	2,000 lpm	132	48

* End user price for complete printer with control unit; announced quantity 1 price or monthly lease price times 35.

** Number of printable characters

***Adaptation of the ANelex Model 4-1000

was dawning, and GE was leading the pack. It is a story of entrepreneurship within one of the world's largest manufacturing companies. It is a story of vision, of how one executive anticipated a court decision that was to fuel demand for a new generation of printer products. It is a story of electromechanical innovation.

It is a story that follows the classic skunk works scenario. This time it's a team of engineers locked away – figuratively speaking – not in a garage, but in a former Studebaker car dealership building some miles from the parent GE plant in Waynesboro, VA.

The assignment: Design a new generation of teleprinters. Start with a clean slate; take advantage of the latest advances in data communications and electronics; worry about time but not about money.

The result: the GE TermiNet 300. According to one report, it was Dow Jones that encouraged GE to develop a faster teleprinter and that placed the first major orders. When installations got underway, Dow Jones decided the new printer's 30 cps performance was such a leap ahead they had to slow it down so their customers with the new TermiNets would not get access to trading information sooner than those with older model teleprinters.

The TermiNet 300 and its later progeny achieved significant market share during the decade of the 70s. GE's first product, the TermiNet 300 was the most successful in terms of shipments. It was introduced in 1969 and remained in production until the end of 1975, with a total of 65,100 shipped.

A couple of key engineers who were part of the original GE printer project were happy to share a quick, insider's overview of those early, creative days. First, Bob Garcia:

"It grew out of the GE Specialty Control Department. Back in the 1950s the corporate pendulum was swinging from centralization to decentralization. After they reshuffled operating groups there were some odds and ends left which they didn't know what to do with. So in 1955 they formed the Specialty Control Department in Waynesboro. In the 1960s I was working for GE on NASA contracts in Daytona Beach, Florida, but those contracts began drying up so I went up to the Waynesboro operation and joined the printer development team.

"The impetus had come from Dr. Rader, a GE VP who was formerly with Sperry Rand. It came from the top. Rader just said, 'OK, we'll do this. Don't worry about money.' At this GE location there were the right kind of engineers, optical fencing specialists and the like. So the team was set up a few miles out of town."

Allen Surber, another veteran of the project, now an independent consulting engineer, emphasizes the significance of the Carterphone decision of 1968. Surber joined GE in that year and went to work in the printer skunk works. "We had some great people there, including Kirk Snell who directed the program, and Cliff Jones, who I see as the real pioneer in developing the TermiNet printer.

"Dr. Lou Rader, a VP of GE, had seen an opportunity opening up from the legal challenges being made to the AT&T monopoly. Up to that time all telephones operated at 110 baud and you needed a black box between you and Bell if you wanted to tie anything in. Carterphone had developed a 300 baud modem but was locked out of the Bell network. They challenged AT&T and in 1966 the FCC decided in Carter's favor. The Carterphone Decision opened up the Bell network. When we introduced our first product, the TermiNet 300 in 1969, it was the only 300 baud terminal. That's one of the main reasons we succeeded. Timing. Before that time no one could use that speed, no one needed that speed before.

"The impact of our project there went well beyond the printers themselves. It was the advent of higher communications speeds that were to drive the development of digital printers through the 1970s and beyond. Digital printer progress during those years was driven from the teleprinter side."

Innovations. The initial TermiNet product was specified to be a high reliability, full-character teleprinter that would make maximum use of electronics, with higher cost more than offset by significantly higher speed. The primary target market was the huge base of Teletype Model 28, 33 and 35 users. These printers, as well as the Model 37 which Teletype was developing at the same time, were rated at 10 cps. GE's teleprinter, along with the somewhat later DECwriters, precipitated Teletype's gradual decline and eventual disappearance as a printer power.

The new printer incorporated a number of significant innovations. One, according to Garcia (who was in charge of logic design for the program), is that it was the first successful commercial product to use LSI MOS integrated circuits.

In its basic architecture, the TermiNets are actually small, table-top, belt line printers. Being a new concept, as always, there were quite a few hurdles.

The keyboard discarded traditional electromechanical switches in favor of electromagnetics. They used ferrite cores which closed an appropriate circuit when passed through openings in the keyboard PC board.

Unlike typical on-the-fly line printers of the time, the TermiNets were front printers. The result was quieter operation than contemporary teleprinters, and better print quality than most line printers. A set of flexible print "fingers" made of maraging Swedish steel were carried across the print line by a constantly moving belt.

To meet punishing speed and flexibility demands, proprietary material was developed for the belt. It was a urethane based elastomer reinforced with fine stainless steel thread (and later) Kevlar fibers. Static build-up between the non-conductive belt and the drive and idle pulleys was a problem, solved by coating the inner side of the belt with a water-based graphite coating.

Synchronization of the print hammers was an interesting challenge. Each had to strike the selected character finger as it zoomed by, and then retract fast enough to avoid ripping the next type finger out of its belt socket. After doing some pioneering high speed photography, the designers decided to bevel the leading edge of the print fingers to gain 0.010-inch clearance. There were also innovative hammerbank design, drive circuits, and control logic the details of which are beyond the ken of the non-engineer reader (or this writer). Suffice to say that the basic concept allowed various TermiNet models to eventually print at speeds up to 340 lpm.

Bob Garcia today, and Garcia and Allen Surber back then. Allen Surber (seated), described the scene: "On the left you can see part of the TermiNet 300 development model on the desktop and in the left foreground is a Tektronix oscilloscope. The time frame would be about 1969." (Photos courtesy of those pictured.)

Compared with the targeted Teletypes, the TermiNets were sophisticated and at around $4,000 apiece ($18,780 today!), more than double the cost of the Model 33 Teletype.

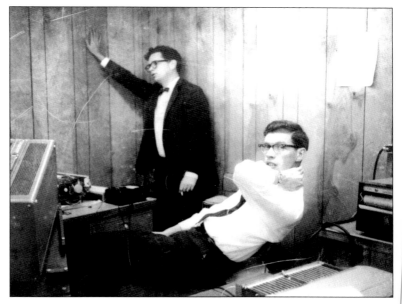

But the product's track record of deliveries testified that its performance and capability more than offset the cost differential for many users, true to Dr. Rader's original product strategy.

GE DCPD From Then to Now. The TermiNet printer family was manufactured by GE's Data Communications Products Department in

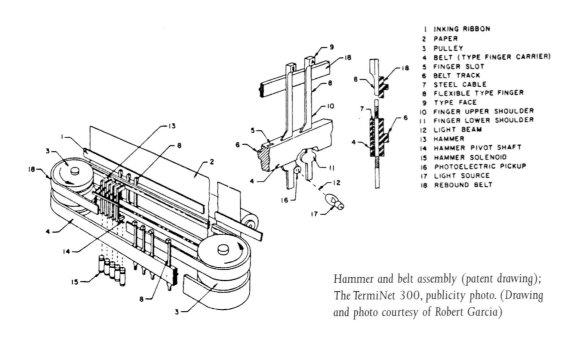

```
1   INKING RIBBON
2   PAPER
3   PULLEY
4   BELT (TYPE FINGER CARRIER)
5   FINGER SLOT
6   BELT TRACK
7   STEEL CABLE
8   FLEXIBLE TYPE FINGER
9   TYPE FACE
10  FINGER UPPER SHOULDER
11  FINGER LOWER SHOULDER
12  LIGHT BEAM
13  HAMMER
14  HAMMER PIVOT SHAFT
15  HAMMER SOLENOID
16  PHOTOELECTRIC PICKUP
17  LIGHT SOURCE
18  REBOUND BELT
```

*Hammer and belt assembly (patent drawing);
The TermiNet 300, publicity photo. (Drawing
and photo courtesy of Robert Garcia)*

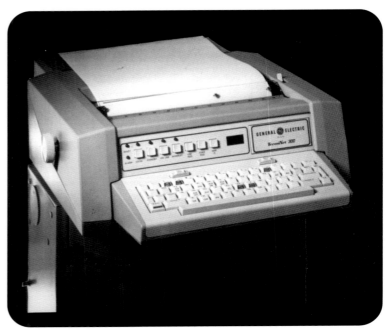

GE DCPD Development Summary

1965 Product Planning	1967 Model Prototypes	1968 LSI Development

Product Line

Year Introduced	Model	Print Speed	Printer Type	Lifetime Volume
1969	TermiNet 300	30 cps	teleprinter	65,100
1972	TermiNet 1200	120 cps	teleprinter	25,800
1974	TermiNet 120	120 cps	lineprinter	9,800
1975	TermiNet 340	340 cps	lineprinter	33,330
1976	TermiNet 1232	120 cps	teleprinter	6,000
1977	TermiNet 510	340 cps	lineprinter	3,400

Waynesboro until the end of 1984. In the mid-1970s DCPD developed and fielded a new family of impact dot matrix printers, the 30 cps TermiNet 30. As of 1975 the printer plant had completed an expansion to 500,000 square feet and employed over 3,000 workers. Their highly centralized, integrated manufacturing was the opposite of the norm today for U. S. manufacturers. In-house CAD was used to design printed circuits and around two-thirds of the boards used were from automated, in-house production.

The original TermiNets were superseded by dot matrix and other technologies in the early 1980s. In an effort to diversify beyond teleprinters into general purpose output printers and specialized printers, the name TermiNet was discontinued around 1983.

How has this business fared since then?

As the 1970s wore one, despite the success of some of the TermiNets, entropy appeared to be setting in. Their "development summary" shows diminishing market success with the

products that succeeded the TermiNet 300 (see box). By today's standards, in fact, all the numbers look somewhat insignificant. But remember, this was long before the days of the commodity printer. They were relatively high ticket peripherals for the then still young computer industry.

Yet it was clearly not a growing business. In the words of one report, "Under General Electric's wing, many felt that the printer operation had not been allowed to aggressively pursue future market demand, leaving the company with dated products and a somewhat stodgy corporate image as a division of one of the world's largest companies."[*]

In 1983 GE sold its printer business in a leveraged buyout. Genicom Corporation was born. The result was new energy. Over the next few years products were diversified, dealer/distributor sales expanded, new executives recruited, most production moved to Mexico, and a European subsidiary, Genicom International, set up. The company also experienced major financial challenges. They went public to retire the debt incurred in the buy-out in 1986 and the same year acquired the assets of Centronics for $75 million. A bold move. But in retrospect, maybe not a smart one. What they got did not justify the debt burden, red ink flowed, and their per share stock prices sank.

There was a reprieve early in the 90s. Bob Garcia described his company in 1992 as "the largest independent domestic printer manufacturer today."[†] By that time Hitachi had acquired Dataproducts, and Lexmark and Tektronix had perhaps not quite caught up. Hewlett-Packard was doing a much larger volume in printers, but was not a pure printer company. They acquired what was left of the Texas Instruments printer business in 1996 and Digital Equipment's the following year.

After several profitable years, in 1998 they reported a loss of close to $2 million on sales of $450 million. Not to worry, Genicom said, the fundamentals are strong. In 1998 Gartner Group/Dataquest ranked them as fourth in overall U.S. shipments in the "midrange" printer market. But just two years later, in April, 2000, Genicom declared bankruptcy. As of this writing it looks like the company will go on the block and perhaps disappear into the product line of a competitor.

NON-IMPACT PRINTERS

During 60s it became more apparent than ever that digital printing would have to go non-impact to support the galloping advances being made in computers and communications. There were some big investments, but very little success in reducing that theory to practice.

One successful type of non-impact printer at the time was the microfilm printer. Microfilm was exposed from a CRT. The data could be archived as film, with the option of printing to hardcopy on-line or off-line. The Datagraphics division of Stromberg Carlson was one of the more successful participants with well over a hundred systems installed. 3M Company's Series F Electron Beam Recorder went 20,000 to 30,000 lpm, cost a bit under $100,000, and used a proprietary 3M dry film development system. Other printers of this type were fielded by Kodak, Control Data, Benson-Lehner, California Computer Products, and IBM.

Direct to paper printers (as opposed to requiring a film intermediary) were less successful.

Radiation, Inc. in 1963 fielded the 31,000 lpm electric discharge printer featured in Chapter 2 (Speed, Specsmanship, and Money). But like the Rank Xerox Xeronic of the 50s, it was ahead of its time. In fact, considering the downside of electric burnoff technology (image quality, paper cost, environmental negatives), its time was destined never to come.

NCR was more successful with its Model 260 thermal printer, a low-end device that found a market as a telecommunications

[*] *1985 Printout Annual*, Datek Information Services, Inc., July, 1985

[†] Garcia, Raimundo R., "And Then There Was General Electric" (unpublished article)

The fastest non-impact printers of the 60s were CRT-to-microfilm systems such as this Stromberg DatagraphiX Model 4360. (Stromberg Carlson publicity photo, from DRA archives.)

printer primarily for the military. The printer used heat sensitive paper developed and manufactured by NCR. But print quality and media concerns limited the market and the product line had a short life.

Ink jet was pioneered by Teletype, which introduced the electrostatic pull Inktronic teleprinter in 1966. The machine had 40 jets, each of which covered two print positions. The Inktronic offered break-through speed at 120 cps, but cost, image quality, and reliability were among the negatives that doomed this product.

During the 1960s it was primarily A. B. Dick Company that managed to field commercially successful non-impact printers. Part of the company's success was choosing the right application. Non-impact printing at the time simply could not deliver the image quality needed for office and other general-purpose applications. But where speed was paramount, it was a realistic fit.

A. B. Dick's electrostatic Videograph, first delivered in 1960, is viewed as the first commercially successful non-impact printer (excluding the microfilm printers). According to Wieselman and Tomash, the Videograph printed address labels on a single 3-inch strip of dielectric paper at around 14,000 lpm (36 labels per second). Character generation was by a specialized CRT with a matrix of fine wires embedded in the faceplate. Early models used dry toner; later, liquid. The systems sold for between $250,000 and $500,000 − pricey, but for large circulation, time-sensitive

mailings, it was for around twenty years the only system that would do the job. Time-Life was the first customer and by the late 1970s the system was used by a good percentage of periodical mailers throughout the country.

"Unfortunately, there were only a limited number of large mailers in the world who needed that capability, so the market got saturated. Over the twenty-year life of the product, I believe around 25 systems were shipped." That was Al Johnson, a retired electrical engineer who in the late 60s was a product line manager at Sylvania's Electronic Tube Division in Seneca Falls, NY.

According to Johnson, development of the tube was a joint project among Sylvania, Corning Glass, and A. B Dick Co. The concept was to conduct a charge pattern with minimal diffusion onto the dielectric paper label stock that was pulled past the faceplate at around 40 inches per second.

Sylvania developed and manufactured a variety of specialized CTRs for printing and recording applications. Two types were fiber-optic faceplate CRTs and electrostatic charge tubes with embedded wires. A. B. Dick used the latter. Tubes available at that time used soft glass and #4 alloy wires which suffered too much charge diffusion. Hence the Corning effort was to develop a hard glass/tungsten wire system. This allowed for much better high speed printing capability. The printing tubes for A.B. Dick contained a 0.25" x 2.75" array of tungsten wires 0.005" in diameter on a 150 per inch grid. As Corning achieved better manufacturing capability, some later tubes were increased to 250 wires per inch in both axes. These wires had to be perfectly flush to both the inner and outer surfaces of the hard-glass plate, and the paper had to be in contact with the face of the tube. For several years these machines were running 24 hours a day to maintain the flow of address labels.

Although the addressing market was limited, A. B. Dick Company did find other applications for the technology including full-

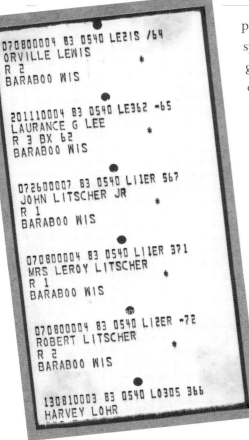

Videograph test labels dating from the pre-ZIP code era of the 1960s; the non-impact print obviously ages well. (Archived label samples courtesy of Al Johnson.)

page fax printers and high speed graphics printers for governmental agencies. One of the most interesting applications was a camera-printer system that clicked on when freight trains passed by to make a visual record of every freight-car.

A few years after launching the Videograph, A. B. Dick Company committed further to non-impact technology, licensing the Sweet continuous ink jet patent in 1963. Their first Sweet technology product was the Videojet 9600 in 1968. Priced at $7,500, the machine quietly printed 132 print positions across page-size documents at 250 cps. It was faster than the impact dot matrix printers of the time, but also more expensive. In the end, for office page printing, the Videojet could not compete.

But they didn't give up. From then on, A. B. Dick Company began to focus on industrial applications for their technology, with much better luck. One of the first products printed codes on beverage cans moving past the print head at up to 2,000 cans per minute. Introduced in 1973, A. B. Dick Company claims their Model 9000 was the world's first industrial ink jet ID system.

Running hard in that direction, they changed the name of their IDP Division to Videojet Systems International in 1980. In time, the tail became the dog and Videojet and A. B. Dick Company went their separate ways. In 1993 Videojet "united operations" with Cheshire, a major competitor, and in 1998 they acquired Marsh Company, a leader in large character marking and coding. Now owned by The General Electric Company, p.l.c. of London, England, they recently were renamed Marconi Data Systems.

The moral of the story? One might be, If you can't fit the technology to the application, fit the application to the technology. Another might be, Unless you have the resources of an IBM or an HP, find a comfortable specialty rather than trying to be all things for all people. By keeping focused, Videojet (as Marconi) is still alive and apparently well after forty years. Parent A. B. Dick Company was founded back in 1884 and is still with us. Myriad computer and printer companies have come and gone in much less time. This brings us to this decade's "defining company."

DATAPRODUCTS CORPORATION: AN EXERCISE IN ENTREPRENEURSHIP

Ironically, Dataproducts originally had its sights set primarily on storage, not printers. The printers were something of an accident. Yet in the eyes of observers twenty years later, it was printers that made the company.

We bring you the story primarily as oral history by founder Erwin Tomash. The founding and nurturing of the company through its adolescence, through his eyes and memory, looks like an exercise in flexibility, perseverance and leadership. The main focus seems to be building the company. The product is secondary. As an example of the interplay between invention and enterprise, the Dataproducts story seems heavily weighted toward enterprise. It is a story that might also be entitled, No Guts, No Glory. It is a glimpse into the mind of a classic entrepreneur.

The first business plan was on a restaurant placemat. In March, 1962, Erwin Tomash and Bill Mozena met for lunch at the Bel Air Sands Hotel and scribbled this five-year plan. The original placemat was later framed and hung for display at Dataproducts' corporate headquarters. (Courtesy of Hitachi Koki Imaging Solutions)

Note: This account is excerpted (with some later edits by Tomash) from an interview on May 15, 1983, in Los Angeles, CA with Arthur L. Norberg, Professor of the History of Technology at the University of Minnesota, who at the time directed the Charles Babbage Institute. The interview is part of the Institute's archived oral history collection (OH60) and is used here with permission (for information on the Babbage Institute, see Bibliography). All the illustrations in this section are courtesy of Hitachi Koki Imaging Solutions, made possible with the assistance of Paul Weiser, Joseph Ryan, and Sharon Abdallah.

The entire interview carries Erwin Tomash from his youth as the child of hard working immigrants to the founding and early years of Dataproducts. The story he tells is a fascinating saga. The path to Dataproducts began with efforts by the film industry to develop a coin box pay TV system in the late 1940's to counter the threat they envisioned from TV. Computer science had not yet invented the kind of processor they believed was needed for such an application. Memory was a bottleneck, so Paramount financed International Telemeter to develop the technology and as a result of a chance airport conversation, Tomash joined the new company and soon became its president. Paramount sold the company to Ampex in 1960 and this made Tomash unhappy. Italic text is Norberg, Roman text is Tomash.

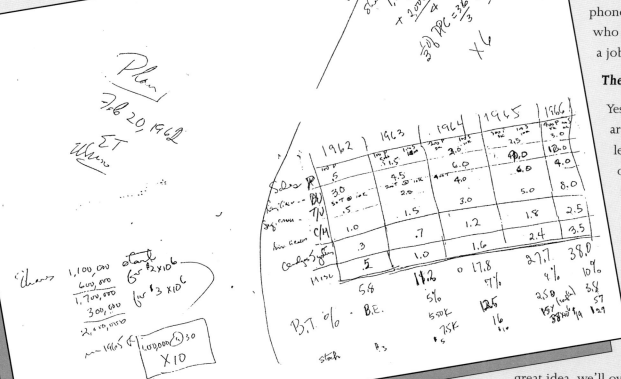

TOMASH: I didn't want to move. I guess I really didn't want to work for someone else. At the time, I had no idea about starting Dataproducts. So I announced my resignation. The moment I left, the phone started to ring. It ranged from investors who had deals, to people who wanted me to take a job.

These were people you knew?

Yes, people I knew. At the same time, Ray Stuart-Williams and Graham Tyson [former colleagues, still at Ampex], suggested that we ought to start a company. It was really Raymond who said, "Everybody's waiting for you to start your company. We all want to come."

I really had no idea of my future. I was going to take my time and then see what I wanted to do. But I talked to some of the others and I found that what Raymond said was true. The moment I said, "Let's start a company," they said, "Yes, great idea, we'll own more of it this time."

The key people were Raymond and Graham. Another one was Cliff Helms ... a classmate of mine at Minnesota and he had worked at ERA. He's a very good engineer. He started as a project engineer and he rapidly rose to chief engineer. Raymond was more the systems conceiver. Raymond was an engineer like Pres Eckert is; very broad, very, very fast, very good at solutions in the development sense and very poor, really, at making products. Helms is a basic thinker; he really spends his time trying to understand the basics, the components. Once he gets that right, he feels that you can build all sorts of machines, big ones and little ones. Cliff and Raymond made a good team. They didn't always agree. Indeed, it was very seldom that they agreed, but they had genuine respect for each other.

So, in the first months of '62, we started to plan Dataproducts. Raymond, Graham and Cliff were all employed at Ampex, so we couldn't include them. But Bill Mozena [at the time CFO of TMI, helped found Dataproducts and later became the first CFO of Amdahl Computer] had already quit. So he and I could noodle and draw up the plans and think about what we were going to do. Our choices were to start from scratch, or to try to save time by buying a company. We decided to do the latter and we looked at a number of things.

One that popped up that was most interesting was the Telex situation in St. Paul. The moment he heard that I quit, Bill Drake [an engineer, a founder of Control Data, who later formed a venture capital firm that backed Telex] had called. I had been an investor. I knew about Telex. They were a hearing aid company and they also had bought a portable phonograph company. They also made earphones for telephone headsets and things like that. They also had a group that had broken off from Univac to develop a disk file in St. Paul. It was called the Telex Data Systems Division. Another group that they had backed was headed by Sam Erwin in

Detroit. He had an idea for a printer. That was part of this same Data Systems division. They had a little ten or twelve man group in Detroit.

[The St. Paul group] had a contract from General Electric for disc files but couldn't deliver. The system didn't work. It was still in development. They had tried two or three managers and all had failed. Telex was losing a lot of money on this contract, about $100,000, $150,000 a month was going into it. So the moment he heard I quit, Bill called and said, "You really ought to come up and run Telex. The big future is this disk thing. All this other stuff is not important . . ."

I just said, "Look, I'm not moving to St. Paul and I don't want to run all of Telex. Let us buy the disk file. That's the kind of business I do know something about. I think we can bring it to market. We'll raise some new money and that will stop the bleeding at Telex." They didn't want to sell and the time dragged on.

Bud [Arnold] Ryden [Chairman of Telex, previously a founder of Control Data] is the one who came up with the idea of a spinoff. That is, to set up a holding company, which owned the Data System division, then sell us an interest in it, and then send the shares out to the Telex shareholders as a dividend. So that there would then be Telex and the new company. If you were a shareholder of Telex, you'd then have two pieces of paper. That was Ryden's idea. He was a financial man and very creative that way. That solved the problem I had, which was, if Telex were to remain a major shareholder, they would dominate, even though it wasn't part of Telex. The spinoff satisfied that.

So the next thing we had to do was raise our money.

I made a lot of visits to sources. One was an investment banker that I had met through Paramount. He was at Lehman

Erwin Tomash, 1982
(Courtesy of Hitachi Koki
Imaging Solutions)

The moment I said, "Let's start a company," they said, "Yes, great idea, we'll own more of it this time."

Our choices were to start from scratch, or try to save time by buying a company. We decided to do the latter and we looked at a number of things.

Another group that [Telex] had backed was headed by Sam Erwin in Detroit. He had an idea for a printer.

That's how Dataproducts was formed. Dataproducts ended up a public company the day it was born, because as a spinoff it had all the Telex shareholders.

Brothers, who were the Paramount banker, his name was Bill Osbourne. He's still a partner at Lehman Brothers. He came out and heard our story. By now we had a plan and a pitch about what we wanted to do. He said, "We'll raise your money for you." Lehman Brothers agreed to raise 3 million dollars for us. This was going to be enough money to take over the Telex division, finish the development, and deliver to GE. Encouraged by that, Tom Clausen at the Bank of America agreed to loan us short-term funds until all this could happen. [At the time Clausen was a young VP at BofA. He went on to later become President/CEO, and later headed the World Bank.] By now it was the middle of March and we went ahead and negotiated our deal with Telex. Our people only got 25 percent of the company and the Telex shareholders got 75 percent. We agreed to finance it after April 1. So that was how the thing was set up. That's how Dataproducts was formed. Dataproducts ended up a public company the day it was born, because as a spinoff it had all the Telex shareholders.

We started in temporary space [on Wilshire Boulevard] until we could find some space. Graham Tyson set about looking and in Culver City he found a building where a company had gone bankrupt. It had furniture in it; it had offices. It was a little shabby and so on, but we could move in fairly quickly and cheaply. At this point, after April 1, we're getting money for the payroll from Tom Clausen. We'd put in our own money, which was about one quarter of a million dollars, and we had all the assets in St. Paul – a building, we owned the building, and we had the group up there. But of course, they're spending money like crazy. So you know, it's a lot of money going out the door. We had a payroll in Los Angeles, a payroll in St. Paul of 80 people and a 10 man group in Detroit. So that was April 1st.

You know, I was naïve enough not to worry about details. I didn't know there was a union problem. They struck. The GE people telegraphed us the 5th or 6th of April canceling their contract for non-performance. It was exciting.

Excitement. We kind of sprang full blown, including a contract from GE. It was kind of fun, but not much, because April 2nd we had a strike in St. Paul. You know, I was naive enough not to worry

about details. I didn't know there was a union problem. They struck. The GE people telegraphed us the 5th or 6th of April canceling their contract for non-performance. It was exciting. So that's how we got the company started.

Well, don't leave the strike! How did the strike get settled then? Were they after more money?

The union was very concerned that all this reorganization was some kind of complex management ploy to break the union or something. [They] settled in about 3 weeks. In the meantime, I went to Phoenix – that's the headquarters of the GE Computer Division – to find out the depth of our problem there. That's where the prior contacts and the long association worked, because coming from Telemeter Magnetics, we had a good reputation with GE. We knew the marketing people. We knew the purchasing people. We knew the manufacturing people. The general manager was a new one – GE kept changing. We bought some time. They agreed to reinstate the contract, if we showed some performance.

So Mozena and I and Ray Stuart-Williams went to St. Paul and I sent Helms and Tyson to Detroit.

That first evening Helms and Tyson called [from the printer start-up in Detroit] and said, "There's very little here." They were just getting ready to ship their first printer. Tyson asked them how long they'd run it and they replied that it hasn't run at all, but that it's overdue. So Tyson made them run it.

It lasted about 15 minutes.

By the third day, I was still up in St. Paul; Helms and Tyson phoned, "Look, there's nothing here. There's one young engineer that's really pretty good. If we could get him, he knows something. The rest of the people aren't competent. The printer design is terrible. There's no hope for it. But there is a pretty good idea here. They got an idea for a moving coil actuator, which really looks pretty good." Helms said, "I don't really see how to make the actuator so it won't fly apart, but still it has a controlled arrival

time and that lets it print a straight line – if we could ever figure out how to make it."

So we concluded that first weekend that we would shut down in Detroit. We hoped to move the one engineer and Helms and Tyson would take over on the printer project in Los Angeles. Raymond and I and Mozena would stay in St. Paul where there were also all kinds of problems. To help save the GE contract, everyone else in Los Angeles would come up there and we'd try to help straighten out the disk file program.

The man that was running Detroit was a fellow named Sam Erwin. Sam didn't like the fact that we shut down in Detroit. He didn't talk to me for a few years, because we had shut it down so peremptorily. But, we just had too many troubles at once, not to move fast. All this time we were borrowing money at the bank. Lehman Brothers' was saying they were going to raise our $3 million.

So, in St. Paul, we were working away and getting a handle on who and what was useful and what wasn't. We found no manufacturing organization and we realized we had to build that up. The engineering development leader, a fellow named Don Sampson, was pretty good, but he didn't have any electronics strengths, so we gave him Wieselman. We had to try to find out what the costs were, because in addition to this contract, there was a big option in the GE contract for a lot more machines. I needed to know whether we didn't want GE to reinstate their order. Perhaps it would be better to let them cancel the whole thing. I sure didn't want them exercising this option until I had figured out what things were costing us.

So that's what went on in the month of April. In May, there was a big stock market break. The market dropped 20 points in one day. Then, in subsequent days continued to retreat. Those were very, very sharp drops – 20 points on a Dow Jones of about 500, not 20 points on 1200.

After a couple of days of hemming and hawing, the Lehman Brothers' people simply backed out on our financing. They were polite and they didn't want to say they were backing out, but after explaining all about the market conditions and so on, they said they would continue to try, but they could no longer guarantee us any money. They never did raise any money for us.

By now the Bank of America had agreed to loan us $400,000 and our $250,000 was in. We really had to scurry. We ultimately raised one million and a half dollars and gave up as much equity for that as we would have for the $3 million. We got a third of it from the Bank of America SBIC, a third of it from Continental Capital, which was a new SBIC in San Francisco, and one third of it from Greater Washington Industrial Investments in Washington D.C. All these participants came from very old connections. So by the time we got it all together, it was July. These things take time. We owed the bank $800,000 by July.

But by July, we had made progress on the printer and we had the disk file product straightened out and we were starting to ship to GE. We'd brought in a manager to St. Paul, a manufacturing man, and we started running it as a business. In that same period, Helms had really designed the moving coil hammer and outlined the process for its manufacture. That hammer built the company. He just went at that step by step to isolate and identify the problems and figure out how best to solve them. In a very innovative way, he concentrated on getting a hammer mechanism that was both accurate in its flight time and extremely reliable. That gadget, which we patented, and with a larger number of variations – lighter ones and heavier ones and thicker ones and thinner ones and so on – has built Dataproducts.

But for the first couple of years, we lived off the disk file business with GE and others. We sold them to ICL. We sold them to Ferranti and then to ICL. We sold a few to Japan. We sold some to RCA. The disk file business maintained the company until the printer could come on stream.

They were just getting ready to ship their first printer. Tyson asked them how long they'd run it and they replied that it hasn't run at all, but that it's overdue. So Tyson made them run it. It lasted about 15 minutes.

But, we just had too many troubles at once, not to move fast. All this time we were borrowing money from the bank.

After a couple of days of hemming and hawing, the Lehman Brothers' people simply backed out on our financing.

In that same period, Helms had designed the moving coil hammer and outlined the process for its manufacture. That hammer built the company.

The Informatics Hedge. In the same period from January to March, after my resignation from Ampex was announced, one of the people who called was Walter Bauer [Ph.D. mathematician, previously had managed the computer department of Ramo Wooldridge]. He was ready to start a software company and wanted to talk about raising money. So I said, "Let's get together to talk about it, because I'm doing the same thing. I'm starting to raise money." After that meeting I said, "Why don't you throw in with us? You know, software and hardware are going to be linked inexorably, and it all fits together. We have to raise money anyway. So why don't you think of your software activities as part of Dataproducts?"

At first, he wasn't completely sold on that. He'd like to have his own company, wanted his own company. But it also had some advantages in his mind. In the end, the two other key people from Bunker-Ramo who were joining him thought it over and they finally agreed. So on April 1st, when Dataproducts was organized, it also had a 100% owned subsidiary called Informatics. It was already in the software business.

The Data Products 3300 Printer was the first commercial low-cost line printer. Introduced in 1963, the printer's superior reliability and print quality was attributed in large part to the "revolutionary" Mark I hammer. (Courtesy of Hitachi Koki Imaging Solutions)

Now I don't understand that because of my lack of background in economics and setting up corporations, etc. What was the advantage to Walt, first of all, being capitalized in this way?

He didn't have to raise any money and he spread his risk. He wasn't sure he could raise the money. I was going to do it. He put in his savings, $20,000 or so as I recall, he and his key people, bought Dataproducts shares. He didn't have to raise the half million dollars or so needed to run Informatics. We also would give him administrative service: finance personnel, space, insurance, etc.

Therefore you were promising to raise the capital needed. What was the advantage to you?

I thought of Dataproducts as a company which would serve the computer industry with both peripherals and services. I saw the software market — and so did he at that time — as selling software to computer systems builders. That was also who we expected to sell peripherals to. The idea of a software product sold to the user didn't come along for really another at least five years. People who rented or bought computers expected to develop their own software, or did it in a cooperative manner, or expected it to come free. It was bundled, you know. What was available came free with the price of the equipment.

Then was Informatics providing the software to Dataproducts?

We didn't have need for any at the time. We were strictly hardware. We were just building printing devices and memory devices and we really didn't have any software. No, Informatics started right off doing contract software work.

By the way, Walt, as he reflects on this, points out that he was hedging his bet too. That is, when you're starting one of these things, you're not really certain that you're going to get these orders and that you're going to succeed and build a company. Even if he was less successful than he hoped, Dataproducts hardware might be more successful and therefore the hardware side would still give value to his investment. And we, I suppose, had the same thing going for us.

But some of the hardware products didn't do so well like the disk files, is that correct?

Disk files didn't in the long run. They did very well in the beginning. They carried the company, including Informatics. Paid for the printer development. At the outset, they did. But we weren't able to keep up with the pace of development later on.

I see, why not?

Because we didn't have enough money, didn't have the resources. IBM really started to step up the pace with replaceable disks and other models, and big files and little files. Our little group couldn't keep up technologically.

So what happened when you realized that you couldn't keep up?

Well, we had to get out of the disk business. What occurred was something like this. Informatics really succeeded quite quickly. The first year they had some problems, and Walt felt some insecurity because it wasn't going well. But, by the second year, it was nicely profitable. Then by the third year, it was growing. In that same three-year period, we had skimmed the cream of the previous investment by Telex. We'd cleaned up the disk file design and had them working. Our shipments were profitable. But, IBM was introducing new models and we were already starting to fall behind. Our printer program was not going along very well yet. So the hardware side wasn't doing great just as Informatics was doing very well.

When that happens, there's a natural tendency for the subsidiary to say, "Who needs you? You're holding us back." It happens over and over again. To satisfy them, we deliberately took the step that we knew would separate us in the long run. We sold them some shares in Informatics and gave them some options in Informatics. Once you do that, the door is open. The next step, now that they have shares in Informatics, is to set a value on their holdings. What value do the shares have if there's no market? So soon there is pressure to take Informatics public. Once you've done that you now have two sets of shareholders whose interests might or might not be congruent, and inevitably the two companies diverge. It's a particularly difficult people problem. We had it to a degree with Data Card. But it's particularly difficult in software where the business is so people dependent. With hardware, when you differ with

the management on this point, you can say it's not going to be separate. The choices are for management to agree or leave and go do something else. You own the product and its customer base. In the software business when you've lost the people, you know, it's like a university without a faculty.

You may have buildings, but you don't have a university. Walt knew that we needed him. Well anyway, that was what happened with the Informatics/Dataproducts relationship.

It took us about a year to develop a printer that was reliable mechanically. And it was 18 months before we had the electronics in good shape and the styling set. Our competition was a company called ANelex in Boston; they had what printer business there was.

The market in the early '60s is not the market today. The market was the big systems companies. It was GE and RCA and Honeywell. Then there were some smaller OEMs. Collins Radio took a few and so on. There were no mini-computer makers yet. The whole terminal business wasn't there. You either sold to those few key accounts or you didn't sell to anybody. So our printer business was very slow in coming.

We did pretty well with the new accounts as they emerged. We sold to SDS (Scientific Data Systems), then later on we sold to DEC, which became our biggest customer. After that, we sold to Hewlett-Packard. Step-by-step. But in the meantime, after we finished developing the product, we really tried to sell the majors: Honeywell, Burroughs, Univac, RCA and GE. ANelex, to keep the business, cut the price, because they didn't have the product.

In 1968 Data Products reported sales up 77% and earnings up 300%. Heartened by six years of steady growth, the company acquired a tract of land in Woodland Hills, California and Erwin Tomash helped break ground for a new headquarters complex. (Courtesy of Hitachi Koki Imaging Solutions)

Later on, it [printers] was a great business but, if we hadn't had the disk file and Informatics, we wouldn't have been around to enjoy that.

Typical.

And they cut the price to where they finally went bankrupt. And they were sold in bankruptcy to Mohawk. Later on, it [printers] was a great business but, if we hadn't had the disk file and Informatics, we wouldn't have been around to enjoy that. So the printer was slow in acceptance and slow in reaching volume. We finally introduced a small tabletop 80 column printer for under $10,000

dollars, really under $7,000, and that product finally got us going with the DEC's and the HP's. Those are big prices now, but then those were breakthroughs. When we started, Anelex was getting $35,000 for 600-line-a minute printers and they worked pretty well but required daily maintenance. Ours was $15,000 and needed monthly maintenance. When they cut their price, they had to cut it down to $8000.

They were keeping us out, but they ruined their company.

Technology: "The Hammer Saved the Company." Another inside view of what made Dataproducts happen is on the technology side. In the words of founder Tomash, "The hammer saved the company." It wasn't all entrepreneurship.

Richard Lee Forman in 1980 completed an exhaustive evaluation of the first ten years of Dataproducts, an unpublished paper entitled, "Tales in Peripheral Enterprise: The Rise of Dataproducts Corporation in the Computer Industry, 1962-1972." (See Bibliography for complete citation.) One interesting fragment of this 495-page document describes the development of the second generation Mark IV hammer. The earlier hammer worked just fine, or at least quite a bit better than the competition's. But Cliff Helms wanted to improve it. Here is Forman's account:

Upon completing development of the Mark II hammer for the RO-280 and 4300 printers in 1965, Helms continued his thinking on advancing hammer design. One of the ideas that occurred to him was placing the hammer on top of the actuating voice coil rather than balancing the coil on the hammer. The coil would be imbedded within a thin, insulated plastic strip, with the hammer or striking piece bonded on top. The electrical current would be conducted through short, cross-flexure springs, instead of long-leaf

The Mark IV hammer. (Courtesy of Hitachi Koki Imaging Solutions)

springs, inserted inside the strip and connected to the coil. This would create a smaller, lighter, more compact hammer which would be easier to control and more durable.

It would have increased reliability because the vibration experienced by the Mark II, an elongated device, eventually caused wear within the metal joints where the leaf springs connected to the hammer. Vibration frequently caused the springs to snap and the joints to fail. The substitution of two short stiff flex springs minimized the vibration. The leaf springs frequently failed because of the rigidity of the metal joints. The flex springs, however, would be inserted in plastic, a more flexible material that could absorb the vibration.

The catch in building the above impact hammer, the Mark IV, was to find some means to literally bond metal (the flex springs) to plastic (the insulation strip containing the voice coil).

At that time, current state of the art technology mandated that the two substances be joined by encapsulating metal parts within a plastic joint. Encapsulation unfortunately created rather rigid and brittle joints. Strong vibration eventually would cause the plastic surrounding the metal springs to crack.

Helms became aware of the growing technology of epoxy resins, particularly used in the aerospace industry to bond metal parts to each other. If epoxy could be used to glue metal to metal and plastic to plastic, why could not metal and plastic be glued together in a lasting bond?

Initiating his search for the right epoxy, Helms began making inquiries of epoxy manufacturers for a suitable product. He received a universal answer from the leading producers that it was impossible to bond metal to plastic. They further informed him that no epoxy existed that could cling to both substances equally, and that all industry research efforts to create an epoxy to glue metal to plastic in a lasting bond had failed. The only means by which the two materials could be joined was by encapsulation.

Convinced that nothing was impossible and that there are always logical reasons for failure, Helms refused to accept industry answers to his inquiries. Instead, he began conducting laboratory tests of his own on the one hundred most common epoxy resins available on the market.

The effort was worth it.

Close study of the glue manufacturer's procedures and inspection of the bonding process revealed the bonds broke down due to a lack of thorough preparation and cleaning of the two surfaces prior to application of the resin. Tests of the top one hundred epoxies revealed twelve were sufficiently capable of holding metal and plastic together. In addition, Helms further discovered that the best and most flexible bond for absorbing vibration was one which used two glues – one primarily for metal and one primarily for plastic – and stuck the two separately costed surfaces together, actually bonding glue to glue.

The result of the experimental process was the Mark IV impact hammer. The Mark IV, more than any other component, product or technical achievement of Dataproducts, is perhaps the greatest single reason for the company's success as a printer manufacturer. Being the most important component in impact printers, the Mark IV results in products of high performance (the hammer is capable of 1,800 lines per minute), high reliability and long life (as its springs and joints do not break). It is the secret of the company's success. The secret of the Mark IV is the glue which holds it together.

Epilogue:
The Rocky Road to the Present

In 1971, after nine years at the helm, Erwin Tomash asked Graham Tyson to take over as president of the company, Tyson accepted, and a new era began. The end of the prior era was even more tumultuous than the early years. Their diskfile business was ailing and some acquisitions were proving hard to digest. And there were too many acquisitions and new business ventures, including even a personnel agency business, Staff Dynamics. Author Forman notes that at least one school of business theory says companies that try to expand horizontally before they have achieved solid vertical integration in their core product line tend to run into trouble. Dataproducts was trying to be an all around peripheral vendor with four product lines. They simply did not have the resources to build all four.

By 1970 it was clear it was printers that were carrying the company. However, in the printer market competitive pressures were increasing. And the

Dataproducts switched from drum to band technology with the BP series in the early 1980s. Pictured here is the first unit of the second iteration BP2000 to ship. Seeing their 2000 lpm baby off are Patrick Maher, BP2000 Product Manager, Dave Leighton, Asembler II Lead, and John Edling, Manager Engineering II of the High Speed Printer Division. (Courtesy of Hitachi Koki Imaging Solutions.)

ground was shifting in the market which was still primarily the huge base of IBM computer users. IBM was taking action to head off costly incursions by the plug compatible peripheral equipment vendors.

IBM's new strategy, as described by Forman, included three

Hard Times at Dataproducts
Source; Dataproducts Annual Reports

$US, Millions

YEAR	REVENUES	NET INCOME
1985	$471.8	$27.7
1986	353.8	(26.7)
1987	338.7	8.9
1988	345.2	(20.2)
1989	353.3	3.8

tactics that were effective in maintaining user loyalty. They began making small improvements every few months, thereby creating a moving target for the plug compatible vendors. They developed equipment interfaces which were harder to emulate. And they began embedding more of the peripheral control functions within their mainframe computers.

In the midst of these and other pressures, the Dataproducts Line Printer Division managed by Graham Tyson remained profitable, a factor that no doubt led Tomash to recruit Tyson as his successor. The next few years were a time of belt-tightening, layoffs, and consolidating operations with emphasis on the printer business.

The result, as described by Forman, "is a maturing corporation which . . . attempts to plan its growth on business realities [rather than] on entrepreneurial instincts and desires alone." Within a few years profitability had been restored and by the company's twentieth anniversary, Dataproducts was solidly positioned as the world's largest independent printer manufacturer, second in volume only to IBM.

During the decade of the 70s, Dataproducts began to build an identity as a printer company, plowing resources into broadening its product line to include low-end line printers, serial matrix printers, and thermal printers (the latter based on Olivetti mechanisms). A line of impact band printers was developed which gradually replaced the original drum printers. Military printers began to supplant other products in their Telecommunications Division plant which was renamed Dataproducts New England in 1978. The following year they acquired the Plessey daisywheel printer operation in Irvine, CA to serve anticipated demand in the rapidly heating word processing market. Of its non-printer businesses, the original core memory segment continued to be profitable during much of the decade.

By decade-end, Dataproducts was clearly identified as a printer company, even though their name was not well known because of their emphasis on the OEM market. Their customers put their own labels on the printers. Their reported engineering and development investments were significant, but not at the level that would be expected today. In 1980, for example, they reported they had increased their engineering and development investment 42% over the previous year to $13.4 million, a year when their revenues were $180 million.

They had the vision even though their level of investment in the future was a bit cautious. Their vision was emblazoned on the cover of their 1978 Annual Report: "Another record year. And a big step forward in our continuing preparation for the future, when printers will serve the needs of people more extensively than typewriters do today."

They predicted the digital printer revolution that was destined to blossom throughout commerce and society. But although positioned as a healthy company with interesting prospects, they failed to harvest the full fruits.

The early 80s continued prosperous for Dataproducts. How-

ever by mid-decade, the challenges of rapidly changing printer market structure and new technologies began to impact the company (see box). Technology was shifting toward non-impact. (Alas, no hammers.) A promising order for matrix printers from IBM fizzled due to both demand and production problems. They joined with Exxon to develop solid ink jet. They fielded laser page printers based on print engines from Toshiba which by 1989 represented more than 25% of their revenues. They set their sights on developing controllers for non-impact print engines over a wide performance range. They fielded a new impact line matrix product and the response was said to exceed expectations.

But it was a new ball game.

By early 1989 Dataproducts was widely labeled as "beleaguered." All that year the company fought off the takeover efforts by a consortium of at least eight parties who called themselves DPC Acquisition Partners. It was a scene Tomash would have found "exciting." To save its independence the company undertook a major restructuring and sold off assets including even its showcase 21.5 acre corporate headquarters in Woodland Hills, CA. What remained of domestic manufacturing was shifted overseas. Despite this, DPCAP in the fall initiated an unfriendly tender offer. The stockholders remained loyal, the dreaded "partners" finally gave up, and in April, 1990, in the words of Printout newsletter editor Charles LeCompte, "an exhausted Dataproducts collapsed into the golden embrace of Hitachi."*

The name "Dataproducts" however, survived well into the 90s, to be finally eclipsed by Hitachi Koki Imaging Solutions (HiKIS) in April, 1999. With the announcement of the name change, the company also described a strategy to focus on independent copier dealers as a primary channel for their digital copiers and printers all the way up to the high volume, on-demand DDP-70 "Digital Document Publisher" system.

* *Printout*, the Datek Printer Report, May, 1990 issue

What's in a Name?
What are the Lessons?

Very few company names have survived through all five decades to date. But there have been many successes. Dataproducts, which among the independents led the industry for a decade or more, is certainly one of them. Like many of the industry leaders, the company went through a series of lifecycle passages and then, as part of a giant conglomerate, lost its individual identity.

This raises the questions, just what is success, and how important is corporate identity?

Is success to invest, invent, grow, and lead a market? To provide the better mousetrap and in the process build a thriving business family of workers and investors, then morph into obscurity? That might be seen as the short story of Dataproducts and, yes, it is a success story. Still, the loss of a name is not nothing. A name is more than a word. It is, or at least should be, imbued with something of the spirit of the named. The disappearance of a name is a loss, no matter how logical and synergistic the acquisition.

What are some of the lessons of the Dataproducts story? Two consultants at I T Strategies, Inc., Marco Boer and Mark Hanley, late in 1999 volunteered opinions on some strategic missteps they believed the Dataproducts experience illustrates.

Overall, they saw Dataproducts as a company that might have been blinded by success. It grew rapidly to a

The Dataproducts LB Series Band Printers, with speeds from 350 to 1,500 lpm, were a centerpiece of the company's product line in the 1980s; the state-of-the-art Model 8500 is a current matrix printer rated at up to 780 cps; the LX-455, now available in two configurations, is a dot matrix line printer. Printing three lines simultaneously and nick-named "The Rebel," performance ranges from 105 lpm to 600 lpm depending on the selected print quality and number of columns printed. (Photos courtesy of Hitachi Koki Imaging Solutions)

critical size that called for a break, a new strategy. Instead, they made primarily product moves, but these were somewhat half-hearted, or too late, or too reactive. Among them:

Page printers. Dataproducts saw the impending tectonic shift toward non-impact technologies and responded in two ways. For a mid-range page printer, they adapted a laser printer from Fuji Xerox. Their first major customer was Apple, which sold the product as the Laserwriter 810. "But what happened," according to Boer, "is that HP came along and priced them out of the market and they did not have a strategy, or perhaps not the resources, to react. Printers had become a very high stakes game."

Solid Ink Jet. Creative. But they didn't read the market right. As recounted by Boer:

> They knew they had to get into non-impact, and they focussed mostly on ink jet, acquiring Exxon ink jet patents for certain markets. Owning these patents is still believed to give them around $5 million per year in royalties. The SI 480, introduced in 1986, was a beautiful machine with stunning print quality. Monochrome. But it never sold to any extent. One problem was that it needed hours of warm-up time. Another was that it was too expensive. In 1990 the Jolt printer was introduced, a color printer. Because of its raised image, one customer sold it as a Braille printer. Now, there's some irony: a color printer sold as a Braille printer!
>
> By the time it became apparent solid ink jet would not fly in the broad market, it was too late to get into developing and marketing their own laser printer. At the same time,

they became engaged in a draining legal battle with Tektronix and Apple/Spectra. Dataproducts won. Apple decided not to make the solid ink jet printer after spending $10 million dollars on it. Tektronix pays continuing royalties. But rather than fight the legal battle, it might have been better to let Apple popularize solid ink jet than to fight a defensive legal battle. If they had let Apple make the solid ink jet printer, it might have legitimized the technology very quickly.

Protecting your assets is a natural reflex. But not necessarily the way to go.

Supplies. The story, they said, seems to overlook Dataproducts' diversification into supplies. This included ribbons, toner, and ink jet supplies. This was apparently profitable and, according to Boer, the supplies operations were probably the most attractive part of their business for Hitachi Koki. They wanted the brand name and the distribution channels.

Current Product Mix. As of late 1999, Dataproducts' revenues were believed to break down more or less as follows – $180M printers, $60M supplies, and $15 M components (solid ink jet, including royalties). It is expected, with their currently announced shift toward the color copier distributors and non-hardware solutions, their business will significantly change over the next few years.

This is not the end of the Dataproducts story. As HiKIS, the company that changed the industry in the 1960s is still with us and may well indeed succeed in reinventing itself.

In the 1970s the daisywheel was hot, driven by and helping to drive the growing word processing market. This new letter quality technology, first fielded by Diablo in 1970, doubled the speed of existing LQ printers.

CHAPTER 5: THE SEVENTIES

THE SETTING This decade saw the ferment and idealism of the Sixties slowly subside. In 1972 President Nixon made his historic visit to China, beginning a process of normalization of relations. But the same year the presidential Watergate scandal erupted, leading Nixon in 1974 to achieve the distinction of being the first ever U.S. president to resign.

It was a decade of "stagflation," of a sluggish economy combined with high inflation, attributed by some to residual debt from the Vietnam War, by others to the OPEC oil embargo. The USA and Soviet Union worked on defusing the nuclear "balance of terror" with the SALT treaty. The first Earth Day celebration in 1975 launched the environmental movement, a need underscored later in the decade by several major oil spills and the nation's biggest nuclear reactor accident to date at Three Mile Island. In 1978 California voters approved Proposition 13, the beginning of an anti-tax trend in many states. This presaged the Republican years to come and a growing bias toward private rather than public wealth.

During the 70s computers infiltrated almost every facet of business and government. Computer vendors sprang up like mushrooms, and the number of printer suppliers to serve them tripled to at least 93 companies worldwide (or well over a hundred, counting those fielding various specialized printers). Intel produced the first computer on a chip, and a few years later the first microcomputers appeared based on the Intel 8-bit 8080. Another step toward the Internet was the 1977 merger of the ARPAnet with other governmental and academic nets to create a single network of networks. The first word processor, the Wang 1200, appeared in 1971. Word processing was destined to blossom in the 1980's, enabled by and a major market for affordable, letter quality printers. Stephen Jobs and Steve Wozniak came out with their Apple I computer in 1976. Bar codes, first used in the railroad industry in the 60s, became ubiquitous in the 70s when the grocery industry adopted it as their standard UPC in 1973.

Standing at the center of our fifty-year history, this was the printer industry's pivotal decade. Shipments, which earlier had been tabulated in the thousands, swelled to the hundreds of thousands. The vendor roster multiplied. New printer concepts, new materials, microelectronics, and automated production brought down printer prices more than 50% in five years and opened up at least a small consumer market. Around 1975 Centronics began to supply retailer Tandy speedy but stripped-down matrix printers which they could sell for $1,995. By the end of the decade, Tandy broke the $1,000 barrier with a Centronics model priced at $995.

Three main printer technology and industry trends explored in this chapter are the swing to dots, the emergence of non-impact technology at the high-end, and globalization of the industry.

For the glory days of dots, we focus on Centronics and Printronix.

For the emergence of non-impact printing, we focus on Xerox, IBM and Siemens.

And as an example of the globalization of the industry, we visit Okidata.

CENTRONICS AND THE INEXORABLE LOGIC OF DIGITAL TO DOTS

The logic of printing digital data with dots is a no-brainer. Digital data – binary, on-off patterns – are virtual dots. If you convert them to codes that are linked to a fixed set of fully formed characters, you are restricted to that set of characters. Asian and many other non-Roman character sets are difficult and graphics out of the question. Looking at the laws of physics, moving relatively heavy, raised metal type elements imposes printing speed limits and electromechanical costs that are hard to overcome.

As noted earlier, various impact dot printers appeared in the 1940s and 1950s. But not for general purpose, page-size printers.

This had to await the 1970s. At the dawn of the decade, in April, 1970, IBM fielded the Model 2213 serial dot matrix printer as part of the IBM 2770 Data Communications System.

But it is Centronics Data Computer that is most often seen as the originator of the general-purpose serial impact matrix printer. This is only in part because they repeatedly made this claim. In reality, Centronics was the first *independent* company to develop and field this ground-breaking technology and bring it to market on an OEM basis, thereby making it available to all.

It is said that the brightest stars burn out the fastest. The Centronics saga is a perfect rags to riches to rags story. Founded in 1971 by mercurial entrepreneur Robert Howard, Centronics began a revolution with its first printer, the 101. The product took off like a rocket, propelling company revenues to the $25 million mark by1973. The company established a close relationship with Brother Industries in Japan, which fabricated the print mechanisms (the print heads were made by Centronics in NH) in the early printers, and by 1979, revenues surpassed $100 million.

This meteoric rise through the printer industry ranks in the 1970s was followed by management shake-ups and declining profits. In response, in 1982, Howard decided to sell his controlling interest in Centronics to Control Data and move on to other enterprises.

Centronics was just one phase of the Bob Howard career. But one that left a substantial mark in the printer industry during its fifteen year history. It is nearly impossible to discuss impact printing without mentioning the company's name. After all these years, the name Centronics still lives on as the de facto standard parallel printer interface in much of the world of PCs. An early commitment to speed, cost-effectiveness and convenience proved a huge success during its early years.

Following this logic, it is certainly appropriate to highlight Centronics as a defining company of the decade and visit with founder Robert Howard to hear in his words how he did it.

Need-Seeking: Robert Howard
Tells the Centronics Story

We spoke with Robert Howard in September of 1999. At the time, he was at his second home in St. Jean-Cap Ferrat, France overlooking the Mediterranean Sea. His other home is in New York City, in midtown Manhattan. He was born and raised in New York, and still spends much of his time there. In the course of his long business careers, by his count, he founded 25 companies. He was a recent recipient of the prestigious Cary Award presented by Rochester Institute of Technology for his contribution to the graphic arts. NOTE: Some material from secondary sources has been woven into this dialogue with the approval of Howard. Italic text is John Webster for DRA; Roman is Bob.

How did you got interested in electronics and eventually in printers? Did you get your education in engineering?

My undergraduate education started at Columbia University school of engineering in New York City. I started as a full curricu-

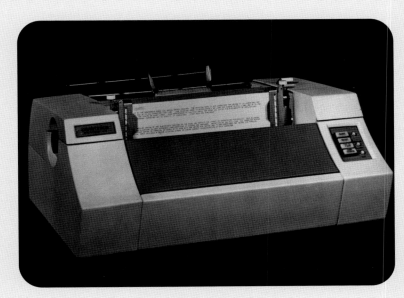

The Centronics 101, a printer that changed the industry.
(Photo courtesy of Robert Howard)

lum student, and continued going to school at night for the better part of ten years, while working during the day at Sperry Gyroscope Company and Sylvania Electric Company. My first employer, Sperry, could not hire a sufficient number of degreed engineers so that they then started to hire people who had engineering training and experience as engineers. While I worked there I continued to take nonmetric-related classes at Columbia. Later, I took other very specific classes at MIT [Massachusetts Institute of Technology in Cambridge], UNH [University of New Hampshire in Durham] and others that I cannot remember.

Actually, it was at MIT that I met Dr. [An] Wang when I was a student of his in about 1950. He was a professor of physics and among other things, an inventor of magnetic core memory, which really allowed computers to be used on a widespread basis. His invention was a read/write memory which gave computers greater memory capacity, making them faster, and thus cheaper.

I understand you developed your first printer when you designed and built computer systems for Las Vega casinos. This became the first Centronics product. What did these computer systems do exactly?

The first products of Centronics were remote terminals connected to a central processor, including a number of printers in the pit of each gambling game. Only single transistors were available at that time, and they were then used only by the military. We bought up transistors by the barrel-full that were rejected by the military. The computer systems consisted of the central processor with a terminal, custom keyboard, and printer attached. I don't believe there was a limit to how many keyboards could be attached to the processor, but they usually had anywhere from 20 to 50 terminals. One terminal was located at each gambling table. Every time a transaction was completed, it would be entered into the computer. Some casinos required receipts to be output, and to do that a local printer was situated in each pit. The pit is the con-

Robert Howard

The first products of Centronics were remote terminals connected to a central processor, including a number of printers in the pit of each gambling game.

trol area for the games. A pit might have included ten blackjack tables or four crap tables. The printers would print out receipts, in addition to general analyses of transactions.

Dr. Wang and I jointly developed this system and it's a patent that I currently own. That is the only patent framed and hanging on my office wall in Manhattan. My patent activities and participation go back to the mid '40s.

How many patents do you currently own?

I don't know how many I have. There were many patents that were developed into other patents. As an example, Presstek currently holds 58 patents, I believe, and has over 100 others pending. They all grew out of my ideas of imaging plates on a press.

My philosophy in developing ideas and companies to implement these ideas is this: it is not my motivation to invent things but rather to find industrial needs, or things that can be done better, faster and cheaper. This goes back to my cable and television company and back to my employment as an engineer.

Why did you decide to base Centronics in Hudson, NH?

Originally, Centronics was a part of Wang Labs. Then Wang Labs started to grow, and Centronics started to grow as well. Dr. Wang didn't want anything to do with gambling. He was a very straight guy, and he wanted no part of the gambling business. So we came to a financial agreement: I left Wang Labs and took seven people with me for Centronics. I was the president and chairman of the board. I also took a young man named Prentice Robinson, who became the head guy. Prentice was terrific, highly qualified and very innovative. At that time, I owned an industrial building on Long Island, NY. The former tenant manufactured radios, but went bankrupt. It still had all the electronics manufacturing equipment. I brought Prentice down there, and he asked what I wanted him to do there. I said, "I want you to set up Centronics here." He said, "I wouldn't move down here for anything!"

At the same time, southern New Hampshire was the best place in the U.S. for manufacturing because there's no state tax, no income tax, and no business tax. Prentice wanted Wang to move out of Massachusetts, which at that time was called "Taxachusetts." I said, "Where should we be?" "Hudson, New Hampshire," he said. He told me it's about 50 minutes from Boston's Logan airport and it's a great place to start a business. So I gave him a check and told him to rent a space in Hudson. He found a second-floor space in a wood residence building on Library Street, and that was the first Centronics headquarters.

Of course, Prentice was right. It was the best thing I ever did. The people in New Hampshire are like America used to be. In addition, it was (and still is) a technical hot-bed.

As time went on, we acquired a factory in Ireland, and worked with Brother International's production facility in Japan. At our peak, we were the largest independent computer printer manufacturer in the world, with 2,500 to 3,000 people in Hudson, and 6,000 worldwide in our manufacturing facilities and service centers.

Can you talk a bit about the history of the first Centronics matrix printer?

We came to the conclusion that the two technologies available at the time — Teletype and the large, cumbersome, expensive line printer for computer printed output — were too far apart. One was too slow and too unreliable; the other was reliable enough, but too costly and physically too large. There needed to be something in between and I came up with the idea of a matrix printer.

Now, I had seen matrix printing done by an NCR thermal printer that made dots on special paper with heat. That was no good because it was not impact. If we could make the dots with a wire instead of a little heater, we could make multiple copies. Actually, the original idea was to make the dots on paper through a ribbon with a bullet of ultrasonic sound. That didn't work and

My philosophy in developing ideas and companies to implement these ideas is this: it is not my motivation to invent things but rather to find industrial needs, or things that can be done better, faster and cheaper.

Dr. Wang didn't want anything to do with gambling. He was a very straight guy, and he wanted no part of the gambling business.

then I came up with the idea of striking the ribbon with a wire driven by a solenoid. It had to be ten times faster than any existing solenoid. This design, my original design, was not improved until about 20 years later.

I did some calculations and found I could print at about 165 characters per second, which was 10 or 20 times faster than the Teletype. We built it at Centronics, and once we saw what we had, I decided that's a better business than the casino business.

We very quickly saw that there was a great demand for printers in this price/performance category. Executives and engineers from IBM, NCR, Control Data, Burroughs, and others at that time said no one would accept matrix printing because it's not fully formed characters, but I said if it's fast enough and reliable enough and is priced right, the poor print quality will be acceptable.

Who manufactured your dot-matrix print head?

We manufactured our own print head and electronics at Centronics in New Hampshire as well as in our factory in Ireland. Brother International built some of the printer mechanics. In manufacturing, you have to know what you can and can't do, and to build a print mechanism takes lots of in-house know-how, and you can't do it quickly. Brother was the largest sewing machine manufacturer in Japan, and a sewing machine is a lot like a printer in its mechanics, its moving parts and electronics. Their sewing machines were extremely reliable. So I knew they knew how to build mechanisms that would be reliable. Brother also built and sold a line of typewriters. They were excited about building our printer mechanics, and so they started making them for us. They would send the mechanics to us in New Hampshire where we manufactured the print head and electronics.

What led you to Brother?

Dealing with Brother was actually very easy because the president of Brother International, which was their international marketing

arm, was a good personal friend of mine. We played golf together and he lived in my neighborhood. Additionally Brother was anxious to get into computer-related products, Centronics gave them that opportunity.

They agreed to invest $3.5 million into tooling and the pre-production costs in exchange for a royalty-free license for Japan, because they were anxious to get into this kind of business. We signed a 10-year contract, and after 10 years we renewed it for 10 years. It worked very well for us and today Brother is a multi-billion dollar enterprise principally in electronic office equipment

Can you tell me more abut the famous Centronics interface?

The Centronics interface was worked out by Prentice Robinson, Dr. Wang and myself. At the time, there was no way for a computer to communicate with more than one type of system-dedicated printer. As a result, every printer was different. Every printer had to be custom-made for the computer, usually via the computer manufacturer, and no two were the same. So we said, "Let's make a standard interface, give it to the world free, and put an end to having peripherals furnished only by the computer manufacturers."

As you may have noticed, the interface has a very large plug. Dr. Wang had 20,000 spare plugs in inventory that he gave to me. He had invented the first scientific calculator, which was equal in size to a desktop computer of today. The plug went from the keyboard to the main board, the calculator. He had all these plugs left over from that, and that's why the plug is so big today on the interface. Every computer made still uses that big plug on it's Centronics interface. The standard interface was the solution to the problem. The plug was simply the implementation and Dr. Wang gave me all the plugs I needed free of charge. Otherwise, if we had to buy these terminal plugs, we would have used much smaller ones.

The size was no problem in those days. I never dreamed it would end up on every computer, every type made, thirty years

Of course, Prentice was right. It was the best thing I ever did. The people in New Hampshire are like America used to be.

Executives and engineers from IBM, NCR, Control Data, Burroughs and others at that time said no one would accept matrix printing because it's not fully formed characters, but I said if it's fast enough and reliable enough and is priced right, the poor print quality will be acceptable.

later. The earliest users were the early digital computer makers: UNIVAC, DEC, and Data General were a few of them. Of course, Wang Labs was a user on its very early digital computer.

I've heard about your development of a low-cost mini-printer that ran into problems. What was that all about?

In 1978 I felt that the personal computer would become a small business computer, and we had to prepare for it with a printer that had the proper price and performance because in the beginning of the computer on a chip, the printer would cost more than the computer. We engaged in an R&D effort because we said that this had to be a matrix printer that would do all the things that the $2,000 retail model 779 did for no more than $500 retail.

We developed a product in 1979 that was introduced early in 1980. It was called the Mini-Printer Model 770. It was the first of what everyone knows of today as a small matrix printer with a small head and very low-cost architecture, thus a low selling price.

Unfortunately, our production quickly went up to over 200 a day and we found that in a small percentage of these machines in the field, under certain conditions, the print head would be fired by the special microprocessor continuously and then go up in smoke. This was a custom microprocessor problem. The microprocessor would go out of control and keep firing the head until it started to burn. We had our own uniquely developed microprocessor in that printer, but unfortunately, in order to meet our production demands, our vendor made a change without telling us in an effort to increase his yield. This design change required substantial testing, and as a result we had an insurmountable problem. The only solution was for us to call back all the machines we had shipped. In addition we had to stop production for almost a year to make the necessary changes, to develop a microprocessor with another vendor, and to test it properly. This permitted the Japanese manufacturers to get into the matrix printer business and to fill the void that we had left. We opened the doors

wide for all other competitors to take this market away from Centronics. We never did recapture this rather important market.

How did Centronics chapter come to a close for you?

I sold out to Control Data Corporation.

What did they do with the company?

They took a piece of jewelry (Centronics) and crushed it with their inadequate management. There were some agreements and products, and they had major contracts with companies like Nixdorf and IBM, but Control Data couldn't come up with anything new. In that business, if you don't make your own products obsolete someone else will. You have to have the wisdom to know when and how to obsolete your products so another company doesn't do it before you.

What did you do after you sold Centronics in 1982?

I started PH Research and began developing thermal solid ink jet technology.

How did that work?

The ink is a solid at room temperature, and you have to melt it to jet it, and then when it hits the paper it freezes to become a solid again. The texture of the paper then doesn't matter. At the time, there was no other color ink jet technology that was quiet, printed on any paper, had brilliant colors, and was reliable and very fast. This was the first printer using this technology,

How did you develop it?

I went to a plastics factory owned by a fellow that was a neighbor and a friend. He was a scientist, and we discussed how to develop a plastic with a very low melting point that would also have a high level of surface tension, similar to that of water The color pigment had a very small particle size. It melts and gets spit out from the crystal gun to form dots. My friend gave us formulas for the plas-

---

So we said, "Let's make a standard interface, give it to the world free, and put an end to having peripherals furnished only by the computer manufacturers."

Unfortunately, our production quickly went up to over 200 a day and we found that in a small percentage of these machines in the field, under certain conditions, the print head would be fired by the special microprocessor continuously and then go up in smoke.

tic inks, but this was only a starting point. One of the formulas ended up working and we finished it off at RH Research.

What does the formula determine?

It allows you to heat the plastic so it behaves like ink in an ink jet print head. It had to be able to be melted so it could be jetted, and not cool to a solid in the one inch of air space before it hits the paper. And the ink had to have a great deal of surface tension so it would form round balls right after it was jetted. It had to gather and become round before hitting the paper. The surface tension was a big problem to overcome. We also developed a piezoelectric crystal to do the propulsion of the ink drops.

When I had the basic technology working I started Howtek in 1984, and gave all the know-how and patents to them. The result was the PixelMaster.

What about the litigation tussle between Howtek and Data Products? Can you talk a bit about that?

Dataproducts nearly ruined Howtek. We were sitting on 3,000 PixelMaster orders per month. We were tooled and set up to produce the printer in Japan. But the Data Products litigation delayed everything for two whole years. During that time, the Japanese yen dropped in value from $2.45 to $1.25, which doubled the price of the printer for Howtek. Here in the U.S. Apple, and many other companies would have been interested in it if that hadn't happened.

That caused the beginning of many years of restructuring at Howtek. That iteration of the PixelMaster was a "tweener". It was too expensive for general purpose computer color printout, and not good enough for graphic arts proofing. That could only be fixed with high-volume production, which was never possible at the increased price. We produced thousands rather than tens of thousands. They were produced in Japan by Juki. Like Brother, they made sewing machines. Juki bought shares in Howtek back then, at least 15 years ago.

What was your impetus for starting Presstek?

I founded Presstek in 1987 based on my idea to eliminate the time and substantial cost in the preparation of high-quality lithograph printing plates by digitally imaging the printing plates in daylight directly on the press.

Dick Williams is now Chairman of Presstek, right? How did you meet him?

Bob Howard, proud parent, showing off the original Presstek prototype, November, 1987 (Courtesy of Robert Howard)

In the early days of Centronics, IBM was a potential customer. Their people had seen our manufacturing and wanted a tour of our development labs. We had some unique systems, like a model shop that worked all night. If an engineer needed a part, he'd lay it out, draw it, and we would manufacture the part overnight so it was ready in the morning.

When IBM asked to tour our manufacturing facility, I said, "Okay but I also want to see your facility." They agreed. When I got to their printer manufacturing and engineering facility, I found that the only person there who knew anything about printers was Dick Williams. So I gave him an offer he couldn't refuse and he became Vice President of Engineering at Centronics. He's the most creative engineering/executive I have ever known in my entire life.

So he was worth your initial investment?

He was worth it a thousand times over.

They took a piece of jewelry (Centronics) and crushed it with their inadequate management.

Centronics was serious about laser printing – unfortunately, too early. These snapshots are glimpses of parts of a lab laser printer prototype at Centronics in 1977: the seamless drum, the developer tray (removed from the machine), and the mechanical drive. To the right of the developer tray is the liquid toner bottle and on the left, under a vented cover, the pump that stirs the developer. (Photos courtesy of Robert Howard)

Observations

We have two threads here. The interview is a revealing glimpse into the personality of one of the industry's most prolific inventor-entrepreneurs. Second is an inside story of how Centronics spear-headed the dot matrix revolution of the 1970s.

The personality. With his record of starting well over a dozen companies, Howard has surely demonstrated a certain genius. Some of his assets are obvious from the interview. There is pragmatism and maybe a bit of luck: free plugs from Wang and surplus transistors from the military. A neighbor who was the connection between Centronics and Brother. A friend who came up with solid ink formulas for RH Research. Later in this chapter Gary Starkweather tells about his own improvisations as he worked to develop the first high speed laser printer.

Other assets include an air of sincerity, bordering on passion. He comes through as a believer, not only in his ideas, but also in people. "Prentice was terrific, highly qualified." "Dick Williams was worth it a thousand times over." He likes to talk about his own accomplishments, but is also quick to share credit with others. He is a people person as well as an idea person.

At RIT's Cary Award celebration, Howard mentioned that he has talked with Dick Williams (at least by phone) every day for 25 years. There was testimony to his commitment. "He is willing to take huge financial risk to achieve a goal."

Associates shared some Bob Howard's favorite sayings:

"Even a blind squirrel finds an acorn once in a while."

"There are three kinds of people: those who make things happen, those who watch things happen, and those who wonder what is happening."

A secret to achievement is to move forward "one small step at a time."

Inventor, passionate promoter, pragmatist, need-seeker, team-builder – all proven to be useful attributes for the entrepreneur of the 70s. There were plenty of missteps. And Howard is quick to blame. But he somehow usually managed to land on his feet.

The Company. Some see the reversal of the company's fortunes beginning with the introduction of the 700 Series when they decided to move all manufacturing in-house. This caused quality control problems that were so severe that the company's manufacturing had to be completely shut down for six weeks in 1980, which contributed to the eventual demise of the 700 Series.

Another misstep was the Quietwriter. In response to Japanese dot matrix competition beginning in the 1980s, Centronics decided to do something completely different. They spent millions of R & D dollars in an effort to perfect a non-impact stylus printer based on technology acquired from Olivetti in 1982. This printer garnered perhaps more public interest than any other printer project at the time, even though it never made it to market following its demonstration in 1979.

This was not their first such gamble. Earlier, in the mid-1970s, in a program to diversify, they worked with Canon on a low-cost laser printer program. The effort, according to Howard, was premature, but he later saw it was the seed that led Canon to field a commercial low-end laser product four years later.

Ongoing financial woes prompted Howard's decision to sell his 45 percent interest in the company to Control Data. Under the agreement, Centronics was merged with Computer Peripherals

Inc. (CPI), and Howard agreed to step down as Chairman. After the merger CPI's line of medium and high-speed band and train printers were added to the Centronics product portfolio.

Moves made to regain profitability included drastic staff cutbacks and closing plants in Puerto Rico, Ireland, and the CPI plant in Rochester, Michigan. And indeed, the company managed some profitable quarters. 1984 saw Centronics take strides toward broadening its customer base.

Trilog was acquired, an Irvine, California-based manufacturer of line matrix printers with speeds ranging from 150 to 300 lpm. Then they acquired Advanced Terminals Inc., a sheet-feeder company. At the time, the annuity of service, supplies, spares, and parts kept flowing, accounting for around 30% of their revenues..

Centronics also restructured its manufacturing in the mid-1980s. This included building a final assembly and test site in the U.S., and then introduction of an Americanized version of kanban, the Japanese production philosophy that reduces inventory of parts, products and material by following a just-in-time manufacturing model. These moves generated some profitable quarters, but the overall direction was down. Except for their work with Canon in the 1970s, Centronics never made a serious attempt to add non-impact page printers to its product line and they were swimming against the tide.

In 1986 the company's final product rolled off the assembly line. Called the Tempest series, the printers were 800 and 1200 lpm Linewriters configured for security and targeted at government, banking, and brokerage printing.

The Centronics story came to end in February of 1987, when Genicom purchased all of Centronics' printer assets for $75 million. Printer products bearing the Centronics name would no longer be available. But we still have that widely used Centronics interface and plug. This legacy, not to mention their achievements, keeps the company name alive long after it has faded from the face of a printer product.

LINE MATRIX PRINTING: ROBERT KLEIST AND PRINTRONIX

In the fiercely competitive computer industry, any company that can last the better part of three decades, remain independent, and keep at the top of its organizational chart the same person that founded it in a Southern California garage must be doing something right.

Just as Centronics pioneered serial dot matrix printing, Printronix pioneered line matrix printing. A challenge for Printronix right at the start was the recession plaguing the U.S. just when founder Robert Kleist and his engineering team were trying to get established.

Fortune magazine highlighted Kleist and his company in a December 1977 feature about entrepreneurs who "beat the odds," founding technology companies which stood out in terms of growth and profitability. According to the magazine, Kleist and his associates put up an initial $670,000, and then raised an additional $550,000 by selling licenses abroad. Later, with the firm in the black, two venture capitalists invested $280,000. In the late 1970s, they attracted an additional $1 million from two other venture capitalists that would be used to support the company's expansion through the coming year. Commitment and persistence paid off.

Keeping a sharp focus appears to have been a second winning asset. Printronix's first printer was the P300, a 300 lpm line matrix printer introduced in 1974. They added a 600 lpm and a 150 lpm version, but they did not broaden their product line until 1981 with the MVP, a lower speed, dual-mode matrix line printer. As of the mid-1980s, they claimed their share of the non-captive line matrix market was close to 80%.

During the mid-80s Printronix broadened its product line further via two acquisitions. The results were mixed. Acquiring Data Printer in 1984 got them into full-character line printers and for a time this was a financial plus. Anadex, with its line of serial dot matrix printers, was acquired in 1985 and turned out to be a financial drag. Also around the same time Printronix got into a patent infringement dispute with C. Itoh/Citizen and Mannssmann Tally. Although the lawsuits were eventually settled out of court, both were considered victories for Printronix. More recent alternative technology moves include thermal and laser.

Lately Printronix has shown resourcefulness in making strategic moves beyond hardware products. Several initiatives are aimed at customers' printer management needs, leveraging the Web. One move is a partnership between SAP, the enterprise resource planning (ERP) software vendor, and the company's own PrintNet Plus software which allows large user organizations to centralize printer management.

The strategy is to work with various ERP software suppliers to develop middleware to tie printer management with online global management of a business. SAP, for example, has hooks that provide for things like label printing. Printronix now supplies proprietary middleware that lets SAP provide data on devices in the network, viewable on a Web page. No one else offers this. Ralph Gabai of Printronix explained it this way:

"Our PrintNet Plus product combines a network interface card with direct Ethernet communications (10- or 100BaseT) complemented by the Printronix Print Manager software. The latter is a Windows NT program that monitors our printers, which can be located anywhere on the network. Each printer can be assigned an HTML page, which lets administrators remotely monitor and control it from anywhere in the world. For example, if your company has fifteen locations worldwide, and you want to change a compliance format, instead of sending service guys to each location, you can do it remotely from one console. You can configure and control printers located anywhere in the world."

All this seems to be working for Printronix. In late 1999 Kleist cut the ribbon on a new corporate and manufacturing facility situated on 12 acres in Irvine, California. The 186,000 square-foot building was designed as an open environment to encourage collaboration and team participation for the company's 520 full-time employees at that location. The site adds to the company's existing facilities in Irvine, Singapore and Holland.

The Printronix P Series (left) was the company's sole product from 1974 until joined by the 150 lpm MVP in 1981. (Courtesy of Printronix, Inc.)

Oral History:
Robert Kleist Tells the Printronix Story

This Printronix story makes for an interesting contrast to Centronics. Centronics today is mostly just a memory. Printronix never became a supernova in terms of growth or sales volume, but it is still shining. Bob Howard has started at least three companies since Centronics. Founder Bob Kleist still leads his company. Talking with him is a different kind of experience — he is clearly a nuts and bolts kind of guy compared with the mercurial Mr. Howard. This interview was conducted in September, 1999 by John Webster for Digiprint Research Associates. Italic text is John; Roman text, Bob.

What led you to start Printronix?

I was one of the founders of the company, along with Gordon Barrus from Dataproducts, whom I describe as the architect of our printer technology. Gordon and I worked at Ampex Computer Products in the early 1960s, and Dataproducts was started by people from that Ampex division. Gordon was one of the founders of Dataproducts and experienced in the design of line printers. We got together to start Printronix in 1974 because I understood the emerging market requirements for minicomputer line printers and Gordon understood the technology to fulfill that need.

My interest was that I had been in the systems business, and I saw a need for a reliable, lower-cost 300-line per minute printer designed for minicomputers. Three hundred lines per minute was the sweet spot for speed. At the time Dataproducts was the major supplier of these printers. They were the 800-pound gorilla in this market.

The two key things we strove for were reliability and the ability to print graphics, as well as words and numbers. Affordability was also extremely important. To achieve the reliability we wanted, we used half the number of moving parts, compared to

existing technology in 1974. For Example, typically a line printer used 132 hammers. We used 44. All of a sudden, we could argue that it's more reliable and in this case it would actually be true!

What challenges did you face when you launched Printronix?

The year 1974 was a very difficult year to start a company in the U.S due to the recession. Getting our venture capital normally would have been easy. We had experienced management, and I'd been successful with Pertec. In fact when we first started to approach people, it was easy, but then every handshake fell apart as venture capital people began to withdraw from any startup. So Printronix financed itself. We started in a garage in El Segundo, and then moved to our first facility, in Irvine, California. It took two years for the company to become profitable, and from there, we could easily raise venture capital. Then we went public in June of 1979. We were one of the earliest computer companies to do that after the recession.

I remember when a reporter who was researching how difficult it was to get venture capital in the mid-70s asked me "How many venture capital people did you contact?" I said "A lot." I decided to count the number since I kept records. I contacted 79 total. None of them invested except a friend who's rich now and retired.

What was the competition like at that time?

Most of the big systems companies, including IBM, HP, and Unisys, designed and built their own printers. We really couldn't enter this marketplace when these large companies supplied their own printers. So we took a different approach, the 44-hammers I mentioned earlier. We introduced a printer that used half the moving parts of the printers on the market, which made it more reliable and lower cost to manufacture and sell. We also wanted to let users print graphics. At the time, almost all the printers could print upper and lower case

Robert Kleist

All of a sudden, we could argue that it's more reliable and in this case it would actually be true!

I contacted 79 [venture capital people] total. None of them invested except a friend who's rich now and retired.

characters, but no graphics. To market these printers, we focused on distribution and minicomputer manufacturers. At the same time, the need for bar code and label printing was beginning to emerge in industrial applications, and that needed the graphics capability of line matrix technology. That's why Printronix

From top to bottom: Yes, Printronix really did get its start literally in a garage (1974). Today, P5000 line printers in final assembly, Derian Building plant. Hammer bank inspection. The Printronix stored-energy hammer technology, as explained by Kleist, meant "the electronics were relatively simple, but the mechanics had to be very accurate." Photos courtesy of Printronix, Inc.

has focused more on industrial applications. What Printronix brought to market was line matrix printing.

What was the first product to come out of Printronix?

Our first product was the P300 introduced in 1975. There are still thousands of them in use today. We marketed the P300 as the most reliable printer available, and as a plug-and-play product that could actually print graphics. The problem was, most people didn't know what to do with the graphics printing capability. We first sold them through minicomputer systems integrators whose customers were paper mills, which needed control systems that printed bar codes and labels. That was the only industry that had standardized on bar codes in 1975.

What we could offer was the Printronix interface, I/O compatibility, parallel and serial communication for all the major computer manufacturers, and we had the graphics capability.

What was the graphics language you developed?

Intelligent Graphics Printing, or IGP. We needed it because our first printer had dot-addressable graphics. That meant the application software programmer had to do a lot of work to program their software to each dot. That can mean a lot of dots and a lot of programming. IGP was originally oriented toward printing bar codes and labels. It tells where to put the corners of a label or the lines in a bar code and where you want the barcode to be placed. This saves the programmer from having to do it. The IGP language became pervasive in the early 1980s and we continued to maintain it and enhance it. And like most firmware, we can embellish it for specific OEM customers. We got software vendors to notice it and it is now embedded in many enterprise networks around the world. There are also many WYSIWYG packages, page layout programs, that have IGP in them.

What other challenges did you face?

In the late 1980s, the minicomputer companies began to disap-

pear, taken over by powerful desktop computers. When the mini-computer companies went out of business, we were faced with a major challenge. The only difference between this period and the beginning of the company was now we had money. But we also had a drop in sales of 40% and several years of unprofitable operation.

At that point, we turned to world-class manufacturing techniques. In other words, if it doesn't add value, don't do it. We implemented just-in-time inventory management, lean manufacturing, all those things. As a result, today, we have many facilities that in 24 hours can turn around a custom product shipped from any of our four factories – Irvine, Singapore, Holland, and Memphis. We are able to do this just-in-time inventory management and work closely with the supply chain. The whole principle behind just-in-time is fast response and less inventory.

One thing this shift in manufacturing did was reduce inventory by four to one. If we extrapolate that to today, we're selling double the volume but we have just one-fourth the inventory. On top of that, instead of a 30- to 60-day delivery cycle, we can deliver in 24 hours.

With that cash surplus from the reduction in inventory, we developed our fifth-generation line impact printers. We invested in our core business – line matrix – because we saw that was an ongoing need, since it was the most versatile and lowest cost-per-page technology. We thought, "If we're successful, the major printer companies like IBM, DEC, HP, Unisys, Tandem, Siemens, and Bull might take notice." And sure enough, all those companies began to outsource their line printers to Printronix.

What does your product line look like now?

The line printer is still our core business and it's the cash cow

At the same time, the need for bar code and label printing was beginning to emerge in industrial applications, and that needed the graphics capability of line matrix technology. That's why Printronix has focused more on industrial applications.

that lets us get into laser and thermal technologies. One reason we pursued laser print technology is that about ten years ago we found users who required printing on plastic cards, like credit cards or IDs, and other non-traditional media. When people think of laser printers they think of an HP office printer. But our customers are different: they're industrial users who have continuous-form printing requirements. For example, they want to print on anything from label stock to plastic credit cards and they don't want the print to scrape off. That requirement was behind the evolution of our L5000 family. This line of laser printers now accounts for fourteen percent of our sales.

We got into thermal printing eight years ago, and it's now about two percent of our business. Our strategy is to design and make thermal printers so that they become a second business comparable in size to our line matrix printers. We're focused on the high end of the bar code and labels market. That segment for thermal printers is already bigger than the line matrix market and it's growing faster. And it's the same customer base, so we will continue with the same customers, just selling them thermal printers as well as line matrix printers.

The future of these printers is on the Network, with a direct connection to enterprise software such as SAP R/3. With Printronix System Architecture [PSA™], all three print technologies can be managed remotely in these enterprise networks and will operate from the same application software.

We believe you have been running your company for longer than anyone else in the industry. Any plans to retire?

As far as my own future, I want to stay with Printronix and see thermal printing become our second core business. I can't play golf, so I can't retire!

We marketed the P300 as the most reliable printer available, and as a plug-and-play product that could actually print graphics. The problem was, most people didn't know what to do with the graphics printing capability.

Using the money that used to go toward inventory, we doubled our R&D budget and advanced our line matrix technology.

NON-IMPACT EMERGES AT THE HIGH END

While serial matrix printing was sweeping through the WP and later the PC printer market, an interesting battle was brewing in the high end. The last half of the 70s decade saw a number of great leaps forward in high-end, production printing, defined as speeds of at least 5,000 lpm. For two decades centralized data processing printing had been dominated by impact line printers. Their speed ceiling was reached at 2,000 to 3,000 (or in one case 5,000) lpm. They did the job, but their limitations in the face of expanding computer technology and applications were becoming ever more apparent.

They were slow. While computing speeds were growing exponentially, impact printer speeds were more or less stuck at around 2,000 lpm, only about three times the 600 lpm of the 1950s drum printers.

They could print only on continuous, pinfeed paper with its attendant post-processing burden.

They were noisy.

There was heavy operator overhead. Changing forms required frequent manual intervention.

Fonts were limited, print quality far from typographic, and graphics crude at best. Forms had to be pre-printed on specialized web-to-web presses and were expensive.

Computer centers had to inventory a wide variety of custom printed and stock forms. Paper savings of around 50% were part of the promise of high speed non-impact printers, not to mention the ability to change semi-variable data (the "form") on the fly.

With line printers, color was out of the question.

Non-impact technology seemed like the way to go, but the journey was filled with casualties.

The Rank Xeronic high speed electrophotographic page printer in 1960 worked, but only a few were placed. The Radia-tion 690 electric discharge printer in 1963 churned through rolls of coated paper at 30,000 lpm but few, if any, were sold. Several companies fielded digital CRT-microfilm printers during the 1960s. The prototype Stromberg-Carlson Datagraphix 3400 printed from a microfilm intermediary onto zinc oxide paper with liquid toner at 5200 pages per hour. A 6000 lpm page printer for the military using a fiber optic faceplate CRT was developed and released by Litton Datalog Division in 1968. It never became commercial. Electroprint introduced the 8,000 lpm EP-100 in 1973 and licensed Oki Electric for Japan. There may have been installations overseas, but placements in the USA were minimal or non-existent. In 1973 Uppster announced the electrophotographic Model II, licensed Hitachi, then faded from view.

But over the next few years a set of very significant high speed production printers appeared which finally did have market-changing impact.

The Xerox 1200, announced in 1973 was a prelude. This electrophotographic cut sheet printer offered 4,000 lpm on plain paper for around $150,000. At that speed, it was not quite a "super printer." But it did demonstrate it could replace impact line printers.

Then a year later Honeywell launched its Honeywell Page Printer, a true super printer at 12,000 lpm. Quite a few were believed to have been shipped over the next five years, but despite electromechanical simplicity and reliability, the need for somewhat expensive, coated paper led to its eventual demise.

The real competition over the next decade was between three vendors of watershed high speed production printers: IBM, Siemens (now Océ), and Xerox. Each advanced the state of the art in various ways. Two of the companies had been household names in the USA and around the world for decades.

The third, Siemens, was less well-known in North America. Their bold development and launch of these high-end products

was the beginning of a journey that has brought them to the status of a major competitor with the better known giants. Their 1996 acquisition by Océ to form Océ Printing Systems (OPS) gave them a broad product line and positioned them as a major digital printing player. Better known overseas, in the USA they are beginning to achieve the visibility their technologies, their commitment to this market, and their resources warrant.

IBM announced its 3800 laser electrophotographic page printer in April, 1975. The first installation was at F. W. Woolworth's North American data processing center in July, 1976 where it replaced six IBM 1403 impact line printers. The 3800 continued to be manufactured for over a decade. The last unit was shipped in April, 1989 to a bank in Yugoslavia. IBM claims to have shipped a total of 8,000 3800's and that as of October, 1994, 3,000 were still in use around the world.

Siemens came next, announcing their ND2 high speed laser electrophotographic page printer in June of 1976. It was demonstrated the following year at the Hannover Fair in Germany and the National Computer Conference in the USA. Deliveries began in 1978 with a version designated 3352 for users of high-end Siemens computer systems. This was the beginning of a continuously evolving series of high-end production printers, each new generation embodying significant enhancements to the state of the art. As of 1994 Siemens claimed to have shipped 10,000 high-end production printers worldwide.

Xerox, surprisingly, was third. They had the advantage of a ready-made marking engine, the 9200 copier. Their competitors,

both IBM and Siemens, had to develop their marking engines more or less from scratch. Xerox also had some field experience with their earlier Xerox 1200 printer. Their major hurdle was apparently not the marking engine, but rather the electronics and software. Another was that they did not have a customer base of computer users. Both Siemens and IBM could leverage the intelligence of their own computer systems with which their printers were primarily designed to work. Xerox had to build a lot more intelligence into its printer.

IBM 3800

Considering they were starting almost from scratch with the marking engine, IBM's development and launch of the 3800 makes for an intriguing episode in the IBM vs. Xerox competitive mini-drama of the 1970s and 1980s. IBM was one of the many

Basic Specs, 1970s High End Electrophotographic Production Page Printers

	IBM 3800	Siemens ND2/3352	Xerox 9700
Speed	32 ips, 210 ppm 13,360 lpm at 8 lines/inch	29-1/6 ips, "over" 200 ppm 14,000 lpm at 8 lines/inch	20 ips, 120 ppm "up to" 18,000 lpm
Technology	laser electrophotography	laser electrophotography	laser electrophotography
Horizontal Resolution	180 dpi (original model; later upgraded to 240)	240 dpi; reduced to 180 dpi for IBM compatibility	300 dpi
Paper	continuous, flat-pack pinfeed 6.5 to 17-7/8 inch wide	continuous, flat-pack pinfeed 6.5 to 15.8 inch wide basis weights 60 to 100 g/m2	sheets, 11 x 8.5 inch only basis weights 16# to 110#
Format Line length Line spacing Character spacing	up to 204 characters 10, 12, or 15 per inch 6, 8, or 12 lines/inch	136, 163, or 204 characters 10, 12, or 15 per inch 6, 8, or 12 lines/inch	up to 150 characters 4 to 30 per inch 3 to 18 lines/inch
Price	~$300,000	~$255,000	~$290,000
First Deliveries	July 1976	early 1978	late 1978

Source: Digiprint Research Associates

companies that turned down Chester Carlson when he came knocking on their door peddling his xerography invention in the early 1940s. Later they tried to play catch-up, introducing the IBM Copier I in 1970, followed by the higher throughput II and III.

Xerox's assertion that IBM had infringed on some of their key patents became more or less moot in the wake of the FTC's 1974 antimonopoly decision against Xerox. According to one source, Xerox at the time owned an estimated 85% of the worldwide plain paper copier market. Within a decade, their share had eroded to around 40%.[*] This was due to a number of forces. Monopoly fears at the time were stronger than they are today. The market was changing. Applied science and technology advances were giving developers more choices. Finally, Xerox was seen as mired in a period of indecisiveness.

Xerox had licensed some of its xerography technology to IBM for computer-related applications. This no doubt fed into the development of the 3800, but IBM certainly did not have any ready-made, high volume copier that could serve as a marking engine for a 210 ppm high speed printer.

On the surface, the imaging technology in the 3800 looks basically like standard xerography. But there were interesting twists. The photoconductor is a flexible belt, fed from spools inside a large drum. This concept greatly lengthened the life of the photoconductor, but, as always, there was a downside. Unlike the competing printers, this innovation imposed a non-print area at a fixed point on the drum's periphery. The drum circumference was 79 inches with a 2-inch slot for the photoconductor, leaving a net of 77 inches for printing. This allowed for printing seven 11-inch images on the continuous, flat pack paper. Then the paper had to be halted to let the gap in the drum to pass by. So the net speed was somewhat less than the 32 ips drum surface velocity.

There was provision to overlay a "form" image by means of a strobe flash through a photographically prepared negative. The

xenon flash lamp with a pulse duration of just 125 μs was fast enough so vertical image smear was minimal. For the variable information, the light source was a 40 mW helium-neon laser beam, scanned in the usual way, by a rotating, faceted mirror. In the original 3800 this provided a horizontal resolution of 180 dpi coordinated with the paper speed for a vertical resolution of 144 dpi. The 5.5 mil printed dots were said to provide good, or at least good enough print resolution. In the Model 3, introduced in 1982, resolution was enhanced to 240 dpi. After initially experimenting with liquid toner, IBM ended up with dry.

The 3800, like its competitors, was a big machine. The paper path was over 14 feet for the standard continuous feed, and over 18 feet with the optional burster-trimmer-stacker feature. The machine's long product life and long life in the field, mentioned above, is testimony to superb engineering. The targeted print volume was two to three million feet per month. For a product used in the field for ten years, that adds up to a lot of pages. It also adds up to a good revenue stream for IBM. Many machines were leased. The original basic monthly charge was $7,344. For customers purchasing the 3800, the original monthly maintenance agreement was $455 per month. Whether leased or purchased, there was also a tick charge which at the time of introduction in 1975 was $2.30 per thousand feet.

Two years after the 3800 was announced, IBM introduced its next major non-impact printer, the 6670 "Information Distributor." This was actually a printer/copier, a sheetfed machine that competed more directly with offerings from Xerox. IBM claimed the 6670 offered a combination of features unmatched by anyone else at the time. Documents could be printed or copied on one side or two sides at up to 36 copies per minute. It was also described as offering text processing, communications, and a selection of resident fonts that could be intermixed on the same page. Print quality, at 240 x 240 dpi, was letter quality.

Another product that should be recognized from IBM in the

[*] *Xerox American Samurai*, Jacobson & Hillkirk, MacMillan, 1986

1970s is the 6640 ink jet document printer. Using deflected, continuous technology, this printer was seen as a wonder in its day with its high print quality and 92 cps rated speed. Initial versions were aimed at word processing printing serving clusters of IBM magnetic card Selectrics. It was introduced in 1976 and manufactured into the early 1980s. However, it was pricey at around $20,000 and its performance a bit moody. According to one informed estimate, only around 10,000 were built. Nevertheless, it was a break-through product which no doubt helped pave the way to user acceptance of ink jet printing for the office in the decade to come.

These three 1970s products, in retrospect, might be seen as the swan song of IBM's remarkable string of innovative, in-house developed print engines which dominated computer output printing for three decades. Perhaps they were anticipating the increasing decentralization of computer power and printing. Perhaps they were simply making some business decisions as to where they could reap the maximum return from their R&D and development investments for strategic growth.

On the page printer front, beginning in the 1980's they began to increase their reliance on print engines developed and built by other companies. Two early examples were the 3820, a 20 ppm page printer sourced from Minolta and introduced in 1981, and the 1984 12 ppm 3812 based on the Kentek engine. After years of upgrades to the 3800, IBM launched a successor, the 3900. It was built around a print engine from Hitachi, which has continued to be the basis for their current high-end page printer offerings. The IBM-Hitachi relationship is strong and apparently mutually satisfying, analogous to the HP-Canon relationship in low end page printers. Other IBM print engine partnerships have included Printronix for impact line printers, Kodak for one of its high-end production printers, Xeikon for on-demand color, and Fuji Xerox and Canon for low end page printers.

Abandoning print engine development and manufacture did not mean IBM was abandoning the business of designing and integrating a continuous stream of printer systems. On the controller and software front, they continued to innovate. Advanced Function Printing (AFP), introduced in the mid-1980s, became an important PDL standard. An IBM printer guide published in 1988 listed 75 current models ranging from low-end serial dot matrix, daisywheel, and ink jet printers on up to the 215 ppm 3800-3 page printer.

But in the 1990s, their commitment to printers seemed to become increasingly ambivalent. In 1991 they spun off their lower end printer businesses based in Lexington, KY, giving birth to Lexmark. With that agreement, they committed to not compete in printers performing below 29 ppm for five years. To serve their customer base at this end of the market, they distributed Lexmark printers. However, once this agreement expired, they began working this market again. In 1996 they introduced a line of network printers rated at 12 to 24 ppm for reseller distribution. In 1992 they consolidated their mid- and high-end printer businesses under an entity called Pennant Systems, only to re-embrace the operation as IBM Printing Systems Company (PSC) in 1995. But in 1998 it was reliably reported they were looking for a buyer for PSC, an exploration that was terminated when there were no takers at the price IBM envisioned. For now, IBM plods profitably along as a somewhat reluctant industry leader.

Three IBM groundbreaking products of the 1970s (from top to bottom) were the 3800; the 46/40 (later called the 6640) office ink jet printer, 1976; and in 1979, the sheet-oriented 6670 Information Distributor. (3800 and 46/40: IBM publicity photos, DRA archives; 6670 courtesy of IBM archives.)

Gary Starkweather

Xerox 9700 and the Role of Gary Starkweather

The story of the Xerox 9700 brings us to the fabled Xerox Palo Alto Research Center (PARC), which was established in 1970. When the 9700 was introduced in 1977 it was hailed by Xerox as the first major product to be "enabled" by PARC research. The key person behind this program at the Center was Gary K. Starkweather, often described as the father of laser electrophotographic printing.

The following excerpts from a 1997 interview of Starkweather by Frederick Su tells a bit of the 9700 story as he lived it.[*]

Why don't you talk a bit about the imaging development at PARC?

I came to Palo Alto in 1971 to build a laser printer. It took a lot of memory to store the images and provide font storage. We built a prototype in approximately a year and managed to prove out the fundamentals of the scanner and xerographic exposure technology.

By 1973 to 1974, we had a printer in general service. By 1976, we had built a unit that could be reproduced and routinely used. Hundreds of people [at PARC] were doing their daily work using these devices.

How closely is the laser printer related to the office copier first developed by Chester Carlson?

We were able to use the copier technology by replacing the lens that imaged the copy sheet to the photoconductor with a laser beam scanner that exposed the photoconductor. A laser scanner is just a pointwise method of exposing the photoconductor. The same toners and developer packages used for copying products could be used with the laser scanner systems we developed.

* Reprinted with permission from *OE Reports*, a publication of SPIE, the International Society for Optical Engineering, Bellingham, WA USA, November 1997 issue. At the time of the interview Starkweather was Imaging Architect for the Windows NT platform at Microsoft.

Other folks, like IBM, had to create their own electrophotographic technology specifically for the printer they built. This required a much heavier investment.

Didn't the patent on xerography expire after 17 years?

There were a lot of upgrades that came along that kept the patent pretty much in a fresh state. There were new additions and capabilities and features that came along in the xerographic system. . . you certainly could build a 1959 copier, but who'd want to?

Anything else you want to say about the development of the laser printer?

It was more fun than anyone had a right to expect. Xerox was very generous in its support of research. There is one person that needs to be noted in all of this history and that is Jack Goldman, who was head of research at Xerox. The Palo Alto Research Center was his idea. He deserves enormous credit for being so visionary and putting it where he put it and for setting the goals that he did. I think there is over 50 billion dollars' worth of business that's been generated from the technology that came out of PARC.

Why have laser printer prices dropped so much?

It's because of the economies of scale coupled with high reliability. The xerographic technology that is used in them is driven by both the copier and the printer markets. The economies of scale are enormous when you look at somebody like Canon that may produce perhaps 50,000 or more of these things a month. With economies of scale of that size, for example, the scanning mechanism, which in the early devices cost several thousand dollars, are now 50 bucks. By and large, the cost of the scanning mechanism and the cost of the printing technology inside the machine have come down by sheer force of production volume, and also by very clever engineering.

Starkweather comes through in this interview as perhaps a bit self-effacing, choosing not to spotlight his contributions or the way he had to crusade to convince Xerox management that a laser light beam could be used as the way to place variable images on the photoconductor. His original assignments, after all, were just to develop lens and illuminating systems for copiers. In a 1999 interview with Mike Zeis of Blackstone Research Associates, he displayed his vision of "print unchained" and something of the passion that he no doubt needed to sell the concept.

When asked about resistance to the concept of bit-addressable printing, Starkweather responded, "A common question was, 'Does anybody need this thing?' It took a while for people to realize that the laser printer really gave them a new freedom, because that quality of printing was previously available only at print shops. That was a liberating concept. But some, especially in marketing, wondered whether people needed the flexibility."[*]

When PARC was established he worked at the Xerox Webster Research Laboratory. In order not to drain that site of research scientists Xerox had a policy of not allowing anyone to transfer from that location to Palo Alto. To get permission to transfer, Starkweather reportedly took the risky path of jumping over his immediate superiors and appealed to higher management, and succeeded.[†]

Even though PARC was a world class research facility, much of the actual development has been described as looking like "garage shop" improvisation (which seems not unusual; cf. Robert Howard's group using surplus components to develop the dot matrix printer and Centronics interface). A key advance in Starkweather's laser printer development was the lens system to correct deviations in the light path. An early prototype was said to use a precision-tooled aluminum faceted mirror that cost $10,000. Starkweather used some war surplus lenses from a catalog to correct light beam deviations, thereby making it possible to use a $100 mass produced mirror.

The 9700 went public at the National Computer Conference in the spring of 1977 and customer shipments began in 1978. The optical system was in concept simple, the spinning polygon mirror deflecting the laser beam across the photoconductor which was a flexible belt which rotated around a set of three rollers at 20 inches per second. The marking technique was termed "charged area xerographic development," meaning that the laser beam created the image by discharging all the non-image areas. The resolution of 300 x 300 dpi translates to 90,000 bits per square inch or 36% more than the dot density of the competing Siemens machine and over three times more than the initial version of the IBM 3800. In addition, the 9700 could be programmed to integrate the "form" with the variable information; there was no optical mask form overlay feature. Around 800 page images could be stored on the 9700's hard disk. All this meant a huge amount of intelligence had to be built into the printer.

The success of the 9700 and later high-end production printers during the past two decades does not appear to have sharpened Xerox's focus in the direction of printers, at least not until recently. They began opening retail stores in 1980, but reportedly sold off 43 of their 54 stores in 1983. The same year they diversified into insurance, acquiring Crum and Forster, Inc. They reorganized under the new The Document Company slogan in 1994, the same year they farmed out the operation of their worldwide computer and communications network to Electronic Data Systems under a $3.2 billion contract.

In their published historical highlights, the emphasis is organizational rather than product landmarks. It is noted that in 1976 the last Xerox 914 copier order was taken, and in 1988, they

---

* *Color Business Report,* July, 1999, published by Blackstone Research Associates, Uxbridge, MA.

† This and other Starkweather lore that follows is from the book, *Dealers of Lightning: Xerox PARC and the Dawn of the Computer Age* by Michael Hiltzik.

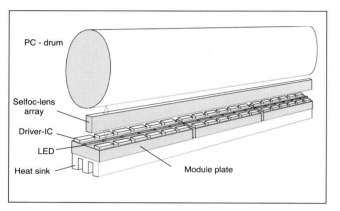

shipped their two millionth copier. In the 1990's they worked to broaden their printer offerings. At the low end, there were partnerships first with Fuji Xerox and later with Olivetti for ink jet printers. They acquired full ownership of Delphax in 1997, partnered with Scitex to develop color digital printers, then in 1999 acquired the printer operations of Tektronix. The 9700 family continued to be upgraded, and as of this writing the Xerox Production Systems Group has maintained its position as one of the three major players in production printing systems. Finally, in the 1990s their major printer coup was the DocuTech, the machine said to create the on-demand segment of the printer industry. The story of the DocuTech is the centerpiece of Chapter 7, The

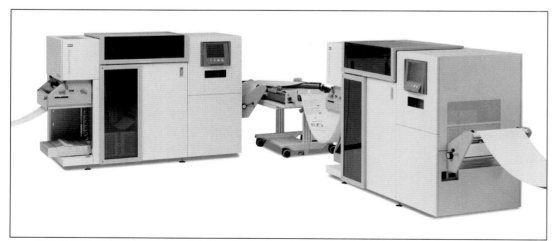

From top to bottom: The original 1978 Siemens ND2, using a laser light source, was fast for its time at 200 ppm. High performance LED print bar developed for new printer series in the middle 1980s. One of Océ's current high end offerings, the 700 ppm DemandStream 8000 web. (images courtesy of Océ Printing Systems GmbH)

Nineties, when we revisit Xerox and consider their positioning for the decade of the 00s.

Siemens/Océ Printing Systems (OPS)

The Siemens effort that resulted in their ND2 production page printer began with an even cleaner slate than IBM's. IBM at least had entered the copier business with a line of xerographic machines. Siemens had a strong computer operation and long experience in impact printers but little, if any copier technology, xerographic or otherwise.

However, as one of the world's largest electronics companies, Siemens has had a history of innovation and the resources to commercialize them. Communications has been a strong suit. Siemens claims the public telex system they built in Germany in 1933 was the world's first. They have been a leader in teleprinters. The T-100, introduced in 1958, became something of a worldwide standard with 500,000 installed. The later PT-80 was a successful successor. Even as Siemens was beginning to develop the high speed laser printer, they launched what was probably the first successful impulse ink jet printer, designated the PT-80i. Around 20,000 were shipped and the basic device was adapted as a color printer by several OEMs in the USA.

If there is one person who can be said to have piloted the Siemens page printer program through the years, that person would be Manfred Wiedemer. He joined Siemens in 1967 and worked with the team assigned to start development of electrophotographic engines in 1971. He was a major contributor to the initial ND2, later became project leader for the ND3 and the second generation LED printers, and is still there in the position of Director of Technology and Key Components. Dialoguing a bit with Mr. Wiedemer was very helpful in assembling the following brief account of this program at Siemens, including some of the challenges that were addressed.

The printer group within Siemens Datenverarbeitung (= Siemens IT) had been developing impact line printers (drum and train) since the early sixties for use with Siemens computers.

It became obvious in 1970 that the future demand for faster printers with higher print quality could not be satisfied with impact technology. This realization led Siemens in 1972 to begin the development of a non-impact-printer which would be five times faster than the fastest impact printer at this time. It seemed like a realistic goal. The Siemens peripherals plant in Munich, in Mr. Wiedemer's words, seemed predestined for building very complex electrophotographic devices. Their assets included a lot of experience in mechanical engineering, and as part of the Siemens IT group, they had extensive background in controllers and systems integration.

They originally planned for an investment that was believed to be only about 20% of what Xerox spent, and 10% of IBM's development costs for the 3800. As it turned out, they ended up spending much more than this. But it was still a lot less than what IBM and Xerox spent, they believe, although they don't explain how they achieved this.

But even at this investment level, a problem was that their captive market of Siemens mainframe users, which was mostly in Germany, was relatively small. Their projected volume of printers in this market was just 400 over five years. So early on they worked to develop the printers for the American and Japanese market in addition to their own customer base.

Consequently, the first pilot ND2 was installed in 1978 in Japan by Siemens' OEM-partner JemNeac using their Kanji controller. Soon their OEM business grew, with Univac becoming their first partner in the USA. Still their volume was too small to support their development costs.

This forced Wiedemer's decision to jump into the plug compatible business so as to access the huge IBM market, especially in the USA. They developed an IBM-compatible controller based on DEC´s PDP11-34. In a situation somewhat similar to that of Xerox with the 9700, the corporate powers did not seem to share the vision. At Siemens they were particularly skeptical about Wiedemer's move toward the PCM business. It was only after the first huge orders came in from former IBM-customers that Siemens top management approved this controller development. It took time, but Wiedemer finally managed to gain top management support.

In short, the start for Siemens was slow, but accelerated after the first few years. Early customers, besides their own computer customers, included Fujitsu, ICL, and, in the USA, Itel. In 1983 they won StorageTek as an OEM-partner in the US. and from that point in time grew more than 50-100% every year. StorageTek, formerly Storage Technology, had acquired Documation earlier, a major plug compatible vendor of high speed impact band printers.

Siemens claimed a number of advantages over the IBM 3800. One was their solid PC drum which did not impose a vertical non-print area or the need to stop the paper with each revolution. Perhaps the biggest advantage was that they had more compact engines which could print from edge to edge, and which were seen as more user-friendly. The printer was said to take a good bit of the European banking market from IBM.

In 1983 they announced their second laser printer. The ND3 had half the speed of the ND2 and was positioned as the world's first continuous forms printer in the 100 ppm speed class. It also

Manfred Wiedemer, born 1944 in Singen/ Germany, studied automation and precision engineering. He joined the Siemens AG as engineer for impact printers in 1967 and soon became leader of his own laboratory. In 1972, he started the ND2 development and was project leader for the ND3 from 1975 onward. In 1985, he was named head of the whole Siemens AG electrophotographic printer department. Thereafter he spent about a year at the US-branch of Siemens in Boca Raton, FL. In 1992, he took up the position of head of the technology department. When the Océ Group acquired Siemens' digital printing division in 1996 he stayed with Océ in that position.

The Océ Group as of 1999 was made up of the following businesses –

Océ Wide Format Printing Systems. Defined as large format printing and copying, rapidly transitioning from analog to digital. Accounted for € 782 million in revenues for FY 1999.

Océ Document Printing Systems. Office copiers and printers, rapidly transitioning to digital and "complete solutions." Accounted for €1,399 million in revenues for FY 1999.

Océ Production Printing Systems. Medium volume and high volume production printers, both sheet and continuous. Accounted for €657 million in revenues for FY 1999.

Océ Business Groups (Imaging Supplies and Facility Services). Océ Imaging Supplies revenues for FY 1999 were €414 million. Facility Services (Document Management) was Océ's fastest growing business in 1999, gaining almost 35% to €197 million or around 7% of total FY 1999 revenues of €2,838 (~$ 2,860). (Business Groups revenues are included in the revenues of each of the three Printing Systems business segments.)

offered a unique cold fusing technique. Fusing technique is as critical as the imaging system in print quality. It also significantly delimits the range of papers which can be used and, if pre-printed, the inks. In the IBM 3800 the original hot roll fusing temperature was over 200°C, which caused problems by vaporing inks used on preprinted forms and creating thermal induced paper handling difficulties. To ameliorate these effects, IBM soon added a pre-heat platen which allowed them to reduce their hot roll temp to around 180°C.

These sorts of problems provided strong incentive to move to a new fusing technology and Siemens chose a process using solvent vapors which significantly broadened the ND3's paper and ink capability. The new engine was especially suitable for the mailer-market since according to Siemens it could handle every type of paper including label stock.

But, again, every up side has a down side. European companies tend to be very sensitive to environmental considerations and Siemens is no exception. By the middle 80s they became wary of

Current headquarters and plant of Océ Printing Systems, Poing, Germany (Courtesy of Océ Printing Systems GmbH)

the environmental impact of the CFC fusing agent used in the ND3. And at the same time they were motivated to upgrade the print quality, the size, and the cost of the existing engines.

These considerations gave birth to the LED-printer family.

They decided the best way to fulfill their product goals – compact design, ease of use, print quality and so on – would be to replace the laser with an LED-printhead among other things. Another change initiated by Wiedemer was to replace the standalone, single model concept with a family concept. Variations of the basic print engine would offer different speeds and features combined with the economies of using as much as possible the same parts in development and production. Siemens returned to the more conventional hot fuser role for fixing.

This, however, was not an easy sell to their top management and marketing people. It took two years to convince the higher powers to commit resources to a new generation of printers for the 90s with new imaging technology. As is often the case, it looked like management dynamics might stand in the way of innovation and product progress. At the time they were successfully established in the continuous forms printing segment and said to be constantly gaining market share against IBM. Now they were planning to expand into the cut sheet segment where Xerox was well established as the leader. In parallel with the decision-making progress on imaging technology described above, they decided to develop their own cut sheet printer.

In order to accelerate time-to-market, Siemens took over a development group that had experience in cut sheet printers to develop a 50 ppm cut sheet print engine, to be designated the

Siemens 2050. They managed to launch the product after just two years of development, and used only 25% of the planned budget. This was made possible in part by using the same electrophotographic components as the fanfold engines, making the 2050 a very cost effective device. It was fast and cost effective. But maybe introduced a bit too fast since they admit it took a bit longer than expected to fully "stabilize" the printer in the field. However that was only a stage and currently they have a cut-sheet printer family descended from that first product with 55, 75, 110 and 158 ppm print speeds.

In parallel with the cut-sheet effort, they launched in 1989 the first member of the LED-fanfold printers which offered resolution up to 300 dpi across a print width of 17 inches. And around the same time peripherals development and manufacturing was moved from the original Munich location to a vast new headquarters and manufacturing plant in Poing, near Munich. Busy times.

Since then both the continuous and cut sheet printer lines have been expanded and their performance enhanced. New toners and new control engineering upgraded print quality, print width, and speed. Important innovations for the print-on-demand-market included "pinless paper transport", optimized print quality using 600 dpi LED heads, RET-technologies, and other control mechanisms. New color toners opened the way to highlight color printing. Duplexing became available, first with a twin engine configuration, and later from the world's first duplexing, single unit system.

Now, the irony.

Our understanding is that in the early 1990's the notion of the paperless office took hold among Siemens top management. So they began a program to divest print-related businesses. First they sold the Siemens-Hell group to Linotype, then the ink-jet division in Berlin to Kodak. Finally, in 1996, they sold the high performance printer business to Océ. Since then history has shown that the paperless office would not succeed in making paper obsolete. Instead, print has continued to expand and new technologies continue to open up new opportunities for digital printing. In short, Siemens' loss was Océ's gain.

The Océ story is an impressive tale of a business reinventing itself. It was started back in 1877 when the founding father, pharmacist-entrepreneur Lodewijk van der Grinten developed and marketed coloring agents for butter and margarine. The company stayed in this business for 90 years, finally selling it to Unilever in 1970. Three of the founder's sons, educated as chemists, had long since taken over, shifting the business to reprographics. In the 1920s they invented a more fade-resistant blueprint process. A later semi-dry "dyeline" diazo process over the years captured a large share of the European market. In the 1930 they came up with a "screen reflex" process for plain paper copying, essentially a variation of the diazo process that enabled copies to be made from opaque originals. This was successfully commercialized until finally eclipsed by xerography.

Despite projecting what has been described as a "certain Old World dignity," Océ has been aggressive both in their expansion by acquisition and product

Among the OPS innovations in high-end production printing –

•The ND3 of 1983 featured a unique cold fusing system permitting wider ink and paper tolerance (but subsequently discontinued primarily for environmental reasons).

•The following year the ND3 Twin was billed as the first such printer to offer duplex printing or a second color.

•In the 1980s they began working with LED imaging, introducing in 1988 an "LED Plus" technology that offered resolutions of 240, 300, and 600 dpi and the widest print width of any such printer to date, 17-inches, for A3 size document printing.

•In 1995 their second generation LED Pagestream 200 DSC was said to be the first duplexing high speed page printer with a second color from a single engine configuration.

•In 1997 pinless paper feeding was introduced, a feature said to cut paper cost by up to 30% and extend the range of materials and documents that can be handled.

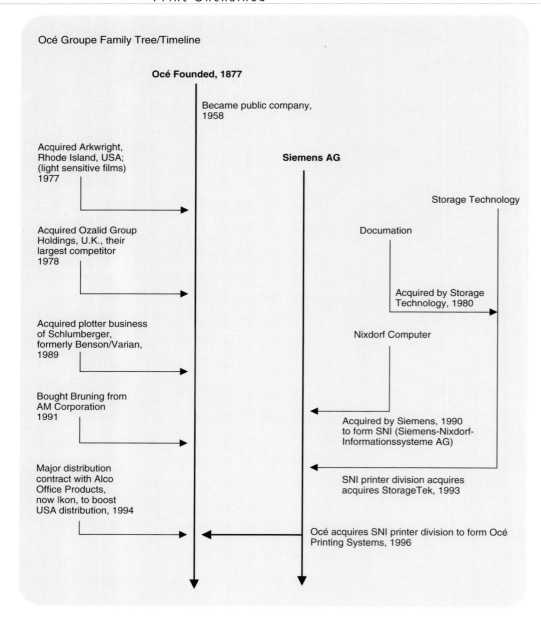

Océ Groupe Family Tree/Timeline

Océ Founded, 1877

Became public company,
1958

Acquired Arkwright,
Rhode Island, USA;
(light sensitive films)
1977

Siemens AG

Storage Technology

Acquired Ozalid Group
Holdings, U.K., their
largest competitor
1978

Documation

Acquired by Storage
Technology, 1980

Acquired plotter business
of Schlumberger,
formerly Benson/Varian,
1989

Nixdorf Computer

Bought Bruning from
AM Corporation
1991

Acquired by Siemens, 1990
to form SNI (Siemens-Nixdorf-
Informationssysteme AG)

Major distribution
contract with Alco
Office Products,
now Ikon, to boost
USA distribution, 1994

SNI printer division acquires
acquires StorageTek, 1993

Océ acquires SNI printer division to form Océ
Printing Systems, 1996

development. Océ Group now encompasses a broad range of print technologies derived from various acquisitions over the past thirty years (see Océ Family Tree/Timeline).

However, it is fair to ask, was this expansion or consolidation?

A historical maturing of the engineering "blueprint" business no doubt worked in Océ's favor. Engineering drawing and CADCAM was becoming increasingly the domain of digitally driven printer/plotters and more recently, wide format digital printers. And, as mentioned above, the illusion of industry maturity, of the paperless office, no doubt oiled the ramp for Océ's spectacular acquisition of Siemens-Nixdorf. The short answer is that for the industries, it was in part consolidation. For Océ, it was expansion.

It is certainly not winner take all, but they have achieved significant market share in each of their three businesses. One of their OEM partners, in fact, became Xerox which in 1995 announced it would be reselling two Siemens continuous forms printers, the 2140 and 2240.

In an era when not only Siemens but other majors seem to be losing their focus on print, Océ continues to bet its future on print. They also go their own way in product development. While most of the other majors including HP, Xerox and IBM treat print engines as an outsourced component, Océ continues to have strong in-house print engine development and manufacturing.

What drives Océ seems parallel with what personally drives a leader such as Manfred Wiedemer. In that sense, the 1996 marriage between Siemens and Océ looks like a good fit.

Asked what motivates him professionally, Mr. Wiedemer offered the following.

For me the driving force behind digital printing is that printout can be personalized and individualized. With all the new channels to access information, including the Internet, opportunities for what I call optimized print are better today than they have ever been. I believe paper remains the most interesting and most easily used medium.

The challenge for the information technology people is to create systems to print the right information, collected from the most current data source, in the best format for people to read and act upon. The old fashioned vehicles for printing cre-

ated much waste because print was not individualized. Our challenge as a printer company is to offer the customer a comprehensive solution including software, hardware and services to produce customized information. Digital printing is the natural link between electronic and printed documents.

Indeed, as of this writing, OPS has taken significant steps toward meeting this challenge. In addition to their continuing hardware performance enhancements, they have implemented a significant shift in the positioning of print capability. The traditional, host-connected PCM approach has largely given way to stand-alone server and network-based systems and, in the print-on-demand segment, what they term "segment specific solutions." In manufacturing, for example, they provide solutions for just-in-time production of printed materials with the print process designed to mesh with the ERP process at the plant location. In book production their digital printers are allowing users to profitably handle jobs that previously were too small for traditional printing presses. In commercial printing, Océ equipment helps implement seamless digital workflow from customer concept to the final printed piece.

Current Status of Production Printing

Despite the apparent ambivalence of IBM and Xerox, production printing today remains a three way race, with OPS claiming to be slowly gaining share. Regarding technology, it is still solidly an electrophotography market.

The three viable alternative technologies are high speed continuous ink jet, magnetography and ionography/charge image deposition (CID). Each of these has found significant specialized markets, but none has come near to a double digit share of the total production printing market.

Each of them also remains pretty much the domain of a single company:

- Ink Jet, the Mead DIJIT system, 1974, acquired by Kodak 1983, renamed Diconix, acquired by Scitex 1993 (but Kodak kept its trade name, Diconix).

- Magnetography, Bull/Nipson, acquired by Xeikon, 1999.

- Ionography, developed by Delphax, which was founded in 1980 to commercialize the technology; Xerox bought a 50% share in 1984, and in 1998 expanded its stake to 100%.

So why, in contrast to electrophotography, have these three technologies remained more or less the dominion of one company? It could be the patent walls. Yet with sufficient incentive and sufficient time, as was the case with electrophotography, there are usually ways around patents. There has been quite a bit of time. All of these alternative technologies were first commercialized almost twenty years ago. One answer, then, might be that the incentive just isn't there at this point. For general purpose production printing various forms of electrophotography seem to be doing the job well enough to date.

Of the three alternate technologies, it could be argued that ionography/CID has exhibited the strongest development dynamic and to have the greatest potential. The original technology has evolved through several generations, and the Xerox acquisition is viewed as a vote of confidence. As of this writing however, Xerox is undergoing yet another corporate downsizing and its Delphax acquisition is reportedly being downsized as well.

In short, in high speed production printing electrophotography still reigns supreme.

OTHER NON-IMPACT PRINTING DEVELOPMENTS

While Canon, Xerox, Siemens and IBM were working to develop and field high-end page printers, a number of other companies in the 1970s were busy at the low-end. Teletype and A. B. Dick had already fielded ink jet printers in the 60s. The following decade saw additional entries.

IBM launched the 46/40 in May, 1976, which the following year was upgraded and renamed the 6640 as a component of the IBM System 6 word processing computer. As covered in Chapter 3, the printer used Sweet continuous ink jet technology, printed at up to 92 cps, was priced at around $24,000, and did not enjoy long term commercial viability.

In May, 1978, Silonics, Inc. introduced the Quietype drop-on-demand ink jet printer. This pioneering piezo printer was speci-fied to run at 180 cps across an 8-inch print line and was competitively priced at around $2,500. Silonics was set up in 1969 under the auspices of System Industries, a disk system company. The development work got expensive and in time outside resources were needed. Konishiroku came to the rescue with a 49% investment in Silonics and at one time reportedly had 25 people involved in the project. Despite this lengthy and expensive development effort, an aggressive launch, and making some customer shipments, the Quietype, too, was short lived.

The same year Siemens introduced the PT-80i, an advanced version of its drop-on-demand PT-80, to the U.S. market at the National Computer Conference. This printer was one of the few modestly successful low-end ink jet printers of the decade, at least in Europe. It was marketed in this country by the Siemens OEM Division in Anaheim, California, but was not able to achieve much visibility as a competitor to the ubiquitous dot matrix printer.

Data Interface fielded the DI-180 magnetographic line printer in 1974. The company was soon acquired by Inforex and the machine was committed to production. However, according to Wieselman and Tomash,[*] problems such as low print quality, unreliability and cost consigned this product to an early demise.

Direct electrostatic printer/plotters appeared in the 1970s. The Versatec 1100A, introduced in 1970 may have been the first such machine. The company was later acquired by Xerox and grew to become a successful vendor of this class of computer output printer. Gould and Varian Associates were also early commercializers of dielectric paper printer/plotters.

Thermal printing did make inroads at the low-end of the market with Texas Instruments and NCR fielding thermal printing terminals. The Texas Instruments Silent 700 series of RO/KSR portable teleprinters enjoyed quite a bit of popularity in the 70s. Shipments by the early 1980s are believed to have been in the range of 200,000 units.

Ink Jet Printing in the 1970s
(General purpose, full page printers)

Manufacturer	Technology	Estimated Installed Base
Teletype Inktronic	electrostatic pull	n/a
Mead DIJIT	continuous (Sweet)	40
A.B. Dick 9600	continuous (Sweet)	250
IBM 6640	continuous (Sweet)	6000
Silonics Quietwriter	piezo, dod	30
Casio	electrostatic pull	7000 (mostly in Japan)
Sharp	continuous (Sweet)	1000 (mostly in Japan)
Siemens	drop-on-demand	2500 (mostly Europe)

Source: data from a presentation by Peter L. Duffield at
The Institute for Graphic Communication, Carmel, California, February 1979.

[*] Tomash, Erwin, and Wieselman, I.L., op. cit.

Then there was the unique electrostatic burn-off miniprinter from SCI systems which actually did find a market and enjoyed a product life of eight years (see the "Gallery," Chapter 2). Introduced in 1977, the SCI Rotary Printer was priced as low as $300 for the mechanism in OEM quantities and churned out 2,200 cps on a strip of aluminized paper.

Finally, at the high-end, Mead Digital Systems launched the DIJIT high speed continuous ink jet system in 1972. This successful system did not compete with the general purpose page printers from Xerox et al., but rather has been developed over succeeding decades as a major vehicle for high speed industrial, advertising, and on-demand document production.

Casio and Sharp in this decade also developed and shipped low-end ink jet printers in Japan. But that was about it for non-impact printing. Ink jet and other non-impact technologies in the 70s were 90% R&D and experimentation, 10% commercial. Bob Howard mentioned working with Canon on a relatively low-cost laser printer project which ran out of money before it resulted in a product. Centronics in the 1970s also invested in continuous ink jet technology. Inside sources let it be known that a 240 cps product would emerge in the early 1980s. The 80s came, but nothing further was heard of this project.

Widespread commercialization of non-impact technology had to await the following decade.

THE DECADE OF THE DAISYWHEEL

In the low end of the market there were also momentous shifts. Even as impact matrix printing was wooing data processing users away from the ubiquitous Teletype, a new breed of printer spawned several major industry vendors. "Word Processing," today virtually forgotten, emerged and grew into a mini-boom in the 1970s. This merger of programmable processors with electronic typewriters, pioneered by Wang Labs and IBM, presaged the personal computer of the 80s. In terms of application, it might be seen as the high water mark of another almost forgotten function, "typing." The word processor was the ultimate evolution of this function beyond the electronic typewriter.

Bob Howard had it right when he recognized a huge, latent hunger for cost-effective speed regardless of print quality. But in parallel with that sea change, the demand for the "letter quality" output that was traditional in the office lived on.

The first to take this capability to its highest level was Diablo Systems. The California company was founded in 1969 by former managers at Singer to build disk drives. David Lee is recognized as the prime mover who took Diablo in another direction by developing the daisywheel printer. The first daisywheel printer went 30 cps, twice the speed of the dominant IBM Selectric. Later speeds peaked at around 80 cps. Xerox acquired Diablo in 1972 and David Lee left to found Qume, which was Diablo's major competitor through the 70s.

The daisywheel market, thanks to word processing, was hot. ITT acquired Qume in 1978. In that year, according to one source, 122,900 units were shipped at an average end user price of $3,100 which amounts to a market approaching $400 million.[6]

As with impact matrix printing, lower cost machines from Japan eroded the domination of Diablo and Qume in the early 1980s. In the OEM market there were NEC, C. Itoh, Ricoh, and Brother. Then along came still others with end-user daisywheel printers priced below $1,000, among them Juki, Brother, and Silver Reed. The daisywheel market remained vital until the later 1980s when PCs began to replace word processors and letter-quality non-impact printers emerged. By that time most of the daisywheel vendors attempted to diversify into impact matrix printers and non-impact, most with minimal success.

This brings us to globalization, the third major trend of this pivotal decade.

* Tomash and Wieselman, ibid.

Statistically Correct Text

Specifying the speed of a serial printer seems simple enough. Characters per second, right? But in the serial printer wars of the 70s and 80s the daisywheel manufacturers decided "cps" left too much room for specsmanship. The speed of daisywheel printers is partially dependent upon how far the wheel needed to move to position given characters.

The effort to agree on benchmark text led to Claude Shannon, a mathematician apparently well-known, at least in some circles, for his classic work entitled "Mathematical Theory of Communication" (University of Illinois Press).

Another Shannon piece published around 1950 in the Bell System Technical Journal is entitled, "Prediction and Entropy of Printed English." This is believed to be the source of the benchmark "Shannon Text." The content was based on a mathematical analysis of the frequency of various characters in printed English. It didn't make everybody happy (where are numbers and punctuation characters, for instance?) but it did win a degree of acceptance and was used by Diablo and many other manufacturers.

For those interested in the mathematical theory of language, or simply adding to their store of useless information, here is the Shannon Text used for daisywheel printers in the 1980s:

The head and in frontal attack on an english writer that the character of this point is therefore another method for the letters that the time of who ever told the problem for an unexpected.

GLOBALIZATION AND THE OKIDATA STORY

A study by Quantum Science in 1980 looked back through the decade at the Japanese "invasion" of the U.S. computer industry. According to the study, the "information equipment market" at the time had been designated a key target by the industry in Japan. The major firms were beginning to have a visible U.S. presence, especially for low cost, high volume products. Stevenson predicted a progression from OEM sales to joint ventures to direct sales. Time proved him right.

At the beginning of the decade, Stevenson placed Japan's imports of computers at close to ¥100 billion and exports almost negligible. By the end of the decade, imports were around ¥150 billion, and exports had swelled to around ¥80 billion. And, most significantly, around 75% of this volume was peripherals, including, of course, printers.[*]

Actually, by that time, this trend was already well along, including joint ventures in computer peripherals. Okidata[†] serves as a significant, early example of globalization in the printer industry. Currently Oki Data is also notable as one of the few impact printer companies from the 1970s not only still in business, but appar-

Bernard Herman

ently reasonably healthy and making a transition from one core printer technology to a diversified quiver of printer products.

Dennis Flanagan, currently President and CEO of Okidata and his predecessor, Bernard Herman, late in 1999 talked with John Webster about Okidata's birth, growth, and survival strategies. Here are some highlights from the story pretty much as they told it, beginning with Herman.

"[David] Nettleton came into the picture because he was an entrepreneur, plain and simple. He had helped RCA establish itself as a telecommunications company, and he had become involved with similar projects at several British companies. At the same time, Oki Electric in Tokyo wanted to market its products in the U.S. Among other things, they were a low-volume, high-cost producer of specialized telecommunications switches for the banking industry, as well as for companies like AT&T. But their products were specifically tailored to the Japanese market. For example, they made a dot-matrix printer for the Kanji character set. So the products weren't quite ready for the U.S."

Herman, who held the top post at Okidata from 1979 to 1992, said the partnership wasn't always smooth sailing, but that the benefits of shared technology development, world class manufacturing and marketing resources greatly outweighed the occasional rough spots. One early challenge faced by the two companies was integration of their marketing programs.

Because Oki Electric had been aiming its products at specialty markets such as telecommunications, the Okidata management team's first imperative was to make them more of a mass-market manufacturer.

---

[*] Mirek Stevenson, Chairman, Quantum Science Corporation, in a presentation at The Waldorf Astoria Hotel, New York City, April 7, 1980.

[†] Name changed to Oki Data Americas, Inc. in 1999.

During Okidata's initial year of operation, one of Nettleton's earliest executive decisions was to convince Oki Electric to acquire two U.S. companies that he thought would help get the fledgling start-up off the ground. The target companies were Bridge Data of Philadelphia, a 20-year-old machine shop and manufacturer of 80- and 96-column card readers, and the Computer Magnetics Division of Applied Magnetics, based in Santa Barbara, CA. The latter, subsequently dubbed Okidata West, manufactured Winchester disk drives.

While card readers became an early staple in their product line, entering the disk drive market proved to be a tactical error. That part of its business was unsuccessful, and the company in 1980 dumped the product line to focus strictly on output products. A good move, borne out by their financial results. After a slow start, in the early 80s, Okidata's sales took off.

Initially Okidata set out to become a dot-matrix printer manufacturer to reckon with. The first in their long line of dot matrix printers was a nondescript but functional and economical 9-pin printer called the CP110. After some years of lackluster sales, Herman arrived in 1979 and directed the company through more than a decade of dot matrix printer line expansion and the development of an LED-based page printer family which is today a significant chunk of its business.

Okidata's most successful dot matrix printer family, the Microlines, originated in 1979, initially aimed at the OEM market. After shopping the printer around to the big guns in that printer market, they ended up offering it to Tandy. But the deal fell through when Tandy decided to go with Centronics.

Undaunted, Okidata brought the printer to market under its own name as the Microline 80. The initial product used a bare-bones, 7-pin print head, but has evolved and the Microline series is still sold today.

Herman added that Okidata was able to expand its stable of OEM customers in part due to the retreat of two of its primary dot

matrix competitors, Centronics and Printronix, in the early 1980s. Herman says that Okidata was destined to become a desktop printer manufacturer, and it soon began to flourish as a supplier of printers for PCs. In 1982 they began to sell directly to users, signing on with retailers such as BusinessLand, Hall-Mark, and Computerland, and some major distributors as well.

Counter-Intuition. By the early 1980s, 9-pin print head configurations were mostly relegated to industrial settings, and 18-pin and 24-pin printers were becoming *de rigeur* on the corporate desktop. At the National Computer Conference in 1983 Okidata demonstrated its first 24-pin printer, called the ET8300, under the Pacemark brand name.

Dennis Flanagan, who came onto the scene as marketing chief in 1986, found a company that needed a shot in the arm. Okidata's sales had begun to plateau for two years at around $200 million. Flanagan traces his own digital printing lineage back to Texas Instruments, where he helped manage sales in the company's now-defunct printer division.

Under Flanagan's guidance, they introduced a line of 18-pin color printers, positioned as a low-cost alternative to 24-pin printers which were beginning to hit dealer shelves in growing numbers. Even in an age demanding higher print quality, more pins did not automatically mean more sales. Flanagan explained why the 18-pin print head offered one big advantage over its 24-pin counterparts.

"The problem with a 24-pin configuration was that it required changes to the software. Without modifying the software, a pie-chart, for example, would be printed oblong on a 24-pin printer. Most printers at the time used either the Epson or Centronics protocols, and application developers like Microsoft and Lotus could write their software to work with any printer that used those

Dennis P. Flanagan

Tank Tough: This tank tread display stand, shown here with the Microline 120, was first used quite a while ago and is still popular. Okidata resellers still mention it as one of their most effective promotional ideas. Okidata has gone so far as to trademark "Tank Tough."
(Courtesy of Oki Data America, Inc.)

Okidata Revenues Selected Years, 1974 to 1991	
Fiscal Year	Sales ($ millions)
1974	5
1976	5
1978	11
1980	18
1982	69
1984	211
1985	209
1988	244
1990	306
1961	376

Source: Okidata 20th Anniversary Yearbook, 1972-1992

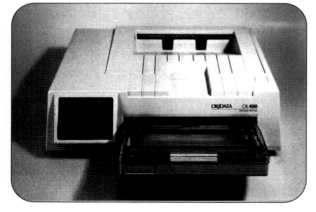

The OL400, introduced in 1990, according to Oki Data was the first toner-based page printer to break the $1000 price point barrier, an achievement undreamed of even a few years earlier. Since then, with the OkiPage 4w (1997), they broke the $200 price point barrier. (Courtesy of Oki Data America, Inc.)

default protocols. But with an 18-pin print head in, the printer would still run like a 9-pin printer, but at a higher resolution, without requiring special drivers."

By 1988, Flanagan asserted that Okidata printers had gained a solid reputation for durability and they unleashed their "Tank Toughness" marketing campaign. Its printers had become popular in harsh industrial environments, where grit and other airborne particles can bring impact printer mechanics to a grinding halt. This toughness image was found to have weight in the office and consumer markets as well.

It wasn't until 1995 that Okidata's dot matrix printer sales hit another plateau then actually began to decline. By then this was an industry-wide trend. Luckily Okidata had been offering LED-based page printers for some time. They chose the LED alternative to the laser, according to Flanagan, because it is reliable (fewer moving parts), compact and fast. Herman, who headed Okidata during its initial foray into the page printer market, said the company was slow to release its first LED product. This was due in part to their delayed implementation of a page description language, which hurt the product's chances early on. "We had an LED printer in the hopper in 1981, he recalled. "We showed it at the National Computer Conference in 1982. But because we didn't understand the application of page description languages like HP did, the printer didn't reach the market as early as it should have."

Better late than never. Okidata's LED printer family has continued to grow along side their dot matrix printers. Their first Postscript-equipped LED printer, the OL830, shipped in August of 1990, the same year they celebrated their first $30 million month in terms of sales. They now also offer a color page printer, the Single Pass Color™ LED Printer.

Okidata has been successful doing with page printers what they earlier did with dot matrix: building new, broader-base markets by lowering the cost of entry threshold. According to Flanagan, now president and CEO, "Dot matrix still makes up the largest dollar contribution to the company. Currently our printer revenue is about 60 percent from dot matrix and about forty percent from LED and other printers. In two years, I predict that we will reverse those figures. Okidata wants to create a larger market in the page printer business than we did in the dot matrix business."

The history of Okidata, in short, comes through as a case study in substance over style. While the company isn't the most flamboyant, its ability to offer low-cost, durable products has made it a success, even up against high-profile giants like HP and Epson. Now, as Oki Data Americas, Inc., the company appears ready to enter the new millennium on a high note as it continues to expand its dot matrix and LED printer lines. Nettleton had predicted the long-range benefits of teaming up with a Japanese counterpart and his foresight produced a company whose printers have become part and parcel in businesses and homes alike.

Okidata was founded in the 1970s, but didn't begin to blossom until the 80s. To do this, they had to maneuver in the shadow of giants as the industry entered a decade of concentration. Okidata managed. So did many others, as we shall see.

HP LaserJet Plus, introduced late in 1985, offered a wider selection of typefaces and other features thanks to PCL 4. (Hewlett-Packard Company. Reprinted with permission.)

CHAPTER 6: THE EIGHTIES

THE SETTING   As in each decade, the 80s had it share of scandal, wars, and economic swings. These were Republican years, bracketed by Ronald Reagan moving into the White House in 1980 and George Bush at the end of the decade. The Bell System was declared a monopoly and broken up into seven "Baby Bells." The Iran-Contra scandal intrigued the nation, but had little impact on the Administration or its policy to oust the left-wing Sandanista government of Nicaragua. It was a decade of mini-wars: the British fought Argentina over the Falkland Islands, Israel invaded Lebanon trying to crush the PLO, The United States invaded Grenada and late in the decade Panama, too, trying to find General Manuel Noriega. There were king-sized environmental accidents including the toxic gas leak from a Union Carbide plant in Bhopal, India which killed 2,000 and injured 150,000; the Chernobyl nuclear accident that radiated the Ukraine and alarmed the world; and the Exxon tanker Valdez which spilled 11 million gallons of crude oil onto the Alaska coastline. Mikhail Gorbachev came to power in the USSR, the Berlin Wall came down, and the end of the Cold War was in sight.

The economy recovered from the sluggish 70s, the stock market twitched with the mini-crash of '87, and people worked, worked, worked.

The democratizing of the computer began with the introduction of the IBM PC in 1980 and the much more people-friendly Apple Macintosh in 1984. Today's Worldwide Web emerged as the number of Internet hosts swelled from 1,000 in 1984 to 10,000 in 1987 to over 100,000 by 1989. Standards groups including CCITT and the industry's Corporation for Open Systems worked to topple the "Tower of Babel" plaguing both vendors and users of facsimile, electronic mail, and computers. The results were mixed. More successful was the introduction of Windows 1.0 by Microsoft in 1985. Its growing acceptance allowed most PCs to talk with each other.

This dynamic setting also changed the face of the computer and printer industries.

Early in the 1980s traditional impact printers continued to produce the lion's share of computer hardcopy output. Market researchers at Dataquest estimated that in 1982, in the North American market, impact printers still represented 90% of the industry's equipment revenues of around $12 billion.*

The performance of impact line printers peaked out around the turn of the decade with Documation fielding a 3,000 lpm band printer. Impact printer prices were down, even as speeds were up. IBM's 4245 Series performed at up to 2,000 lpm, end-user priced at around $35,000. Honeywell began using belt line printers from CII-Honeywell Bull in the 1970's, still manufactured and supplied to their users into the mid-1980s. Speeds ranged from 900 to 1600 lpm, end user priced at $30,000 to $65,000. Among the independents, Dataproducts continued to be a leader with a series of band printers offering 1000, 1500 and 2000 lpm.

Impact matrix and daisywheel printers continued their roll early in the decade. Digital Equipment Corporation was a leader in impact matrix teleprinters with more than 200,000 of its DECwriters shipped by 1982. But the days of low-end impact printers were numbered. This is the decade in which non-impact printing got real. Laser printers became affordable. Ink jet finally became practical and even more affordable

In this chapter we will focus on four themes.

First, page printers, graphics and color have ushered in an electronics-intensive age in which controller technology and software gate printer capabilities. Adobe Systems has been a major independent printer software pioneer. Company co-founder John Warnock will share some glimpses of how he helped make that happen.

Next, an in-depth exploration of the Hewlett-Packard phenomenon. Barely visible as a printer supplier in 1980, HP revolutionized the technology and the industry in a few short years with

ground-breaking laser and ink jet products. It is logical that HP is featured as this decade's defining company.

During the decade there was a lot more going on in non-impact printing than the HP and Canon partnership. A third focus will be the many other non-impact developments including ink jet in Europe and the U.S and thermal printing.

Finally, a look at the major impact matrix vendors who emerged during the 80s to steal the show from Centronics. They were just beginning to enjoy the party when new technologies came along. Epson and a few others managed to make the transition. Most others did not. This episode will close with some perspectives on their strategies and what worked and what did not work.

PDLS, ADOBE, AND JOHN WARNOCK

The most elemental coding system needed to operate a computer or a computer printer is sometimes termed "machine language." The line printer needs to be told when to fire hammers at each print position. The serial matrix printer wants to know which wires to fire when. In the case of the page printer, the challenge is awesome: at 600 dpi, a page printer needs to be told where to put up to 30 million dots per page. If developers of applications were stuck with coming up with programs to drive printers at this level, there would be few applications indeed.

Traditional ASCII, the coding system designed for text, is limited to 128 characters and functional codes. Some larger coding systems are used by IBM and the International Standards Organization. With escape codes, the character/command set can be multiplied a number of times. But the power of non-impact page printers is the capability to mix text with graphics, print a repertoire of fonts, and increasingly, handle color. An "escape code command set" is a step toward handling complex images, but the real answer is an actual, specialized computer language, a page description language or PDL.

---

* Tomash and Wieselman, op. cit.

This is a technology in itself. It is generally accepted that this was the capability that allowed HP, not Canon, to convert the bare-boned CX laser print engine into the LaserJet printer. As page printers proliferated, so did command languages. IBM, Xerox and HP each developed their own, and several independent vendors promoted their systems. Images formatted in one system could not be used by the others. So it was that in the mid-1980's that the "PDL wars" erupted.

The wars did not end with a treaty mediated by the ISO or other standards body. In fact today there is still no official standard. The nearest thing to a *de facto* standard is felt to be Adobe PostScript. This pioneering page description language is seen as an instrumental software product gating desktop publishing and accelerating the proliferation of page printers and graphics applications.

When John Warnock and Charles Geschke co-founded Adobe Systems, Inc. in 1982 they knew they had the key to workable computer desktop publishing, but they never dreamed their innovation would take them so far so fast.

Adobe's PostScript page description language fueled the company's early growth and soon became a world standard. This and later products, including PDF (Portable Document Format) have brought to the computer publishing industry the Holy Grail of more or less true device independence for electronic document distribution and printing. Since then they have built on their core strengths in graphics, publishing and electronic document technology to become the world's second largest desktop publishing software company.

With now over 2,600 people worldwide and annual revenues now in the $1 billion range, Adobe markets a portfolio of more than fifty products including PostScript, PageMaker, Photoshop, Illustrator, Acrobat, FrameMaker, and GoLive. In 1999 they launched InDesign, their self-proclaimed answer to Quark Express. The emphasis early on was licensing printer OEMs and their

technology is currently built into the printing systems of over 55 manufacturers. Well over 300 printing devices are now said to use PostScript, ranging from small desktop printers in the home up to the IBM InfoPrint 4000, described in 1999 as the world's fastest RIP-while-print system at 464 impressions per minute.

Software products are marketed through conventional channels including retailers, various resellers, system integrators, and now directly to users via the Web. In 1994 they began leveraging their financial success with an aggressive program of venture capital limited partnerships with promising entrepreneurs worldwide to support the development of new technologies and services, generate returns for shareholders, and monitor emerging market trends.

John Warnock:
Interview, April, 1999

Dr. Warnock holds degrees in electrical engineering, mathematics, and philosophy from the University of Utah. In recent years, his technology innovation and entrepreneurial success have been widely recognized with awards from numerous organizations including Inc. Magazine, Rochester Institute of Technology, ACM, National Graphics Association, and PC Magazine. He received the Distinguished Alumnus Award from his alma mater. He has been named one of the "Ten Revolutionaries of Computing" by Computer Reseller News.

Warnock moves fast but in April, 1999 I caught up with him and we had a chance to talk a bit about how Adobe got started and some of the unexpected twists that lie behind this entrepreneurial success story. Note: Italic text is Ted; Roman, John.

John Warnock, CEO and Co-Founder, Adobe Systems.

How did you first get interested in computer graphics?

It was back in 1968 when I was getting my doctorate in the Computer Science Department of the University of Utah. Dave Evans was down the hall working on computer graphics under funding from the government, from ARPA. He needed help with one of the problems, and I solved it which got me the Ph.D. the following year.

An interesting thing is that it seems almost all computer graphics came out of the University of Utah. We were all there around the same time, people like Alan Kay, who is now with Disney, and Jim Clark of Silicon Graphics and now Netscape.

Somewhere I read you began working on PostScript — or at least the seeds of PostScript — at Evans and Sutherland Computer Corporation.

Yes, after Utah I worked with a couple of companies around the country. Evans and Sutherland made flight simulators and I went to work for them doing interactive graphics. Then I went to Xerox PARC (Palo Alto Research Center). It was Chuck Geschke, by the way, who hired me, and I brought in a different kind of graphics disciplines to Xerox. At PARC I worked with interactive graphics, but during my four years there, got interested in printing. With the Research Group we developed a language that came to be called Interpress to take instructions from the computer to produce high quality printout.

But that standard got corrupted and they finally put the lid on it. Chuck and I said, "Hey, we put in two years of our lives building it and now Xerox is putting it in mothballs." And we left with the idea that we'd start a business putting together graphics work stations.

About when was this? It seems like in the early 1980s there was a flurry of PDL activity. What was the gating factor?

You're right, there were at least five start-ups around the same time including Interleaf, TexSet, Xyvision, Vutek, and us. What happened was that laser printers were getting cheaper, down to below twenty grand, and Sun Microsystems was coming out with their workstations. It was basically printer-driven, and we got Adobe funded based on that. We build a demo of PostScript at first using a twenty thousand dollar Xerox printer which Digital Equipment lent us. After this demo, there were two critical meetings.

Steve Jobs at Apple just by chance heard that two guys who recently left PARC had demonstrated a page description language. So he came to a demo and he instantly fell in love with it. He said "It's exactly what we need, you've developed the individual aspects we're looking for. Don't go into the workstation business. Work with us and OEM your software to Apple." This was around 1983.

The second critical meeting was when Gordon Bell, Digital Equipment's top engineering manager, came and exclaimed, "Do you realize I've spent $6 million on this problem?" Well, that really changed our work vision. We decided to definitely go into the software business based on PostScript.

At the same time we had started going to Seybold seminars and it looked like no one was thinking about this and we recognized there might be a market. But we knew to be credible we had to connect with a typesetting manufacturer. We first contacted Compugraphic, but they didn't have systems that could be driven as a raster device and we got nowhere. So we went to Linotype and with the Linotron 100 we demonstrated we could drive their device. We wanted to license Helvetica and Times Roman from them, and we got them.

We spent 1984 building the Apple LaserWriter; we did the software, Apple the hardware. When the first LaserWriter was introduced in 1985, the people who knew what it was were just blown away! It caused a really large stir and early on sales were very strong. But soon, with the management changes at Apple when [John] Sculley took over and Jobs left, things slowed down.

But the industry had changed for good. Pagemaker came in. The "desktop publishing" concept was born with the Macintosh,

Laserwriter and Pagemaker launched as a package. The next year, in 1986, we went public.

How does a software company like Adobe maintain its products as proprietary assets?

Secrecy is the main thing that keeps it private. Typefaces used to be held in a propriety way. But with PostScript, for the first time, users had access to scalable, good-looking type.

The real break came when IBM licensed our technology. With that, everyone said PostScript is it, and all the other page description languages are dead. And this led HP to license it, too, so it effectively became the *de facto* standard.

How much of your revenues now come from PostScript?

Just fifteen percent.

What do you see as the future of the digital printing industry over, say, the next ten years?

Well, it's not about selling printers, not about selling the hardware anymore. It already has become more about selling the expendables, the materials.

One more thing: Did you and Chuck envision anything like this when you started Adobe?

There's a quote that has been attributed to me: "Adobe will never be more than fifty employees."

HOW THEY DID IT

as told by Dick Hackborn and Jim Hall

The centerpiece in the 1980s was the emergence of Hewlett-Packard as the dominant player, continuing to the present. At the moment "reinventing the company" is something of a mantra at HP. Their track record over the past twenty years is testimony to the reality of an earlier transformation. They walked their talk and evolved from an instruments company to the printer company.

Dick Hackborn, perhaps more than anyone at HP, is recognized as the guiding light of this transformation. His first major responsibility at HP was integrating their early minicomputer and high-frequency test instrumentation businesses in 1967. In 1971 he was named engineering manager for computer hardware development. In 1979 he moved to printers as general manager and later vice president of the printing business where he was responsible for the development of their laser and ink jet technologies, printer product design, and manufacturing. He retired from the position of Executive VP in 1993 and as of this writing serves on the HP Board and also as a Director of Microsoft Corporation.

Dick Hackborn
(Hewlett-Packard Company. Reprinted with permission. Photo by Steve Castillo.)

PART ONE: Archived History

Based on conversations with Dick Hackborn

This previously unpublished manuscript is on deposit in the Hewlett-Packard Company Archives, used here with permission. These excerpts include observations on how the company's printer programs relate to some of the famous HP "Principles" and perspectives on distribution, product positioning, and leveraging the color capability of ink jet.

The business models HP used to create a laser printer business, on the one hand, and an ink jet printer business, on the other, were quite different. Certainly they are one of the enduring contrasts of the computer industry – our close partnership with Canon on laser printers and its strong competition with Canon on ink jet printing.

But the basic set of principles upon which both these businesses were built are exactly the same principles that led to our initial success in handheld calculators and, indeed, upon which the company was founded:

1) Seek opportunities and develop businesses that allow HP to bring to market products that make a real contribution; our

products have to be breakthrough products, not me-too products.

2) Give responsibility for exploiting the market opportunities to focused, autonomous teams that manage their own business – create effective business entity autonomy. The business entities own their decisions – profit and loss performance, capital investment decisions, R&D programs, marketing/customer focus – everything. The accountability in this structure lies in the third principle.

3) Each entity must pay its own way and reach corresponding profit levels (if not initially, at least in a reasonably short period of time).

The LaserJet Business Model was dictated by some basic realities.

We saw an opportunity to bring a relatively inexpensive, personal computer-connect desktop laser to business users.

This was not an opportunity we thought would exist indefinitely without other entrants. So time-to-market was critical to success.

Because we expected competition fairly quickly, establishing price leadership at the outset – *price leadership that we could sustain* – was also critical.

Finally, we believed a continuous stream of innovative products, achieved through the rapid introduction of new models, was necessary to maintain market leadership, even when this meant obsoleting our own products before our competitors did.

This meant that, though it was traditional for HP to invent its own technology and then put that in products, in this case we did not have time to do that. We already knew Canon had the technical and manufacturing expertise for the print engines we needed. We already had the working relationship with Canon on our system laser printers, and thus considerable knowledge of this technology's capabilities. We also had an excellent under-

We believed a continuous stream of innovative products, achieved through the rapid introduction of new models, was necessary to maintain market leadership, even when this meant obsoleting our own products before our competitors did.

standing of how users wanted to interact with their printer and how to interface printers to computer systems and the corresponding applications. Finally, we felt we could further contribute to this collaboration by developing the controller electronics that told the marking engine what to do. If our approach here was somewhat novel for HP, it was entirely consistent with the First Principle.

Principles Two and Three. . . . By the early 1980s, the Boise Idaho team had already gained a lot of experience in laser printers based on their own minicomputer system products. It was through their insight that the huge potential of a desktop laser printer was first recognized and subsequently turned into an immensely successful new business for HP. There was no backseat driving from me or Bill Hewlett or any of the corporate executives in Palo Alto. Boise set its own agenda with Canon, negotiated its own contracts, jointly did product definition with Canon, developed its own printer controllers, planned its marketing strategy (including distribution, a not uncontroversial item) and made the business happen.

We have always had within HP ongoing senior management reviews. Boise presented its plans and progress to upper management in the normal course of these reviews as it continued to execute its basic business strategies.

The first LaserJet was introduced in March, 1984. Nothing like it existed previously. Other laser printers had been introduced, but these previous products, whether from HP or another company, did not meet the *complete set* of market requirements for a desktop printer: fast, flexible, high-quality printing; small form factor, and affordability, an *upgradeable* printer command architecture insuring broad support from personal computer software applications, and excellent reliability. The first LaserJet met all these requirements and as a result, created a totally new mainstream office printer market (similar to what handheld calculators had done twelve years before).

Right from the beginning, the Boise business team recognized that the new LaserJet printer was not only going to be extremely popular in medium to large size organizations, but also with smaller companies, and even in the home office.

Rethinking Distribution. It became immediately clear that the historical approach of selling HP printers through our own direct sales channel was not going to work — not only was it insufficient to reach a large portion of the total potential market, but it also had the wrong cost structure. Selling through the reseller (dealer) channel was going to be imperative.

Dealer distribution of printers was something that raised more than one eyebrow within HP. Direct sales of computer systems to large end-user customers was and still is the right way to distribute these products. It's also the right way to distribute the big system printers that attach to larger computers. Since our first desktop laser printer owed its pedigree to our system laser activity and was managed by our system laser business team in Boise, the conventional wisdom was that it should be sold direct. Furthermore, at $3,495, the LaserJet was viewed as too expensive for the dealer distribution channel.

But the Boise team was not following the computer system paradigm. They were tracking with the new PC business. HP already had a small sales force focused on resellers supporting calculators and HP's own early entrant in the PC market. Boise went to this sales force and convinced them to carry the LaserJet too. The tremendous success of the product put an end to internal second-guessing of this strategy fairly quickly. The importance of the decision to distribute the LaserJet through resellers cannot be overemphasized. It was the critical final piece of the overall strategy. It established a fundamental channel strength for HP that has been a major competitive advantage over the past ten years.

More for less. At the time the first LaserJet was introduced, HP already had a five-year vintage chart for subsequent introductions and line extensions. We introduced two more LaserJets (the LaserJet Plus and LaserJet 500) based on the same print engine within eighteen months. But even before the second and third LaserJets were introduced, indeed, even before the first one was introduced, HP was working closely with Canon on the development of the next generation engine which went into the LaserJet II, introduced in March, 1987.

"More for less" became the code phrase in Boise for what each subsequent LaserJet had to offer the market. This mantra reveals the largest lesson learned from our experience with calculators. For many years we continued to introduce increasingly sophisticated [calculator] products, but these line extensions also cost more at a time when our competitors were offering the "same for less." The more sophisticated HP calculators sold to those who needed more advanced capabilities — but a large portion of the marketplace was lost in pursuing this strategy. With LaserJets, each basic platform revision offered greater capability at a lower price than its predecessor.

Two years later we stunned the market by introducing a completely unexpected product. The IIP embodied the other major lesson from our calculator business: a smart company can also introduce a product which makes a breakthrough contribution to the market by offering "less for much, much less." The [LaserJet] IIP had all the print-quality, reliability, paper-handling, font and graphics capabilities of the II, but it was half as big, half as fast, and was priced more than $1,000 less.

We couldn't make enough of them.

And our competitors, who had been reduced to competing with the II just on price (and that not very successfully because we had the lowest cost structure in the industry), suddenly had to face a price-leader product of an entirely different order. We went down-market on them and forced them to spread their R&D and marketing dollars that much thinner.

The importance of the decision to distribute LaserJet through resellers cannot be overemphasized. It was the critical final piece of the overall strategy. It established a fundamental channel strength for HP that has been a major competitive advantage over the past ten years.

Ink Jet and The Principles. This history actually began in the late 1970s with a chance discovery at our corporate labs in Palo Alto. An engineer working at the time on thin-film technology for integrated circuits was testing the response of a thin silicon-based film medium to electrical stimulation. The electricity superheated the medium and droplets of fluid laying under the film were expelled. An idea was born.

The [LaserJet] IIP embodied the other major lesson from our calculator business: a smart company can also introduce a product which makes a breakthrough contribution to the market by offering "less for much, much less."

What if you could finely control these jets of fluid? Large industrial ink jet marking devices already existed, but up to this point it had been accepted that only crude printing of quite large characters for industrial purposes was practical.

Suddenly it looked like this marking technology could be miniaturized – and it had the advantages of requiring very little power and being inherently inexpensive.

Refinement of the technology continued in Palo Alto and Corvallis, Oregon. Again, the advent of the PC made it clear where the market opportunity lay. Only this time there were no other suppliers of this technology and we felt we had sufficient time to develop it completely ourselves and still bring a product to market in time. This was because we believed it would take longer for the personal computer to become pervasive enough to open up the market for a truly personal printer at a very low price – one printer for every PC.

So our ink jet business model reflects the more traditional approach of developing our own technology to bring to market.

Regarding the first principle, ink jet offered us the opportunity to eventually replace the least expensive printers in the market – serial dot matrix impact printers – with products that were superior in every way. Ink jet offered the potential for better print quality, greater font and graphics capabilities, quieter operation, extremely low power consumption and, most important, high-quality, low-cost color.

The second and third principles: We empowered business teams, first in Corvallis and subsequently in Vancouver, Washing-

ton to bring this revolutionary printing technology to the marketplace. Though we felt we had more time to refine it, time-to-market was still considered critical. Corvallis introduced our first ink jet printer, the ThinkJet, both to get to market first and to develop, before anyone else, the customer knowledge we knew would be essential to long-term success. ("ThinkJet" comes from a combination of thermal and ink jet which characterizes the drop-on-demand type of ink jet which HP pioneered.)

The Disposable Cartridge and Color. Right from the beginning we concentrated on the disposable ink jet print cartridge as the best way to satisfy end-user needs for very easy-to-use and low cost personal printers. This was a major breakthrough. However, the ThinkJet's resolution was no better than a dot matrix printer, and it required special paper. It was a success in terms of revenue and profitability (in fact we still make and sell the print cartridges today), but it did not meet mass market demands for high print quality or plain paper. It was still a specialty market.

The market breakthrough came in 1988 with the introduction if the HP DeskJet printer from our Vancouver business entity. It offered plain-paper printing at industry-standard print resolution. It was the least expensive non-impact printer on the market and as a result, had considerably higher sales volume.

Canon and a couple of other companies were beginning to think about entering the ink jet printer business, but none of them had any notion of the value of color at the low end. Ink jet is the technology most adaptable to color printing. Color would allow the DeskJet to offer something no laser printer could offer at a dot matrix price, plus near-laser print quality capabilities no dot matrix printer could match.

Prior to the introduction of the color DeskJet 500C in 1991, color printing was a niche application. Color printers were expensive, purchased only by users whose special needs justified the price. Attempts to sell even less expensive devices did not result in

large volumes of sales because the average user did not want or need a specialty printer. Mainstream printing was and still is mainly black text in standard business communication. The black printing capabilities of inexpensive color printers was inferior, so users were unwilling to trade quality for their primary printing need (text) to gain the most obvious use of color (color graphics merged with black text).

Our market research clearly showed that mainstream users were not looking for color printers. Asked to prioritize their requirements for printers, users consistently put color way down the list. Our competitors were discovering the same thing and the answer this led everyone to was that there was no mass market for inexpensive color printers.

But they were asking the wrong question. "Do you need a color printer?" Always, the answer was, "No."

We asked a different question. "If we satisfied all your black printing requirements and offered you the ability to print in color as well for little or no price penalty, would you buy such a printer?"

The overwhelming response was "Yes!" The market didn't want color printers, but it was very interested in mainstream printers that could print color, too – in short, HP should offer color as a feature.

In 1991, around 360,000 non-impact color printers were sold worldwide by all vendors. We alone sold more than five million color printers in 1994! Today, all DeskJet printers also can print in color; the low-end of the DeskJet line has nearly single-handedly reduced worldwide [impact] dot matrix share to twenty percent (down from 65 percent in 1991). And HP enjoys an 80 percent market share worldwide in color printers, with color moving into the high-end of the mainstream market in the form of the Color LaserJet.

❖

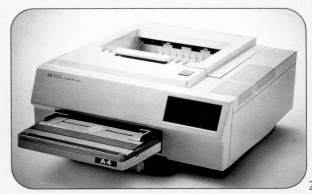

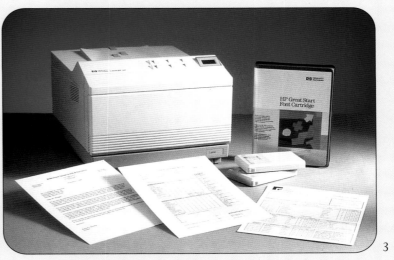

LaserJets in the 1980s: 1. HP's original LaserJet, introduced in 1984 and described as their most successful single product ever. 2. LaserJet II, 1988. 3. LaserJet IIP, successor to the II (1989) offered less for $1,000 less. (Hewlett-Packard Company. Reprinted with Permission.)

PART TWO: Dick Hackborn

An Interview with Alan Bagley, 1996

NOTE: This dialogue is excerpted from a longer interview provided courtesy of Hewlett-Packard Company Archives. Al Bagley is an engineer who had recently retired from HP's Test and Measurement Division. The segments selected are those which focus on some of the strategic decisions guiding the HP printer business, how they were made, HP cost structure, and views on the competition and future.

Bagley's side of the conversation (italics) has minor edits in the interest of continuity. Brackets denote our edits.

Canon and a couple of other companies were beginning to think about entering the ink jet printer business, but none of them had any notion of the value of color at the low end.

So we had these two different business models, low end ink jet and the laser, which would be leveraged from Canon's basic copier technology. Today, the industry sells more printers than they do copiers but at that time, the copiers were king.

Yes, that's funny because you justified getting the engine from Canon because the copier business was so big.

Yes, right. Now it's reversed. We leveraged from them. Our contribution was in other things, like the printer control language, the electronic controller, the distribution channels, support, and so forth.

Dick, as you talk about all this stuff, you have yet to say the words, "consumer products."

Well, right. I think that came along a little later. We started with the first LaserJet, introduced in the spring of 1984. In fact both products, the Thinkjet and the LaserJet, were introduced at that time. Before these introductions, there was a big debate going on. Some parts of HP's computer business thought we should introduce a printer that just worked with HP systems. It should only have an HP-IB interface. And they thought we should sell it through our direct sales force.

Basically, it was left in typically HP style: "OK, well, you're running the show. Just make sure you're right."

But we looked at this thing. We said, "No, this is a PC business model. This is not a minicomputer business model. That means resellers. It means having a standard PC interface, a serial interface, a parallel interface. It means having software that application developers will have drivers for." It was an entirely different business model.

But it wasn't a consumer market yet. We were still talking about the business PC market. We're talking the mid-'80s here.

I know, but what is important is that you broke away from the old model, and it took a bit of fighting City Hall to do it, I believe. You broke away from the idea that we need to have this great, proprietary interface. You said, "No, we're going to serve more than HP computers!" Isn't that true?

Well, it was a battle. It's reputed to be a bigger battle than it really was. It was discussed, but nobody said, "Well, you can't introduce it if you don't do it this way." Basically, it was left in typically HP style: "OK, well, you're running the show. Just make sure you're right."

Yes, that's a helluva lesson right there!

That's right! So the LaserJet didn't really become a consumer business until we introduced the [LaserJet] IIP, which was roughly three years later. [Ink jet] became a consumer product in the late '80s when we introduced a version that [unlike the Thinkjet] used plain paper and was 300 dpi. That's when it hit the consumer area.

These are milestone events in the total HP history, both LaserJets and ink jets occurring quite close together. I think they both very much exemplify the "HP Way" right down the line. They were division product line focused – know your markets, know your technology – and P&L accountability.

Oh, I'll tell you another story, about when we introduced the laser printer. I think Dave would be proud of this!

My boss wanted me to introduce it at $2,999 and I looked and I looked at that and thought. . . gee, no matter how many we sell, we are not going to make money at that price. We could make a good return for the company at just under $3,500. I got another one of those comments: "I would introduce it at $2,999. But it's your decision. If it fails, it's your problem because you introduced it at too high a price at $3,500."

I kept thinking, "Gee, you know, at $2,999 no matter how many we sell, I'm not going to make any money and that's going

to be the ultimate failure. Meanwhile, this thing had so many advantages over daisywheel printers which were selling for $2,500 to $4,000. I can't help but believe that we're going to get at least as many of those customers and they number in the hundreds of thousands, if not millions."

So, you know, you listen to all sides and do some soul-searching but finally you make your decision. We introduced it at $3,495, or whatever it was, and the rest is history. And sure enough, as we got the volumes up and more importantly, as we got lower cost models, we reduced the price and the profitability remained. Today you can buy an HP laser printer for less than $500 and it's better than our first LaserJet.

Moving into consumer products, what did HP have to learn about marketing, since that was not our forte for a long time?

Well, we had to learn — and we're still learning, I think. The first step is getting into the dealer channel and then make those channels available to anybody in the company who feels that they have something that could go through those channels. Don't duplicate them. To be a leader in channels is expensive.

The second thing we learned was that we had to work extremely well with software vendors. We weren't doing the applications or operating systems. There were thousands of people around the world writing PC applications that we wanted to work well with our printers. Being an established computer company gave us more knowledge about how to do this. Canon had a great problem getting started in this area. We made a significant investment with all those independent software vendors out there to get our drivers or have them write drivers for our printers that would work really well.

[We had these two kinds of partners,] sales partners and software application partners. And I might add that you have to earn these people everyday; you don't own them. You don't own your people and I think there are some companies in the computer business today — I think IBM is one of them, at least parts of their organization — that still have this "we own the customer" attitude.

You've got to earn your customers every day and you've got to keep earning them.

The third thing we realized was that our financial model was different, particularly in the laser printer business where we could not mark up a lot from Canon's price. So we had to keep our R&D investment down. I remember people saying in the early days, "Well, your R&D is one or two percent of your sales revenue. When is it going to be [our] typical eight percent?" I'd say, "You don't understand. Our one or two percent is to invent our electronic controller and other things to get the product tested. We don't do Canon's six percent R&D. That comes at a price; that gets transferred to us." We had to have very tight cost structures, but we made significant investments where appropriate in other areas to give us market differentiation. Because of that, we marked up.

I remember when I would go to DEC their managers would say, "Well, how do you do this? How do you sell this laser printer for such a low price? Canon must sell to you at a ridiculous price." They assumed a 2.5 or 3.0 times multiplier [mark up] as in the minicomputer business. But one needed only a 1.5 times multiplier. And I'd say, "Oh, well, you know, they're competitive with us." They'd go beat Canon on the head to get this thing at half the cost we get it for, and they didn't understand. . . . .

Then they'd multiply it.

Right, they'd multiply it because they were applying a minicomputer cost structure against a high volume peripheral. Our ink jet printers are much more representative of a typical HP structure. But in either case, a model fits the business conditions.

So this was our initial dealer channel. Then, after some time, probably starting maybe four or five years ago at the most, we re-

My boss wanted me to introduce it at $2,999 and I looked and I looked at that and thought. . . gee, no matter how many we sell, we are not going to make money at that price.

Don't duplicate them [channels]. To be a leader is expensive.

alized there was yet another channel. This is the mass-volume channel. This is the consumer business you're talking about and we're still in the process of learning.

We've learned a lot.

We've learned that you have to guarantee deliveries, you have to guarantee inventory to these people, you have to be willing to take some of it back, you have to have point of sale material – the thing is sold right out there by walking down the aisle – you have to have a superb order processing system. Now we're moving toward the grocery store/drug store type of distribution. And that's different than the dealer channel.

In the case of one of your software partnerships, I think you started with somebody other than Adobe's Postscript and then switched to Postscript.

Right, good point. That was another key ingredient of both our printer product lines in the beginning, the hardware/software interface. The common one for serial printers at the time was the daisywheel interface. This was clearly inadequate for a laser printer. The alternative would be complex, custom types of software and drivers which would be different for every computer manufacturer.

Inside HP we had been experimenting with a thing we called printer command language (PCL). This was a way to have all of our different line printers, which had different characteristics, require only one software driver in our computer systems. It was a printer kind of command language architecture. So in Boise, they adopted that to go on the laser printers and subsequently for our ink jet printers. So all we had to do was get the independent software vendors to write their software and their drivers in particular to interface with PCL, which our printers understood, and they would work over whole generations.

We always updated it and expanded it but it was always back-ward compatible. It was very fundamental to our early printer success. You could take a new laser printer and run it on your PC with the old driver. It would only run like your old laser printer, but at least it would run! And then, of course, if you had new software drivers it would run to its full capabilities. So we popularized that to the tunes of millions and millions of printers out there and it's been emulated. You know, it's THE standard basically in the industry today.

So, how does that compare with PostScript? I remember Steve Jobs calling me up and saying, "Why aren't you using PostScript? And on and on and so forth." And I'd say, "You don't understand, Steve. It's a market focus. Our market focus is not publishers. Our market focus is mainstream office users and therefore we don't need something as complicated as PostScript. And our printers aren't as expensive, either, by the way." And so we focused on mainstream office printing with PCL.

[Can we look more at the ink jet business, more specifically, at the print head and how it relates to the business model?]

[The head has] changed some since the 1980s. Today it's a mixture of both solid state electronics as well as a silicon substrate with thin film disposition on it that is used for channeling ink through very small channels and into a cavity where there is a resistor.

But it uses semiconductor technology.

Yes, that's part of it. For example, the drivers for picking which nozzle is going to squirt, the multiplexers to drive the resistors, that's all semiconductor. It sits on a silicon substrate. But there is also a lot of technology you would not find on an IC and that's the fluid dynamics. This includes moving the ink inside this head to the proper cavities, keeping out air bubbles, super heating of the ink in those cavities which is what causes it to go out of the nozzle, even the shaping of the nozzle.

... you have to earn these people every day; you don't own them.

So they weren't exactly like the razor with no profit. Both the printers and the print heads are P&L accountable.

They do excellent color, so does HP. Their problem is that their black text is not nearly as good as HPs.

Do we have a pretty good set of patents on that?

There is nobody even close to HP and Canon in this thermal ink jet. We have made a tremendous investment. But it isn't just the patents, or the intellectual property. It's the know-how, how to build these things in high volume. Today, my understanding is that HP produces six thermal-ink jet print heads every second. Somewhere in the world.

How about that! And they sell for $30 list?

Yes, on the average.

That's quite a deal. That must be harking back to the Gillette idea: give them the razor and sell the blades.

Right. It certainly crossed the minds of a lot of the financial analysts who follow HP that we are making much of our money off these heads. They certainly are a big revenue-generating business. But they also require a tremendous amount of assets [so] we've got to have good profit margin so that our return on assets is reasonable.

But we also realized that we could not, we should not expect to make all of our money off of our ink jet heads. . . . that our printer products had to make some money, too. So they weren't exactly like the razor with no profit. Both the printers and the print heads are P&L accountable.

That is probably HP's biggest single business, if I'm not mistaken.

I was told that last quarter it was the biggest business.

[However,] we do make decent profit margins and excellent return on assets with [all] our printers. It's not all supported by our ink jet print cartridges. However, [people] are incorrect when they talk about disposable [LaserJet] toner cartridges. That is probably the most leveraged of all our businesses in that we pass those things through with very little markup because we're competing directly with the manufacturer, in this case, Canon. And we do that because

the customers want the HP support. So we pass it through but we must incur very low expenses doing that because we get very little markup. So it's not a high markup business; it's probably the lowest markup of our [businesses].

[What about ink jet competition, for example, Epson?]

They do excellent color, so does HP. Their problem is that their black text is not nearly as good as HPs. It's not close to looking like an HP LaserJet, the inks are not as black, the dots are not as accurate.

Anyhow, that gets back to one of the fundamental strategic goals of our ink jet business. This was to provide the best possible black text quality, while giving them color for a low incremental cost. So we came at it differently than Epson did and for a while Canon was also different. Canon has since moved more to the HP strategy of having really high quality black text.

But isn't Canon's long range objective photography?

Yes, it is, but I think the two are going to blur. In the future, people who use printers with computers are going to want a good photo-realistic color output and not just the person who is into

HP Ink Jet in the 1980s: The ThinkJet, 1984 (top) 96 cps, limited to black print, required special paper for adequate print quality. The HP DeskJet, 1988: black and color on plain paper. (Hewlett-Packard Company. Reprinted with permission.)

photography. They may be bringing documents across the network that have photographs in them. The graphics may have been generated on a computer versus a photograph. This requires even higher resolution and really excellent color output. So I believe the difference between whether you're doing this for photographs or whether you are doing this as part of your everyday use of a computer on a network will be less significant over time.

PART THREE: Jim Hall
On the Canon Courtship and Birthing the LaserJet

Jim Hall joined HP in 1972 and has managed laser printer development there since the beginning. He is now R&D Manager for LaserJet Business Printing. In December, 1999 I asked him if he could fill in some details on the LaserJet program and he offered the following vignettes. (Note: Italic text is Ted; Roman text, Jim.)

What's the real story behind the Canon-HP connection? The lore is that it grew out of a friendship between the sons of the top executives at Stanford University.

Well, it really started back in the 1960s or early 1970s when HP partnered with Yokagawa to form a division in Japan. Ownership was 49% HP, 51% Yokagawa. That took Bill Hewlett to Japan and while touring around talking with leading business people he met Dr. Mitarai, Sr., the founder of Canon, and they got to know each other.

Later Dr. Mitarai's son Hajimi applied to Stanford for a PhD in their electrical engineering program, and was turned down. His dad was puzzled and called Bill Hewlett and asked him to check out what happened. So Bill called the dean for an explanation, and the dean said he could not see any reason for that decision. So the net-net was that Hajimi was accepted and he was graduated in due time with his Ph.D.

It was a small favor, but this means everything in the Japanese culture.

Later, in 1975, Canon introduced its first laser printer, based on their NP5000 copier. During their first showing in the U.S. they contacted HP and offered us first dibs on licensing the technology. We took the offer and invested a couple of million dollars in the license and were able to introduce our first laser printer, the 2680 in late 1980. The friendship between Dr. Mitarai and Bill Hewlett was the key circumstance that led to the 2680.

Would you call the 2680 a success?

I'd call it an interesting lesson. A technical success, but not a commercial success. It was priced around a hundred thousand dollars, about the same as the HP 3000 minicomputer it was designed for. It embodied very innovative stuff. The print quality was good, and a key feature was that we could print both text and graphics. The software became the basis for our PCL, which came later.

But market reality came down on our heads. We knew we had to get the price down. We also knew it would be a huge R&D investment. So we said, "Let's look at third parties."

By then the Canon LBP-10 laser print engine had been introduced, which was much lower cost but used liquid toner. The toner solvent was like kerosene. The copies would come out smelling like kerosene! So we surveyed the vendors and selected Ricoh with its dry electrophotographic technology. We based our 2684 laser printer, which came out in 1983, on Ricoh. It was a ten thousand dollar machine that went around ten or twelve pages a minute and was not especially reliable.

But going with Ricoh had one important result: it sure got Canon's attention. Canon's Mr. Kitamura took me to dinner when I was in Japan and when I said I am going with Ricoh he spent hours telling me that was the stupidest thing I could do. I told him we went with Ricoh because we wanted a dry system.

The message got through and in just six months Canon was

But going with Ricoh had one important result: it sure got Canon's attention ... The message got through and in just six months Canon was back with a dry system.

back with a dry system. They were worried we wouldn't accept it, and we asked Dick Hackborn to host the meeting where they presented it. It was exactly the engine we wanted, with the only quirk being that it delivered output face up so pages were in reverse order.

Besides being dry, why was it just what you wanted?

The key was reliability. The cartridge concept made that breakthrough possible. The electrophotographic systems at that time were the two component copiers. Their failure mode was not that they just stopped copying. It was more often degraded print quality, and this was hard to trouble-shoot. What do I replace? The cartridge concept avoided all that. It's like putting a new engine in a car every ten thousand miles.

I've heard the naming of the LaserJet was controversial. What was that all about?

Yes, that is a fun story. In R&D we had a block of numbers reserved for us and it should have been called the 2686. That was the next number in line. About the time we were to launch the printer, in March, 1984, HP introduced the Think Jet — a contraction of "Thermal" and "Ink," right?

When our marketing in Boise heard of this, they said, "Hey, that sounds jazzy! Why not 'LaserJet?'" But we engineers all said, "No, there's no 'jet' in this machine. It is lying to the market! The market won't go for it."

Our first few shipments in fact went out as the HP 2686. But our division manager came out on the marketing side and the name got changed. And in retrospect, of course, he and the marketing group were right.

One side benefit of that name we didn't expect is that it works in every language. Since both "laser" and "jet" are new terms, they are usually the same in every language and don't need to be translated.

Was there a channel marketing organization in place for you for the LaserJet launch?

Actually, when we started the project in the spring of '83 our only vision was that it be sold with the HP 3000 minicomputer by our direct sales force. That was the only way we knew at that time. Meanwhile HP had developed the HP 150 personal computer with a touch screen monitor. A small marketing group was put together to try to sell it through the COMDEX channel. As they were getting ready to go to COMDEX to show their computer to the dealers, we begged them to let us tag along. They had a two hour presentation for the dealers and they let us have about ten minutes.

Well, after an hour of the presentation most of the dealers had left. Our ten minutes came at the end. Two minutes into our demonstration of the LaserJet we had to stop because it seemed almost everyone had run out of the room. We wondered what we had done wrong. What they were doing was going to look for their top management because they knew they wanted to sell the LaserJet. Anyway, with all this excitement, we started to get some belief.

Can you fill me in on the software, on the origin of PCL?

At Boise back when we were building impact printers we knew we couldn't have a unique language for every application, so we came up with PCL. It was a hierarchical language, with each level building on the preceding level, a superset of the previous level. But when we started with PCL for the laser printers a horrible argument developed. It looked like we had to choose between the Epson standard which was good for graphics and the Diablo standard which was good for characters.

Luckily we had religious training from our 2680 experience: we knew the future was mixing text and graphics. And now it seemed we were being forced to choose between text and graph-

But we engineers all said, "No, there's no 'jet' in this machine. It is lying to the market! The market won't go for it."

Two minutes into our demonstration of the LaserJet we had to stop because it seemed almost everyone had run out of the room. We wondered what we had done wrong.

ics. PCL at the time could mix text and graphics, but was not written to be compatible with standard personal computer software. We took the gamble to go with PCL. We put together a group to work with the application software vendors and it was a massive undertaking. With the 2680 we saw the power of mixed text and graphics. Without this we never would have endured that pain.

Why didn't you just go to PostScript?

Apple introduced their LaserWriter with PostScript in January 85. The printer was very expensive because PostScript needed such heavy computing: five times the memory, three times the process-

ing. The average office user just did not need all this. PCL was right at that time and was a tremendous advantage for us for many years.

In succeeding years, under our basic "more for less" strategy, we upgraded PCL to levels four and five with vector graphics and scalable fonts. What happened is that eventually PCL and PostScript more or less merged in capability because the cost of memory and processing came down so much. Now the smallest memory chip you can buy will work anything. Our PCL was the page description language that matched mass market needs. It was not overkill. We wrote the standard for the general office user.

With the 2680 we saw the power of mixed text and graphics. Without this we never would have endured that pain.

This was the scene in May, 1976 when a delegation from Hewlett-Packard, including Jim Hall (on right), visited Canon's Shimomaruko facility to lay the groundwork for the cooperation that led to the development of HP's first laser printer. According to Hall, Canon's Chairman, Dr. Mitarai Senior, at the time made his office at this location even though it was not Canon's corporate headquarters (although it is now). It was the original corporate headquarters of Canon, which is why Dr. Mitarai kept his office there. (Courtesy of Hewlett-Packard.)

At right: Here Canon's Mr. Takashi Kitamura, at the time project manager for all Canon Laser Beam Printer projects, near the blackboard, presents. Currently Mr. Kitamura is Senior Managing Director of Canon. (Courtesy of Hewlett-Packard.)

THE HP STORY: A POSTSCRIPT

Oral history is an important vehicle for telling the story. But we could add, *Which* story? It tells as much about the teller as it does about the topic. But that does not rob the history of its worth.

Both the tone and the content of the Dick Hackborn oral history is triumphal. And rightfully so. This became clear when IT Strategies consultants Marco Boer and Mark Hanley were asked to react to the HP printer story as told by Dick Hackborn.

HP's printer business has experienced success beyond their expectations, the consultants believe. Within fifteen years the printer business carried HP from a $5 billion company to today's $19 billion for printers alone. No one predicted this. Then, as consultants must, they asked Why and How? Was this success story testimony to the Principles, to inspired leadership and planning, to deft response to opportunities as they arose, or was HP simply swept along by fortunate timing and some unplanned events?

Most of history — from wars to spectacular business coups, to human ailments — is driven by combinations of factors. The Hewlett-Packard journey to domination of the mainstream printer business is no exception. The story as recounted here by the man who led the charge was experienced by the consultants as testimony to both vision and agility. Among the impressive examples —

There was the decision to develop and market printers as universal products rather than as peripherals to HP computers.

And to follow that logic by successfully building completely new distribution channels.

And in their product development strategy, to learn from their unsuccessful "more for more" calculator experience and switch to not only "more for less," but also "less for much less."

And to maintain a continuous product roll-out, without worrying about obsoleting their own products.

And to win in low-end ink jet by correctly reading the market for color.

Not emphasized by Hackborn were two other important elements of their success. One was their developer program for the LaserJets which drove the HP PCL standard by assuring compatibility with a maximum number of software applications. Another was their customer support program which continues to be enhanced. According to a source at HP's Boise operation, this program "is always a big plus for us and is still winning 'A's' in vendor surveys."

The Canon LBP-CX. Turning to the relationship between Canon and HP, Boer and Hanley noted HP was closer to the market than Canon so was better able to capitalize on the Canon disposable laser cartridge. Canon had the vision of the disposable cartridge as a way to make laser reprographics affordable. They took the risk and made the R&D investment. Hewlett-Packard ran with it, but this does not mean that Canon was unaware of its potential as a printer engine.

It first appeared in the U.S. market as the heart of the Canon Personal Cartridge Copier early in 1983. Shortly thereafter, in May, 1983, at the Tokyo business show, Canon demonstrated their "personal" laser printer. Then, in the fall of '83 it was presented as the Canon LBP-CX in the USA at the COMDEX computer exposition in Las Vegas. The same year evaluation units were made available to page printer vendors in the USA. The LBP-CX was by far the industry's most dramatic leap in hardware price/performance. According to one observer, the least expensive laser printer up to that time was an OEM version of the Xerox 2700 at around $27,000. This compares with the reported OEM price for the CX of under $1,000 in volume.

How did Canon do it? It was partly achieved with automated, mass production. Printer versions of the CX were built on the same assembly lines as the copier versions. More importantly, it

Canon introduces the LPB-CX to the U.S. market at COMDEX, 1983. (Source: Snapshot taken on the show floor, DRA archives.)

was the cartridge concept. Canon traded low hardware cost for high use cost (see Part One, Technology Overview, for details.). This made it possible to field a Canon-based laser printer, in the case of the LaserJet, for around $3,000, a price unimaginable for a laser printer just a few years earlier.

Early in 1984 HP announced its Canon-based LaserJet. But they were not alone. Other OEMs including Imagen and Quality Micro Systems also introduced low end laser printers based on the same engine. Apple's Laserwriter was the major competing Canon-CX-based laser printer for a number of years.

Apple's Laserwriter was much more expensive than the LaserJet, so it was primarily HP that brought laser printing to the masses rather than Canon or Apple. This was a demonstration of the power of being close to the market and also, as Hackborn mentioned, showed the importance of the controller and other components needed to convert a marking engine to a printer product. The Canon LBP-CX was marketed as an end user product by Canon, primarily in Japan with their own proprietary page description software which was best suited for the domestic market. It is understood that by agreement with HP Canon did not field a version of its laser engine with the more universally acceptable PCL for many years.

"HP deserves tremendous credit," Boer observed with a hint of wonder. "They grew faster than Microsoft. Their 'cheaper/bet-

ter/smaller' recipe fueled amazing growth. But a recipe can also be a trap when conditions change. It's tough to make the "cheaper/better" recipe work now when you have a $199 laser printer from Oki Data. How can anyone make it cheaper and still have a return?"

PDL Prevarication. Conditions did change in a way that challenged the HP LaserJet formula. Hackborn mentioned their commitment to their own PCL page description language with the early LaserJets. However during the ensuing years Adobe PostScript became an industry standard and users began demanding it. HP zigzagged a bit looking for an alternative to PostScript, but eventually gave in and soon became Adobe's biggest customer. Their strategy had been to forego adding value to the Canon marking engine and to add value on the software side. But now they found themselves giving up much of this core value in the form of royalties to Adobe. In time PostScript clones appeared, HP switched to Xionics, and Adobe lost its biggest customer. As the printer leader, this seemed to legitimatize the clone-makers in the eyes of the page printer world, fanning fires of acrimony which some observers reported characterized the PDL scene in the late 1980s. The burden of leadership is seldom light.

Hanley, who spends a lot of time in Japan, offered a glimpse of how it looks from that perspective. "They will tell you the story, which is that the company that really led in the laser printer market was Ricoh. But Ricoh made a fundamental design error. Ricoh believed people wanted long lasting devices and they produced a box that was expensive but durable. Canon saw the opposite: that people wanted a low cost device with disposable consumables, and this was the main technology driver. It was a brilliant idea. HP was smart to key into this idea. But I think they need to give their partner Canon credit, too."

Did the company make the market or did the market make the company? In the case of a successful company such as HP, the an-

swer is both; the company and the market are interactive. Vision and responsiveness to opportunities might be seen as different routes to success. But neither can work alone. There needs to be balance. Too much vision can blind a company to opportunities. As told by Hackborn, it appears the vision did not blind them. The "Principles" were tools which supported their success story. For the coming decades, however, new tools will no doubt be needed.

"Today," Boer observed, "the Principles cannot be a substitute for leadership. They need to go hand in hand with strong leadership, which they had in the past under people like Dick Hackborn, Bill Hewlett and David Packard. I feel these leaders exemplified the values now being lost in business. Compensation packages and perks in the past were very modest compared with what is happening today. Now we see their new CEO Carley Fiorina getting a 3-year $90 million compensation package."

However, today this is what it may take to attract an executive of her caliber. As of this writing, it appears their new CEO understands the need to steer HP in a re-creative new direction appropriate to today's information age. In the fall of 1999, Fiorina keynoted the COMDEX computer exposition in conjunction with the introduction of a new HP Logo. "Our new brand will give us a clearer, stronger voice in the marketplace, and the world will get a better picture of us that reflects our true inventiveness," she asserted. She shared her vision of HP once again reinventing itself. This time, in the new "economy of ideas," she intends to make HP known as the company that will make the Internet work for people. The printer business as described by Dick Hackborn looks like a challenging act to follow. But it does give them the momentum and resources for an impressive launch.

For a start, HP announced plans to spend around $200 million in the first year on brand recognition. Will there be fire beneath this smoke? Time will tell whether the giant is still agile.

OTHER NON-IMPACT DEVELOPMENTS

The 80s was the decade in which non-impact stole the show. And there were a lot of players other than HP.

According to one source, between 1984 and 1987 laser printer unit shipments grew from around 60,000 to 400,000. During the same period, ink jet units grew from around 350,000 to 1.3 million (North American shipments). "I think it is safe to call this a revolution," Tom Ashley, at the time a Dataquest senior analyst, was quoted as asserting, "Just two years ago, you couldn't touch a laser printer for less than $17,000. Now the day's in sight that these things will sell for as little as $2,000."[*]

By the second half of the decade ink jet vendors were multiplying. Canon had introduced their own thermal ink jet printers, developed in parallel with HP. Epson was making the biggest splash with its low cost impact matrix printers, but also had ink jet as early as 1987. Other vendors of page-size ink jet machines included Advanced Color Technology, Diconix/Kodak, IBM, and Siemens. Xerox had a major development program based in Texas. Videojet's specialized ink jet line was expanding.

Exxon's entry, fielded in 1983, was perhaps the most impressive office ink jet printer of the decade. Exxon made its multimillion dollar plunge into office automation around 1980. Under the Exxon Office Systems (EOS) umbrella were Qyx, Vydec, Qwip, Zilog (electronic typewriters, word processing and facsimile) and an ink jet development program in Brookfield, CT. The ink jet printer, designated the Exxon 965, was a 32-jet impulse printer which offered letter quality text at a quiet 60 cps across a 13.2-inch print line. In its first manifestation, it was marketed only with Exxon word processing systems. Later, after disappointing sales, an open version was introduced that could be interfaced with the IBM PC and other computers. However, at $4,500 list, Exxon's 965 could not compete with the much cheaper impact matrix printers

[*] *Electronic Business,* May 1, 1985 issue.

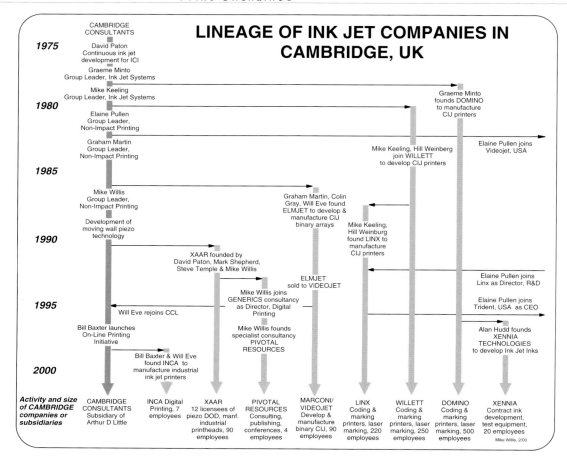

LINEAGE OF INK JET COMPANIES IN CAMBRIDGE, UK

Lineage of the Cambridge, U.K. family of ink jet companies (Courtesy of Pivotal Resources.)

or the faster, low-end laser printers of the time. The operation was a success but the patient did not survive.

Exxon, disappointed in its entire EOS diversification effort, sold it to Harris/Lanier in 1985, except for the Brookfield ink jet operation. But they didn't keep it for long. A few years later they sold it to Dataproducts. Dataproducts liked the proprietary solid ink jet technology Brookfield had developed but allowed employees to acquire the liquid technology. Brookfield became Trident, which continues as a successful vendor of ink jet technology and products for industrial marking and other specialized applications.

Cambridge Consultants, Ltd. (CCL) in the U.K. in the 70s began working with continuous ink jet and seeded a number of ink jet ventures. These include Elmjet (later acquired by Videojet/Marconi), Willett, and Domino, each of which has become significant in the industrial coding and marking area. Around 1985 Mike Willis was named group leader for a drop-on-demand program at CCL which in 1990 was spun off as Xaar. Willis, now consulting with Pivotal Resources, has assembled a comprehensive "lineage" overviewing this family of ink jet enterprises (see chart at left).

Solid ink jet (SIJ) emerged and had its major period of ferment in the 1980s. Like the CCL offspring in the U.K., this also grew into a family of companies. Loosely connected by a network of licensing and personnel migrations, it was a mildly incestuous and sometimes feuding family. The technology held the promise of overcoming some of the limitations inherent in liquid ink jet. The drops "freeze" immediately upon striking the paper, producing vibrant, water-fast, relatively surface-independent color.

The story of this technology is so wonderfully convoluted it begs at least a brief chronology:

1984: Exxon Office Systems (EOS) demonstrates SIJ. The same year Robert Howard incorporates Howtek and hires several key development engineers from EOS. Howard's IPO raises $6.3 million. Ken Fischbeck, EOS SIJ team leader, moves east from Dallas to Creare Innovations.

1985: Dataproducts and EOS form joint SIJ venture. Exxon sells off EOS to Harris/Lanier except for the SIJ development operation.

1986: Creare spins off Spectra under Fischbeck to commercialize solid ink jet. Howtek demonstrates the Pixelmaster color SIJ printer. Dataproducts announces the monochrome SI 480 printer. Dataproducts sues Howtek for patent infringements, winning millions of dollars in royalties, a burden which in time broke the back of the company, according to Bob Howard (although Howtek has survived as a vendor of scanners and related equipment).

1987: Dataproducts buys the EOS SIJ operation, but spins off the SIJ development operation which becomes Trident International. Tektronix demonstrates a color SIJ prototype, developed further by obtaining "key enabling licenses."

1991: Tektronix introduces the first of its successful SIJ Phaser series.

1992: Dataproducts launches its color SIJ "Jolt" printer, but it doesn't fly. Their technology does pay off as Dataproducts/Hitachi successfully markets SIJ components and technology licenses.

Today: SIJ seems to have won a solid niche in specialized applications. The main vendors are Trident, Markem (which acquired Spectra in 1997) and Hitachi Koki. And the Tektronix Phaser series continues to be successful as a color page printer.

The 80s was the decade of thermal transfer and dye sublimation printing. As a technology for general office printing this turned out to be short lived. Pioneered in Japan, thermal transfer did manage to win a following in the USA. One of the more successful products was the Okidata Okimate 20, announced in 1984. The printer offered 80 cps in near letter quality and was priced low for the time at $268. In the brief window of time before low cost color ink jet swept in, thermal transfer was a key technology for economy color in the office. Vendors included Hitachi, Konishiroku, Okidata, Panasonic, Sharp and Toshiba. There were also some thermal transfer line printers from plotter companies including Benson and Calcomp. To hedge their bets, some of the larger vendors also fielded thermal transfer, including HP, IBM (resistive ribbon technology) and Data General.

At the other end of the performance scale, 1980 was the year Delphax systems was established as a joint venture between Dennison Manufacturing Company and Canada Development Corporation. Two years later Delphax shipped its first printer, the 60 ppm, 240 dpi 2460. Since then their ionography/CID technology has continued to be developed and find significant markets, currently as a wholly owned Xerox company. A second important alternative to electrophotographic printing also appeared, magnetography from Bull in France. Magnetography became a viable commercial product which was handled by Bull subsidiary Nipson. Now it is Xeikon, which acquired Nipson in 1999. Despite some unique attributes, the future of these two major "alternative" page printing technologies is by no means certain.

Other, more esoteric page printing technologies were demonstrated in the 80s but have yet to make the scene in general purpose applications. Mead Cycolor, a process of creating high quality color images by crushing dye-filled "cyliths," was invented and simmered away through the 80s with several specialized color products appearing from time to time from licensees and partners. Elcography, the "electro-coagulation" technology from Elcorsy, was invented back in the 70s, with more serious development in the 80s. The process continues to surface from time to time in prototype digital presses. An Elcorsy digital press claiming 1,700 color ppm was demonstrated at DRUPA 2000.

Exxon 965, 1983, the sleek product of what was probably the most gutsy, expensive ink jet development program to date. (Exxon publicity photo, DRA archives.)

MATRIX PRINTING: CHANGE OR DIE

Impact matrix vendors, which rode into the decade dominating the low end, soon found themselves in a life and death struggle. The pioneering U.S. companies including Centronics were being outproduced, under-priced, and outsold by vendors from Japan. Then, in mid-decade, the booming Japanese dot matrix manufacturers found their products were being obsoleted by ink jet technology. It was a case of change or die, first in this country, but soon thereafter on both sides of the Pacific.

Centronics was hit first, and probably the hardest. Late in 1970, after years of a productive partnership with Brother in Japan, the American company decided to manufacture their 700 Series in-house. Observers generally agree this move was bad timing and badly executed as well. This marked the beginning of the decline of that matrix printer pioneer.

Epson is the U.S. subsidiary of Shinshu Seiki Company, Ltd., which in turn is part of the Seikosha group, a precision manufac-

turer best known for watches. In 1980 Epson led the dot matrix invasion. The MX-80 appeared in 1980 quantity-one priced at $650 and performing at 80 cps. A year later the $450 MX-80 appeared, offering the same speed but lower resolution with a 7-pin rather than 9-pin head. Within two years they claimed to have shipped over 100,000 MX-series printers in the U.S. and to own 35% of the low-end printer market. In time sales of the series was well over one million, making it one of the largest-selling printer models ever. It won IBM's stamp of approval when selected as a graphics printer for the PC.

The success was not only due to reliability and the fabled efficiency of Japanese mass production which Epson termed "mechatronics." Marketing was also key. It was unorthodox (for the time) and aggressive. OEM sales were less important than mass distribution through a network of twelve exclusive dealers.

Within a few years the meteoric phase of Epson's growth ended in the face of competition from a host of other Japanese matrix vendors such as Okidata, Panasonic, Star and TEC. Epson managed to continue to lead in volume, but price cutting in the mid-1980s slowed growth.

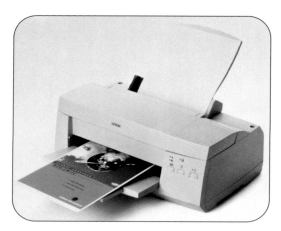

Epson is one of the few major impact dot matrix vendors to make a successful transition to ink jet. They did it by pioneering piezo technology and hyping print quality. This is the Stylus Color 900, introduced in 1999, featuring speed and the smallest drop size to date (3 picoliters) for outstanding 1440 x 720 dpi resolution. (Courtesy of Epson America, Inc.)

IBM's Counterattack. With a good bit of hoopla, IBM announced the dot matrix 4201 Proprinter in 1984. It was billed as a counterattack to the Japanese invasion of the low end printer market. If anyone could do it, it seemed as though IBM could. The design, the manufacturing, the financial resources behind it, and the price/performance specs all looked good. The print head, ironically, was from Epson. But everything else was domestic. It was feature-rich with an ability to handle 230 resident characters, innovative paper handling which could accommodate cut sheets, continuous forms, and envelopes, sophisticated vertical and hori-

zontal spacing, 200 cps draft speed, and a 3-million character ribbon cartridge. The suggested quantity one price was $549. *

In a bold gamble, IBM invested millions of dollars in an automated production facility at its Charlotte, NC plant in a attempt to beat the Japanese at their own game. The Proprinter was one of three major printer product introductions by IBM in mid-decade. In the twilight of line printing, they introduced the 3,000 lpm Model 4248 line printer, and as an alternative to impact dot matrix they introduced the resistive ribbon Quietwriter. Unfortunately, the technology and globalization sea change was even too big even for Big Blue. None of these products, including the feisty Proprinter effort, experienced significant market success over the long term. After this brief resurgence to regain stature as a source of innovative print engine technology, IBM settled back into becoming primarily an OEM packager of outside print engines.

By the end of the decade, the printer landscape had changed remarkably. Most of the larger matrix printer vendors still lived, but their profile was considerably lower. Seeing the handwriting on the wall, they made efforts to segue to other technologies. In the mid-80s Epson was already offering two ink jet printers and a series of direct thermal and thermal transfer printers. Okidata, which we followed through to the present in the last chapter, had begun its diversification. It looks today as though these two companies, Epson and Okidata, have been the most successful in making the transition.

Epson has made the transition to become an ink jet leader. In terms of print quality, with fine-tuned piezo technology and commitment to photo-quality performance, Epson has hit a chord in the marketplace. The Stylus 900 printer jumped ahead of the competition in 1999 with 10 ppm color printing and the industry's smallest drop size at three picoliters. Although still behind HP and Canon in ink jet printer volume, their corporate resources (annual revenues over $6 billion), highly developed ink jet technology and aggressive distribution boded well for their future.

---

* "IBM Fights Back: Comits to Domestic Manfacture of Low-End Impact Printer," *1985 Printout Annual*, Datek Information Services, Inc.

Transitioning Strategies. Gary Bailer, a former executive with Epson competitor Panasonic, described his company's survival strategy. "It was in the 1990-1993 timeframe that vendors began to accept the realization that DM printers were nearing the end of their leading role in the low-end printer market. Our dot matrix volume was $60 million a year. DM was our cash cow. These printers are still in niche markets like doctors offices – you always see a DM printer there. They'd rather use their resources to lease new medical equipment or other equipment to bring in new customers. They need impact for insurance forms. Another niche: automotive work orders. In a low volume application, it doesn't make sense to change the system. Also, dry cleaners. Always a multipart form. POS listers – still a big market. The leaders here: Epson, Star, TEC, and Seikosha."

Citizen, Star, and other dot matrix leaders are gone, Bailer noted. His company, Panasonic, he saw as the last survivor – still hanging in with dot matrix printers, but barely.

"The challenge is how to make the technology transition," Bailer explained. "Do we milk our core technology, do we partner, or do we give up?

"The American executives at these companies warned that their core technology was being obsoleted. We said, 'this is real!' But Panasonic was shipping 100,000 printers a month and didn't want to believe it. By the time they realized it really was real, it was too late! Other companies, in response, tried to outsource, but they often found they were running into conflict since their engine sources tended to be competitors.

"Also, outsourcing is not always the right answer anyway. It is quicker to market, but by the time the products gets to market the vendor may discover they were stuck with an engine that is obsolete. If you are outsourcing your engines, you need a big marketing budget. To succeed you need at least one of three things: controller technology, a state of the art engine, or a lot of money for advertising."

Recalling some of Panasonic's strategic moves, Bailer noted they tended to buck the trends in some ways. "We broke the rules. We shot ourselves in the foot, obsoleting our 9-pin models. We also shot our channels in the foot. We had been going through rep companies and some distributors. We decided we needed to hype our brand name and we pursued the superstore segment, and warehouse clubs, and national distributors. We were the first to get into the superstores and we were exceptionally successful at it, and led the market, at least until HP and Canon came along with their low end ink jet. We eliminated our reps. Oki in contrast, was afraid to eliminate their VAR base."

In looking at non-impact, Bailer recalled that the effort was half-hearted because "Pan" was happy with their existing dot matrix volumes. He said they did pursue monochrome lasers, but had controller problems. Currently they do have a color laser. OEM sales, mostly. Now they are bucking a trend again, concentrating on the fax and copier business, while others are pushing color laser printers to national distributors. "But the margins are low when it is a product retailing for only around $2,000. There is the supplies business annuity which can make up for low margin on the hardware. But to add more value, they are not using that approach. They sell direct to dealers, leveraging the transition going on there from analog to digital, and from monochrome to color. These are the dealers traditionally associated with the NOMDA channel."

Strategies for morphing from a one-product company to a multi-product company are obviously full of pitfalls. As a result dozens, maybe hundreds of companies have come and gone, and the number of survivors keeps dwindling.

But hey, as we look to the 90s and beyond, we see the boundaries of the industry crumbling. Counting companies becomes less relevant than redefining the nature of the industry.

"With ordinary printers, you see the flowers blooming. With our printers, you see the ladies dancing."

Epson hyped the photographic print quality of its piezo-driven ink jet printers with this ad that ran in Time Magazine early in 2000. According to the ad, Epson printers were used to prepare catalogue cover artwork for a textile company. (Reprinted courtesy of Epson America, Inc.)

182

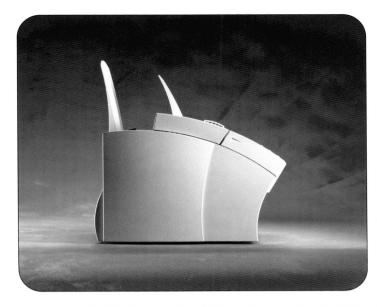

Styling and multiple function capability in a single unit are two 90s themes. This Hewlett-Packard 3100 is an example of both. It was the company's first laser-based multifunction device, introduced in 1998. The compact machine will print, fax, copy, and scan. Pricing, at launch, was $699. (Hewlett-Packard Company. Reprinted with permission.)

CHAPTER 7: THE NINETIES

THE SETTING   The 90s were boom years. People working harder and many getting richer. In the U.S. and the U.K. conservative leaders were replaced by business-friendly centrists Bill Clinton and Tony Blair. In the U.S. unemployment reached record lows and the Dow Jones record highs, cracking 6,000 in 1996, 8,000 in 1998, and 10,000 a year or two later. Stock speculation in dot-com and other high tech companies reached a frenzied pace by the end of the decade.

But there were dark pockets. In the midst of prosperity in the U.S., many were left out. Corporations down-sized, slashing payrolls. New jobs were created, but an increasing percentage was in the low-pay, service sector. The Cold War became a distant memory and the globalization of the world economy accelerated. The interdependence of national economies was highlighted with the passage of the North American Free Trade Agreement, the $20 billion bail-out of the Mexican economy, and the "Asian flu" of 1998 which sent shivers across oceans. Science moved ahead, unraveling the DNA, leading to cures but also to controversial genetic engineering, "Frankenfoods," and cloning. AIDS continued to decimate millions around the world.

More than ever, the line between news and entertainment became blurred. Examples were some great media events: two major show trials, O.J. Simpson for murder and President Clinton for private indiscretions; the death of Princes Diana; tears and twisted bodies in the wake of the Oklahoma City bombing by native-born, anti-government terrorists. Televised coverage of nasty wars – in the Persian Gulf, Bosnia and elsewhere – were piped into the world's living rooms through a filter of meticulous censoring. In contrast, the Rwanda bloodletting in 1994 was virtually ignored until it was much too late.

The World Wide Web snared large percentages of the population. By 1993 the Web had grown to 1.3 million hosts and was carrying 5.5 terabytes per month. *Time* magazine declared 1995 "the year of the Internet." New media piped an increasing flood of information and entertainment into offices and homes. Communication vehicles tended not to compete but rather leverage one another. Print on paper, far from being obsoleted, continued to grow.

Presaging a broadening of attention beyond products, the leading brand and program buzzword of the decade was "solutions."

In the graphic arts, the 1990s saw the typographic industry fade as hundreds of "personal foundries" poured out new type designs. Digital presses took aim at mainstream printing processes with somewhat disappointing results. Heidelberg pioneered on-press platemaking with its Presstek-based GTO-DI. The two major industry associations, GATF and PIA joined forces in 1999.

The digital printing industry echoed the larger currents of technology with its own, continuing consolidation, concentration, and spectacular product price and performance gains.

Hewlett-Packard continued to dominate, especially in the mid- and low-end where growth was the fastest. The industry paused in awe as HP announced shipment of their 30 millionth laser printer in 1998 and their 100 millionth ink jet printer late in 1999. In less than a year, their ink jet printer sales hit 25 million.* Unlike many vendors, HP's ink jet program relied less on outsourcing than on contract manufacturing. For example, in 1999 SCI Systems announced a contract to build HP ink jet printers "on a finished product basis," with SCI, in turn, assigning the program to its new plant across the border in Guadalajara, Mexico.

Xerox worked hard across the entire performance spectrum from digital presses to entry level printers and multi-function devices. Lexmark was born in 1991 and protected with a non-compete from IBM until 1996. Despite the expiration of that shelter, Lexmark continued to be an industry leader through the decade, fielding ultra low cost ink jet printers and, according to one source, ranking second in laser printer sales by 1998. Canon and Epson both carved out significant shares of the pie. Print engine manufacture continued to gravitate to Japan. Other companies began to apply digital printing technologies to various smaller but higher-margin, non-document markets such as textiles, display advertising, wall covering, and fine art reproduction.

In mid-decade, a significant shift in distribution was the rapid growth of discount outlets and mail order/catalog at the expense of the traditional channels such as ComputerLand, MicroAge and NOMDA dealers. It was the decade when multi-function peripherals (MFPs) created a stir. Almost 100 MFP announcements were made in the 1993-1994 timeframe alone. The HP "Mopier" was a move toward capturing more of the analog copier equipment base. The growth of color, networked printers, and MFPs was gated by advances in software and various standards.

Technology trends included the usual: more speed and better print quality for less money. Dots per inch does not equate to print quality necessarily, but is a building block. As the decade wore on, 600 dpi began to replace 300 dpi as a baseline. For digital photography as well as office ink jet, machines appeared by 1999 offering still higher resolution. Epson, HP and others weighed in with 1,200, 1,440 and even 2,400 dpi. Ink jet drop sizes got smaller for sharper edge definition, to 5 picoliters and by the end of the decade, from Epson, to 3 picoliters.

A book could be devoted to any one of these trends. In this chapter we will be selective, focusing on:

- The blossoming of on-demand printing, spearheaded by the Xerox DocuTech, widely viewed as the most important single printer product of the decade;
- Other activity at the high end, including digital presses;
- Printers to the people, printers as consumer products including a look at industrial design and pricing trends;
- A sampling of cutting edge and otherwise significant, imaginative or photogenic applications;
- Transition to the next millennium, introducing the prospect of a redefined printer industry;

*Lyra *Hard Copy Observer*, December, 1999

THE DOCUTECH MAKES A MARKET – AN INTERVIEW WITH CHIP HOLT

The Xerox DocuTech Production Publisher is arguably the most successful printer product of the 1990s. For Xerox, it has been a major revenue generator, an impressive sign that Xerox, sometimes seen as a sleeping giant, has not been sleeping after all. Despite the huge numbers of consumer and office printers flowing from HP and other competitors, the DocuTech is believed to rank first in terms of revenues for a single product during the decade.

The DocuTech furthermore looms as an outstanding example of a product that made a market. It has also been deemed a "change-agent" within Xerox. It both reflects and has supported a quantum shift in Xerox from a 'box' mentality to a 'systems' mentality.

In September, 1999, *Micro Publishing News* named the DocuTech to its MPN Hall of Fame, the publication's choices "for the most important and innovative tools of the last ten years." Their citation reads in part, "If you've ever heard the term 'on-demand printing,' thank the DocuTech. The DocuTech essentially invented the idea of digital or on-demand printing." Two months later it was awarded a Silver by the Industrial Design Society of America in a "Decade of Design" competition sponsored by *Business Week*. The citation read, in part, "The DocuTech is a strong idea supported by outstanding engineering and thoughtful design."

A development program in a company as big as Xerox is obviously the product of a large team. Xerox Senior VP Robert Adams was the strategic initiator of the product. David Kearns, then CEO, became an avid supporter once he was exposed to the product concepts. In time there were many other key managers and supporters such as Paul Allaire, CEO after Kearns, Frank Steenburgh, VP of Marketing for Production Systems, and of course the program team that was assembled to pursue and implement the product concepts.

But looking for one individual who was most important in making it happen, all paths led to Charles ("Chip") Holt, recently retired Corporate VP of Xerox's Joseph C. Wilson Center for Research and Technology. Holt seems an appropriate history-maker upon whom to focus for the 1990s, a decade in which teams and the art of technology management appear to be eclipsing the inventor-entrepreneur in a maturing industry.

Chip is popularly viewed as "father" of the DocuTech. He seems to have earned something close to deity status among many of those who worked with him. Many expected being part of the team would be a one or two-year interlude, but it ended up seeming like a lifetime. Software engineer Tony Frederico, quoted in a Xerox house publication, remarked that the project looked like 200,000 lines of software in 1983, but it ended up more than 1.3 million lines and still growing by 1991. "His leadership skills and charisma made us drive in one direction," Frederico said of Holt. Associates described him as "the only person who could have seen the product through."

Charles ("Chip") Holt, recently retired Corporate VP of Xerox's Joseph C. Wilson Center for Research and Technology

I talked with Chip for a couple hours in October, 1999 on a windy day at his shorefront home near Rochester, NY.

Asked first about his professional background and his path to Xerox, Chip described his previous work developing real-time computers at AAI in Maryland, experience that looks like an appropriate foundation for the intense real time computational demands of high speed electronic reprographics. He developed a state of the art, computer controlled, automatic test system for carrier based aircraft using a preponderance of HP instruments in the late 60's. This activity led Chip to almost accept a position with HP in Palo Alto, CA. They were interested in expanding their product line beyond instruments into systems. But his boss at AAI went to Scientific Data Systems and convinced him to join SDS instead. This looked interesting because SDS had been acquired by Xerox the previous year. Note: Italic text is Ted, Roman text, Chip.

I ended up going to SDS because I thought that Xerox had this great strategic vision of marrying computational technology with reprographics. Peter McCollough in those days was talking about "the architecture of information." Xerox was frightened about the advent of a paperless society. I thought this was going to be a foray into marrying these technologies.

It was a foray, but not the foray I expected. It became evident almost right away there was no strategic vision for the integration. In fact the acquisition of SDS turned out to be a huge disappointment to Xerox. There were cultural mismatches, there was technology underperformance, and it was just a terrible marriage.

What happened first, you going there, or the acquisition?

I was being recruited in 1968-69 while the acquisition was in process. I accepted the offer in 69 and moved out there in 1970. So there I was, watching the deterioration of this relationship between southern California and Rochester with great chagrin. After all, it was not a small thing in the Sixties to take an East Coast person who has the guts to go to California and transport him out there, only to find out why he went there was false.

But although I didn't know it, the pathway to the DocuTech was beginning to open. I had the chance to be project engineer for the SDS Sigma 9 and I took it. The marketing manager for that project was Bob Adams. I got to know Bob very well. It was a coincidental relationship that played heavily in the ultimate destiny of the DocuTech.

The Sigma 9 unfortunately came out when other technologies were beginning to erode the concept of big mainframes and SDS technology was also not the dominant force it had been during the early 60s. In fact, SDS' technology could not keep up with competitors such as IBM, Burroughs, NCR, HIS. These competitors were all in their heyday at that time.

In any case, in reaction to the disappointment of Xerox regarding its acquisition of SDS, and to salvage something from it, they formed a task force to study the emerging electronic needs of Xerox products over time. I leapt at the opportunity to join this task force. I hoped it would be the gateway out of the oblivion I saw for the computer business of SDS because of the cultural mismatch. This task force confirmed that over time the products of Xerox would be becoming increasingly electronic and there would be an advantage to have a division with competency in the development of sophisticated electronic subsystems.

Xerox accepted these conclusions and created a new Electronics Division in which I got a small managerial position with responsibility for the development of microprocessor-based control systems for future reprographics products..

Now, the intent of our microprocessor control systems was to replace the electronic subsystems being built in Rochester by a very capable organization. They had a very strong and talented team of people, but the historic systems they had built were de-

signed using "hard wired" logic circuitry. Although this team would have eventually graduated to microprocessor-based controls, the corporation directed that these designs would emanate from southern California So as luck would have it, I got the job to come back from this hated acquisition that had taken all the profit-sharing performance away from the employees in Rochester – at least that's the way they perceived it – and tell them we were going to replace what they were doing with designs based in Southern California.

So we had another task force to get that resolved and that was really bloody.

Here we were, the turkeys from California, telling this talented team, that we were going to replace their designs with microcontroller-based ones. The theory was that by replacing the hardware that was being designed in Rochester with a microprocessor you'd begin developing reusable platforms and software for the products. And it was a good strategy. So we introduced the first microprocessor-based control system in the 9400 reprographics product – it was a big duplicator. The 9400 was a derivative of the 9200 which was a bet-your-life investment that Xerox made in duplication, going from copiers to duplicators.

The 9200 spawned a whole bunch of duplicating products but also provided the base engine that would give Xerox its second paradigm-shifting event in the marketplace. Our electronic printing was built off the 9200 and this was the 9700 printing product.

My number one interface in Rochester during this time was Don Post, who was also my biggest opponent. We just had an enormously difficult time with one another due to the roles we had been given. Over the years, however, Don and I became very good friends and he ended up running the system development organization under me for DocuTech.

Wait a minute. You said something about a "second paradigm-shifting event?"

Yes, Xerox changed the market three times. I see these as what I call market-changing or market-making events which can occur on the basis of competent true believers railing against the infrastructure of a corporation.

Anyway, I think the three major events were, first, the copier itself, which created a market nobody foresaw.

Second, the 9700 created electronic printing, an investment that was heavily resisted by Xerox.

And finally, the DocuTech created the print-on-demand market. Interestingly, in the period 1973-1979, Bob Adams had become very involved in the emergence of the electronic printing marketplace and was one of the prime movers behind the 9700 and later, of the DocuTech.

So, due to this new arrangement, I started coming back to Rochester with increasing frequency and by 1979 I was asked to relocate to Rochester and learn the mainstream product development process. I was given the 3100 family which historically had been very successful, the so-called low end products such as the 3109, 3450 – all of the low end copiers.

Later I was assigned to the Printer Strategy Extension Project, and that got me re-engaged somewhat with the work that Bob Adams was doing. He was emerging now as a very influential manager within Xerox. In any case, our mission was to come up with truly strategic planning on the application of electronics to reprographics, in short, the question of where the printing business is ultimately going.

Around 1982 Bob took over the Systems Group and had all the systems related products under him, all the printers, all the work stations. It was a big organization. The Company in those days was divided between the core competency products – copiers/reprographics – and the systems products that included printers, the emerging workstations, and all the dreams Xerox had about emerging into a systems company.

I leapt at the opportunity to join this task force. I hoped it would be the gateway out of the oblivion I saw for the computer business of SDS because of the cultural mismatch.

Xerox changed the market three times. I see these as what I call market-changing or market-making events which can occur on the basis of competent true believers railing against the infrastructure of a corporation.

The functions that resulted from an electronic implementation were so dramatically enhanced over the electromechanical copier, that we believed people would be willing to pay the price difference that ensued. This closed the gap.

In 1983 Bob called and asked me to investigate whether the concepts of a product that he had some people brief him on were feasible enough to implement. These were the concepts that in fact turned out to be the basis for the development of the DocuTech.

Can we get a fix on those concepts?

The concept of an electronic reprographics product was that you could replace the photographic process of imaging and the electromechanical process of document handling of a copier with an electronic process of scanning and reproduction. The Company was constantly looking at the cost curves of whether the implementation of those concepts would ever cross the line and be competitive with an electromechanical copier. The answer was that electronics never really would cross the line. However, it would get close enough that it wouldn't matter because the functions that resulted from an electronic implementation were so dramatically enhanced over the electromechanical copier, that we believed people would be willing to pay the price difference that ensued. This closed the gap.

Is this "electronic reprographics" product a synonym for a digital copier?

All of this vocabulary evolved later. Electronic reprographics was really a concept that you were going to build a reprographics machine electronically. And that was all it was going to be. The immediate challenge was the desired design target speed. A one hundred page a minute, high image resolution product requires enormous data rates. Consequently, the computer required to drive the product had to have processing power beyond the state of the art in those days.

The project was to discern whether it was technologically feasible to implement such a system. We didn't have a good understanding beyond a one to one replacement of the conventional reprographics machine. And we were additionally looking for increased reliability. We weren't thinking of all the additional functionality that would ensue and in answer to Bob's question, in May of 1983 we determined the product was feasible even though we knew the technology challenges, especially the computer, would be formidable.

What Bob wanted essentially was to create a business that could handle all the applications on one side of the network and all of the reproduction on the other side of the network. He named the architecture associated with that concept, Distributed Printing and Reprographic Architecture, or DPRA. It was the beginning of the revelation that if you build a product electronically, you could achieve many aspects of the two main businesses Xerox was in – printing and reprographics. You could achieve plug and play architecture. By July of 1983, we confirmed that the new product could be compatible with that architecture.

In this feasibility phase, did you personally look at markets, too?

We were looking at the technological feasibility, although I did have a product planner with me who understood markets pretty well and wound up being the planner for the DocuTech over the course of its development. That was Jack Ratcliffe. He is still there and one of the planners in Production Systems. There was also Don Post. When he discovered that we were going to develop this product, he walked away from his job and joined me.

I remember him. He was the guy you had to go out and give the bad news to, yes?

Right. He ended up being my systems designer on DocuTech. So, by October, 1983 I was transferred from my old job and Bob Adams said we're going to make this program official, we're going to start implementing it. We decided to go with the most aggressive marking engine that Xerox had been building, which wound up being the 5090 product.

We began during 1984 to gain an understanding of what this product concept could do, the seminal features. Some of that was born out of necessity, because to run it at the speeds we had to

run – we were essentially trying to replace photons with electrons and it couldn't be done. So, in other words, how do you replace photons with electrons? How do you pass electrons into a marking system to drive a laser when in fact a flash of a light was what was done in the past, photographically.

The flash is pretty efficient in a lot of ways.

Right, it is. A very high bandwidth. We knew we would have to solve the bandwidth problem architecturally and we needed to think about the features of the product in an integrated systems way. So the principle people in the project virtually locked themselves in a room for a year and started to develop the concepts of the graphical user interface, what the features were and what they were going to look like.

By 1985 we had developed a good understanding of what the five seminal features of the product would ultimately be.

So what are they?

The first was productivity. Our target was to run as fast as a light lens product. But on a light lens product, between every job, you have to take the original out of the document handler and put the new original in and program the machine to run and then cycle down and cycle back up between jobs. We had conceived of a way, because we were acquiring information electronically, to build a print queue of jobs in waiting. That meant once the first job was in and initiated you could simultaneously start acquiring subsequent jobs. We were scanning the data off the documents and putting it in memory, but then also putting into that memory the operator's job instructions. All those instructions along with the data that had been scanned went into the print queue.

We essentially could build a print queue of infinite length and there was no need to cycle down and cycle up between jobs. In those days the operators and managers of the central reproduction departments using products like the 5090 would kill for a couple percentage points of improvements in productivity. Just this feature alone, as I recall, gave a twelve percent jump in the productivity over the equivalent light lens configurations.

Productivity was also improved because the operator could do makeready at the same console while jobs were running out of the back end. The more makeready needed, the greater the productivity savings. In short, we built concurrency into the system.

The second feature was electronic precollation. In conventional light lens machines, for every copy that needs to be made you need to mechanically re-circulate the original. Sometimes the original is a large document with a lot of value put into it, and with mechanical re-circulation the document could sometimes be damaged. This did not make the users happy. With the electronic system the original needs to be scanned just once instead of physically re-circulating it, and the image held in memory.

How many pages could you store?

We had a very large hard drive in the DocuTech, ending up with three 380-megabyte disks which could store hundreds of pages. This let us store latent images of jobs that had been previously put in there that you may have modified. That leads to the capability to merge documents together, all kinds of things that are not available when you have to electromechanically flash images from the original.

The third feature was the ability to do electronic cut and paste. This is the ability to acquire the information through the scanner, store it, and call it back up through the user interface, and modify it. Enlarge it, reduce it, crop it – anything that you could do with scissors, paste, and a camera you could do with electronic cut and paste.

Interestingly, on this feature, there were enormous corporate debates about how you could put such sophisticated features into an environment that traditionally had been run by relatively unsophisticated personnel. Their job was to put paper in and take paper out and keep the toner hoppers filled and anything other than

So the principle people in the project virtually locked themselves in a room for a year and started to develop the concepts of the graphical user interface.

In those days the operators and managers of the central reproduction departments using products like the 5090 would kill for a couple percentage points of improvements in productivity.

Interestingly, on this feature, there were enormous corporate debates about how you could put such sophisticated features into an environment that traditionally had been run by relatively unsophisticated personnel.

that would be a distraction to productivity. The concepts of concurrency and ease of use were lost on the traditional planners and developers of the old reprographics machines. Here with the push of a button the technology behind the product actually made the operations very easy. It turned out that operators became empowered and in fact became more enthusiastic about his job. The new system became a real motivator.

I've heard the theory that if you bring more sophisticated, innovative systems into a market, you face the danger of obsoleting the personnel — and that this means advanced technology needs to find new markets with more sophisticated personnel. It sounds like this has not been your experience with the DocuTech. The success of this product seems to indicate that an existing market can rise to meet the sophistication of new technology.

Actually what happened here is the emergence of a new product concept based on a set of new technologies, incapable of replacing offset presses in their traditional applications. But it did and continues to disrupt the low-end market applications of those presses and created a new market: Print on Demand, which I define as the ability to print current/variable data at the place and time of need.

The fourth key feature was our ability to duplex large, 11 x 17-inch documents. The 5090 engine which we based this product on could duplex other size sheets but it was such a sophisticated engine it didn't have room under its covers to have the staging tray for 11 by 17. Think about the way duplexing is done in the standard reprographics product: first you do all the first side imaging, and store those imaged sheets in a staging tray. Then all the sheets — it could be, say, 200 — are brought out and inverted and imaged on the second side. That requires a lot of physical room for that storage and that was out of the question. Second, think what happens if you get an error in that run. You have to purge the entire job — all the sheets in the staging tray — and you have to start over again.

We hypothesized that because the computer was the source of the image that we didn't have to stage the mechanical duplex path. Instead we could feed the sheets into a dynamic race track seven pages long, for example, and have the computer figure out whether the first or second side needed to be imaged. So, in the dynamic duplex process, you only have seven sheets committed at a time. With the staging trays eliminated we had room to handle 11 x 17 pages, and if you got an error only seven sheets rather than the whole job was purged.

And the fifth major feature, given the fact that we could electronically collate, and electronically duplex even large sheets, we could now, with the touch of a button, do signature imposition to make booklets. What that means is that thanks to 11 x 17 capability, you could put 8-1/2 by 11 originals in the input tray and tell the DocuTech you want to make a signature on 11 x 17. Then the computer automatically figures out how to acquire these 8-1/2 x 11 sheets and impose them to make the booklets. That was an incredible feature because signature imposition management traditionally needed to be done by the highest level people in the repro shop because the manual process was so complex. Now it could be done with just a touch of a button.

And here's one more feature associated with making a book: the thicker the book, the bigger the radius of turn on the sheets, yes? So the computer will figure out which page of the booklet each copy is and incrementally move the image further out to keep the margins of the published booklet consistent no matter what page it is.

But, additionally, all . . . of . . . that . . . stuff could also be networked! And this became the sixth feature, the counterpoint feature. So now DocuTech can be a receiver of information from anywhere in this information age — a compliant product in the information age infrastructure.

Of course in 1984 networking was just emerging. But nonetheless, we had conceived a product that had functionality

beyond simple reprographics. It could be networked and also do printing with the features described, which ultimately enabled it to serve publishing applications..

It was in 1984 when David Kearns, Xerox CEO, started to visit with me to review this product that he believed was enormously important.

When he heard the story he concluded it could be seen as damaging both businesses of the company: printing and reprographics. It was doing things in different ways than the traditional businesses were. And he said, therefore, if that can be done, we had better be the first ones to do it. There was no turning back. So he and Bob Adams tucked us away in an organization that reported to neither of the existing Xerox businesses. This was correct, because the infrastructure of the existing businesses would have warped the outcome of this product in deference to whichever business it reported to. And in fact, as it worked out, as a business DocuTech created a new industry on the basis of both of the existing Xerox businesses.

By the way, print-on-demand is still having enormous effects on the behavior of centralized reproduction and data center printing. I remember as a technology manager the belief that once such things are possible, they will be immediately adapted. Well, the way businesses are run, centralized repro centers are still reluctant to think about their jobs as being sourced from anything other than paper originals. And data centers are still reluctant to think their operations can absorb anything other than computer data streams.

I remember, just after the DocuTech was launched in 1991, giving the keynote address at XPLOR. I told that enormous audience that the concepts of data center printing and centralized reproduction were anachronisms and I couldn't understand the hostile reaction!

You didn't get a lot of applause when you said that?

[Laugh] It took me a long time to appreciate just how much gestalt there was in these two kinds of operations. As we were developing the product we had been insulated from such realities since we were locked away. Toward the end, of course, we did a lot of focus groups with the usual "what ifs" and one way mirrors, asking if you had a product that could do this and this, what would you think about it. There was a lot of market preparation as we got near the launch.

Were you able to forecast the adaptability of your traditional customers to these new systems?

In the beginning I feel our program team did have the belief that the traditional market could not make the shift. But toward the end we mounted an enormous market research and implementation program under Frank Steenburgh in 1988 with the development of what we called Reprographics Advanced Program Management. He selected the cream of the crop from different geographic and vertically integrated industries for training programs. We did a lot of beta site introductions for two years. We also started to develop applications that the DocuTech could solve for identifiable vertical industries such as pharmaceuticals, universities, aerospace, and the like. That was very effective. I think by the time of the launch we had thirty or more production beta sites and they all converted.

Anyway, the rest is history. Within three years of the launch in 1991 the DocuTech had made its mark as a market-changing product. Today, Xerox has shipped 20,000 systems generating revenues of $30 to $35 billion with yearly margins approaching $1 billion. And this, as is the case with any market changing innovation, was a product the existing businesses of Xerox did not want.

We hypothesized that because the computer was the source of the image that we didn't have to stage the mechanical duplex path.

Of course in 1984 networking was just emerging. But nonetheless, we had conceived a product that had functionality beyond simple reprographics.

I told that enormous audience that the concepts of data center printing and centralized reproduction were anachronisms and I couldn't understand the hostile reaction!

People in the market surely wanted it.

But not those enmeshed in the established businesses within the innovating company.

With such a long development cycle, were there issues around software and adapting to changing industry standards?

For sure. Early on, for example, we had chosen Ethernet as our network protocol and Interpress which had been developed at our Palo Alto Research Center by Chuck Geschke and John Warnock. By 1987 the industry had changed a lot because of the advent of the PC. PostScript was taking the industry by storm and there were raging debates at these industry conferences on the pros and cons of Interpress and PostScript. The PARC guys were still bound to Interpress. Geschke and Warnock, meanwhile, had left PARC because of the frustration of not being able to get their ideas marketed.

So the Interpress people were saying their product was the superior composer, that it was faster and all that. But the extant standard, by virtue of sheer volume, was coming to be PostScript. We were getting strong feedback that it didn't matter whether you were better or not. If there was an anomaly between the application and reproduction anywhere in the system and you didn't have the standard component in there like PostScript, you were suspicious. And so the burden of proof would be on you.

Photos courtesy of Xerox Corporation.

Anyway, marketing people came along insisting on Postscript. So here is the engineering team, five or six years into the process, committed to Interpress. Now marketing comes in and starts telling them what to do which was different from the existing commitments. I believe they were correct, but the approach seemed abrasive and it took some time to get through that.

Marketing usually has the last word, I guess. It helps when they're right. Anyway, maybe it's time we should be looking ahead. What are your interests now that you've retired from Xerox?

Well, I'm teaching and getting more consulting requests than I can process.

Is the teaching at RIT?

Yes, in their school of business. One course is on the management of research and technology and the other is on the effects of innovation on the renewal cycle of business. The reason I got interested in the renewal cycle is primarily my first hand experience with the DocuTech. My belief is that Xerox really did have a tempo of changing the marketplace, but that it was not seen as predictable. Yet it seemed to me to have a fairly constant rhythm of every thirteen years, but for unplanned reasons.

The questions are whether that rhythm can be anticipated, what energizes it, and how it might leverage the planning process. I worried about this when I was in product development and then, after I left the DocuTech program and got into research I re-

Today, Xerox has shipped 20,000 systems generating revenues of $30 to $35 billion with yearly margins approaching $1 billion. And this, as is the case with any market changing innovation, was a product the existing businesses of Xerox did not want.

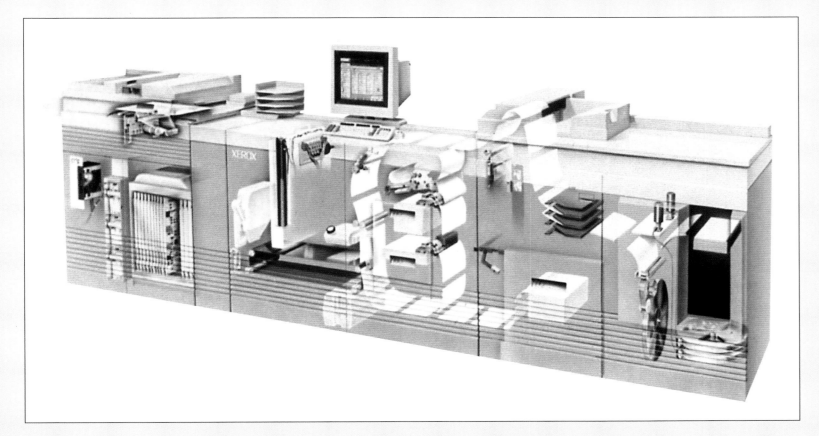

But the extant standard, by virtue of sheer volume, was coming to be PostScript. We were getting strong feedback that it didn't matter whether you were better or not.

ally worried about it. The decade of the 90s, after all, has been preoccupied with downsizing and reduction of investments. It seemed as though all the corporate research and innovation activities were being questioned so I started participating in industry conferences and building rapport with colleagues. This helped sharpen my ideas about these things which I'm pretty confident about now.

One thing I've centered on in my courses is the idea of market-changing events. The pace of development can be hectic, but the renewal cycle seems to be triggered by people I refer to as competent true believers.

Corporate behavior needs to be bi-modal, consisting of defending and attacking the market. Behavior in these modes is completely different.

Defending the market requires productivity, process excellence, certainty, and high attention to customer needs. These are the attributes typically ascribed to the Quality movement.

Attacking the market requires innovation, risk taking, operating with uncertainty, and not paying close attention to your customers. In this regard, products are developed that change the market and give the customers not what they want, but what they will come to *learn* to want. Corporate leadership is required to keep the behavior of the defend mode and the behavior of the attack mode in harmony.

The questions are whether that rhythm can be anticipated, what energizes it, and how it might leverage the planning process.

Most companies today are over-managed and under-led. This results in a heavy emphasis on market defense with huge efforts expended in behalf of productivity and a forgetfulness that the opportunity to be productive was originally created by the behavior in market attack. The imperative for change is forgotten in market defense behavior.

What about the thirteen year cycle you observed at Xerox? Do you think it applies to other corporations as well?

Cycles do, but 13 years is part of the uniqueness of the specific

market you are in. It is something I observed as I started thinking about this rejuvenation process. And yes, other corporations have their own cycles. One of the consultants I've become close to, Marv Patterson, was director of engineering at HP and has written a couple of textbooks. He emphasizes the new product revenue metric they use at HP. They actually keep track of the revenue streams over the life of a product from the time it was introduced. They call it the vintage revenue metric. He has developed some quantitative processes based on the revenue metric that can be used to predict what level of investment is needed to create certain year over year revenue results. And that links quantitatively to different products when the products have different time constants.

Patterson has some plots that show how HP in the 60s and 70s was distinctly different from Hewlett-Packard in the 80s and the 90s in terms of revenues for computers and low end printers. They have radically increased the frequency of their new product introduction. But it's important to understand the industry you are in and be as efficient has you should be for that industry. It is possible to be too efficient. And of course also possible to be too inefficient.

Most of the emphasis is on improving your time to market. But if you have a time constant in your market that is incompatible with the speed at which you go to market you can damage yourself.

In what sense do you use the term "efficient?"

In terms of time, in time to market. For example, if you are on a productivity wave-form and not thinking about changing the market, all of your metrics are associated with product productivity and time is one of the elements.

By "productivity" you mean converting a concept to a product or the productivity of the product for the user?

I'm talking productivity in the financial sense, that is, return on

investment. The conventional idea is that the quicker you can produce revenue and minimize the investment the more productive you are. You can go very wrong with that kind of thinking. On the other hand, if you are advocating change, if you don't support time to market, it can look like you are contradicting yourself.

On the topic of product cycles, they are certainly speeding up. But a lot of new products are just enhanced and renamed old products. Have you thoughts on what constitutes a real product cycle?

That's a good question. People responsible for the product improvement in a company will tout those kinds of metrics when in fact what you are really looking for is what effect the investment in this new product will have on the marketplace as opposed to how fast can it get there. So you get these conversations within a company such as "I have brought out a new product every six months. So why aren't we producing revenue?" And marketing replying, "You are producing products of no value at very high speed."

To rationalize profitable innovation, you have to understand market change rhythm and fit your timing into it.

The DocuTech was obviously a real market-changing event. But the launch was back in 1990. What do you see as the life of that product, and what do you see coming next?

This is the kind of question you have to ask when you start looking at the economics of innovation. You ask, Are the returns adequate? I think it's really getting close to the thirteen years since the last market-changing event. So we ask, if this cycle theory true, where is the energy to create the next change. That is the most fundamental aspect of planning: the awareness that you are subject to or participating in a market-changing rhythm. It is very similar to aggregate product planning where you plot radical

change in emerging markets and try to fill in a spectrum of things you have to invest in even if you don't have a full understanding of all the ingredients.

Is color a possible DocuTech enhancement? I guess that would be positioning it as a digital press. Some people might call it a press now, but I don't.

Well, it's beginning to encroach on a press. My agenda was to get that color equivalent of the DocuTech to market, to continue our encroachment into what is now the offset market. That was a personal strategy which obviously had to be bounced off of the corporate dynamics. The extant business of Xerox is selling consumables that are used in great marking engines and their applications.

That's the perspective now, you say?

That's the business model. Whether the perspective is there or not is another question.

It has obviously been the strategy at the low end with disposable cartridges. I didn't think it was at the high end.

At the high end, even more so. The size and momentum of the materials revenue streams through a product such as the DocuTech is enormous. This includes papers and toner and other supplies, as well as service.

What about a product like the DocuTech producing offset quality color? The technology challenges are enormous. There can be as much information on one sheet of paper in an environment like that as there is in the entire hard drive capacity of the DocuTech. The information content increases exponentially because the square inch density of information. . . . if you put two more bits behind a color pixel it's really two squared. . . the density is so great It can be done with the right technology, but it is a big challenge.

> *Corporate behavior needs to be bi-modal, consisting of defending and attacking the market. . . . . Most companies today are over-managed and under-led.*

> *It is possible to be too efficient. And of course also possible to be too inefficient.*

> *So you get these conversations within a company such as "I have brought out a new product every six months. So why aren't we producing revenue?"*

[With color] there can be as much information on one sheet of paper in an environment like that as there is in the entire hard drive capacity of the DocuTech.

I think the IT function is taking over the world. The data center and central reprographics are really subsets of the IT function these days and IT is trying to find out how to bring these two functions together.

Now here's how I see the opportunity presented in continuing to move up-market with variable data printing encroaching on offset.

I'm not sure of the exact numbers — but worldwide there are something like 4.2 trillion prints made in the printing world every year. And Xerox, with all of its success, makes something like just 8 billion of those 4.2 trillion prints.

So to connect the sources of information with this opportunity, I believe, is the next frontier. Offset will be eroded. Of course people like Heidelberg won't sit around and watch that happen, so they have to have their own plan.

The battleground as I see it is between combining variability with lots of productivity vs. variable printers becoming effective in terms of print quality that matches the traditional press. Companies need to develop their identity and offerings in the information age. How do you flex your current strengths into value added solutions without getting lost along the way? What that really means in relation to the products you offer is, I think a very complex conversation.

This emphasis on "solutions" looks much broader than just printing or reprographics. I wonder if this emphasis on solutions is signaling a move by some of the big vendors away from a major focus on digital printers. For a company making a continued commitment to reprographics, it looks like moving up-market is the way to go.

There are many people who will disagree with that. I'm just saying that is my personal opinion on what is next for Xerox. Xerox has the technologies for such an assault. I see an enormous opportunity to move the concepts up-market.

But again, people have questioned whether the graphic arts industry is adaptable enough to move to these more sophisticated technologies.

My answer to that is that the CRDs [central reprographics departments] and the data centers of the 80s have undergone enormous change as they have come under the influence of the information age forces which were helped along, in their case, by DocuTech. And it's taken them a long time but they are certainly adopting the new technologies that are available. I think the IT function is taking over the world. The data center and central reprographics are really subsets of the IT function these days and IT is trying to find out how to bring these two functions together.

It seems if they are not together now, they will be soon.

Whatever "soon" is. Just a few years ago I was a panelist at a forum with Jeffrey Moore and Benny Landa and a few others at the big OnDemand conference in New York City. Jeffrey asked the audience, which was filled with printers, how many were in the "tornado" of print-on-demand? And, as alert as I feel I am to trends, I was amazed by the response. In his book, he wrote about the early adapters, and then the others being sucked in by a tornado effect. Anyway, the response was less than ten percent and this was nine years into the print-on-demand industry implementation!

Today, Print-on-Demand is a market reality, but it is only a small piece of the dynamics of the information age. The convergence of computation, communication, and information consumption-based technologies is the fuel creating new information age market opportunities. What are the required innovative rhythms? What product opportunities will be seized by companies participating in these markets? These are the strategic questions of the next decade.

PRODUCTION PRINTING AND DIGITAL PRESSES

First, what are we talking about? Definitions in this area can be a bit murky. "Production printing" is sometimes used to cover all high speed page printing, but the notion of "high speed" has evolved over the years. As of 1993 page printers in this category were those rated at from around 60 ppm up to 300 ppm. The Delphax ImageFast 850, according to one tabulation, was an exception, holding the speed record at 850 ppm. Since that time, with twin configurations, several systems appeared which deliver speeds of 700 to well over 1000 ppm.

More specifically, production printing normally means high speed monochrome printing of continuous forms. High speed cut sheet printing, such as the DocuTech, falls into the on-demand category. And high speed *color* printing systems, sheetfed or continuous, tend to be termed digital presses. The line between on-demand printers and digital presses is vague and seems to have to do in part with performance (digital presses are faster, handle larger page sizes, print color, and are more expensive) and in part with the target market.

Production printing with continuous forms is generally business printing from a host computer, an upward evolution of line printing in the corporate data center. An optional feature may be a second, highlight color. On-demand applications tend to be the realm of the quick printer or in-house repro center. And digital presses tend to be marketed (perhaps mistakenly) primarily to commercial printers.

What about variable printing? This defining attribute of digital printing may or may not be relevant, depending on the application. At the low end, digital color printers may compete with color copiers; at the high end, with commercial, offset printing. In both these cases, variable printing capability is not needed, and digital printers will be competing on the basis of cost, turnaround time, and image quality.

Production and On-Demand Printing. During most of the decade this has turned out to be a somewhat slow market, at least compared with the booming small printer scene. A relatively stable circle of vendors fielded equipment at this level. Lists early in the decade generally included eight companies: Dataproducts, Delphax, Kodak, HP, IBM/Pennant, Nipson, Siemens-Nixdorf, and Xerox. Some of these based one or more of their systems on print engines from outside, with Hitachi Koki understood to be the major supplier.

Brand loyalty was high and new sales heavily weighted toward churning, i.e. upgrading existing customers to higher performance systems. Nevertheless, later in the decade there were some newcomers. Océ and Xeikon moved into this market with acquisitions of the Siemens-Nixdorf printer operations and Nipson respectively. Heidelberg launched the DigiSource 9110 in 1999, a 110 ppm on-demand system said to be the first serious competitor to the Xerox DocuTech. Among Heidelberg's OEMs for this engine were Canon, IBM, and Danka.

Performance advances were incremental during the decade. Both speeds and print quality improved with more systems by 1998 offering 600 dpi rather than 300. On-line post-processing advanced, with vendors adding collating and bookbinding capabilities, among them IBM, Océ, Xeikon/Nipson, and Scitex. Also significant were steps taken by the vendors to offer broader solutions such as ways to integrate the systems into their customers' digital workflow.

In 1998 Océ introduced their SRA/2 controller architecture which offers multiple resolution printing to ease the processing of "legacy" applications. The new controller was first offered with the twin-engine PageStream 1060 Plus, which also has pinless form feed as a standard feature. A number of vendors introduced

new software to facilitate print management, among them IBM, Océ and Xerox. Océ's 1998 offering was DMDA (Document Management and Delivery Architecture), software that lets the user assemble, view, and consolidate data from various sources to feed their printers. DigiPath is front-end software developed by Xerox for commercial print shops that lets customers submit and manage print jobs remotely.

Translating variable data theory to practice proved complex, especially for color because of the huge volume of data. However, successful implementations proliferated with the introduction of advanced software and RIPs. Preprocessing and storing the variable images was the common approach. An alliance between Xeikon and Varis produced VariScript, a system which allowed users to compose and RIP at print time.

Growth toward the end of the decade was more in copy volume than equipment sales. Océ claimed growth in both unit shipments and print volume, apparently at the expense of IBM and

After making the market and a decade of domination, the first real competition to the DocuTech appeared in 1999 in the form of this printer, the Heidelberg DigiSource 9110. (Photo courtesy of Heidelberg USA, Inc.)

Xerox. They credited their performance at least in part to their vision of digital printing within a wider context. Manfred Wiedemer of Océ Printing Systems put it this way:

"Our recent story is a shift from a host-connected, PCM approach to standalone server and network-based systems addressing application-specific solutions. With Internet-based commerce gaining importance, one can't view the printing industry in isolation any longer. Ultimately, there will be no demarcation lines between printing, publishing and the Internet. The core demand in today's market is for the integration of digital workflows, customer databases, distribute-and-print concepts, single-source publishing, and electronic media. We and no doubt other digital print providers are preparing for this."

Digital Presses. As noted by Chip Holt earlier, you can be too late to market, but also too early. The digital press vendors learned this the hard way. They were too early, fielding products with high hopes, only to find their customers were not able to make money with their new capabilities. The customers — early installers of digital presses — said they weren't getting the volume because their customers weren't bringing in applications suitable for digital presses.

As the I T Strategies consultants saw this segment in its mid-1990s doldrums, part of the problem was that the digital press vendors were too focused on their existing customer base. "Looking for growth," Mark Hanley noted at the time, "they ask their *existing* customers, which is logical, except that it doesn't work. They have to ask people who are *not* their current customers."

The primary market focus was on short run commercial printers, quick printers and copy shops. In these establishments, users were not taking advantage of the unique attributes of their new digital machines. As voiced by Hanley, "a dark secret people are not willing to admit, is that at all levels most color printers are being used as little more than high speed color copiers."

In this sub-industry one finds a variety of players. First, the "pure" digital press vendors. Here there are just the two established pioneers – Indigo and Xeikon. Xeikon, whose OEM customers include IBM, Xerox, and Agfa (Agfa is also a major investor in Xeikon), is believed to lead somewhat in unit placements. They announced their first digital press, the DCP-1 in 1993, went public in 1996, and as of 1998 claimed 1,448 shipments. Looking ahead, their main challenge is that compared with Xerox, IBM and HP, they are a small company with resources which may be too limited to get to the next level.

Indigo has been more of a loner, both in terms of technology and marketing, at least until recently. The company's family of presses all use their proprietary ElectroInk liquid toner. Liquid ink, in a way, seems appropriate for anything that really wants to call itself a "press." Indigo launched its first digital press in 1993, went public the following year, and began an aggressive direct marketing program rather than seeking out OEM alliances or other partnerships. Their stock, riding on high hopes, was trading at $80 per share at one point. Some seasoned high-level executives were attracted from Xerox, including Frank Steenburgh and

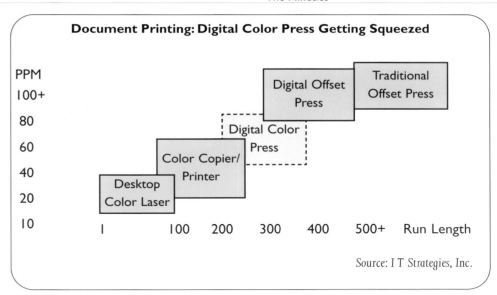

Document Printing: Digital Color Press Getting Squeezed

Source: I T Strategies, Inc.

Wayland Hicks. But when sales were only about 10% of what had been projected, Indigo stock sank to just $4 at one point and the new Xerox execs departed after only about a year.

As of this writing, things are looking up. Indigo has reversed its loner posture and has partnered up for development and/or marketing with a number of companies including A. B. Dick, Gal-

Market Positioning: Color Copiers, Digital Printers, Digital Presses and Conventional Offset Printing.

Termed their "flagship product," Indigo's TurboStream, introduced in 1998, is a six-color press rated at 2000 pages per hour (two-up). Optional features include electronic collation, high definition imaging, and personalization. (Photo courtesy of Indigo America)
Xeikon digital presses such as the DCP/500D offer one-pass duplexing at up to 25 A3-size sheets per minute. (Photo courtesy of Xeikon America.)

lus, Hallmark, HP, KBA Planeta and Datacard. The HP partnership is the most intriguing, although it has been a few years since it was announced and to date not too much has come out of it.

These "pure" digital presses are being squeezed from below and above (see box, page 199). Traditional presses are becoming integrated into the digital work flow thanks to both off-line computer-to-plate and on-press plate imaging. Presstek, the pioneer in "direct imaging," now has competition from Creo/Scitex and Dainippon Screen. Squeezing the digital presses from below are high-end color digital copiers and printers which are going up-market with faster speeds, better print quality, and more sophisticated back- and front-end capabilities.

A second tier of vendors with machines sometimes classed as low end digital presses are Xerox and Canon with high-end color copiers. Xeikon/Nipson, Delphax and Scitex Digital Printing with their "alternative" technologies are also sometimes seen as contenders. In May, 2000 Scitex demonstrated a full-color ink jet digital press that could churn out up to 2,000 ppm (two-up, 18-inch wide webs at 500 feet/min.) at a per page cost comparable to offset. The image quality was variable and did not approach that of offset, but was seen as adequate for some applications.

Finally, there is a group of vendors in an advanced development stage, or who have shipped new entries. Press manufacturer sponsored programs include KBA/Karat and Mitsubishi. Elcorsy with its alternative electro-coagulation technology, surfaces from time to time. NexPress, the Heidelberg-Kodak spinoff, promises a real digital press product in 2001 and Xerox has announced its Futurecolor program intended to take digital presses to a new level with print quality, speed, and running cost closing in on conventional offset.

Placed within the context of the overall digital printer market, digital presses are in their infancy. Annual unit shipments are still believed to be under 1000, and the installed base only a few thousand. Growth is modest. But compared with traditional printing,

digital has compelling attributes: quick turnaround for short run efficiency; seamless integration with the digital data flow; and variable printing capability. Its day is bound to come.

But for whom? Can the tradition-bound commercial printing industry complete the transition` to digital? Some believe today's larger commercial printers are best suited for tomorrow's totally digital workflow. After all, it is they who know the market. But others assert the typical commercial printer will never wake up to the immense opportunities, that it will take new, service-oriented vendors. As of the end of the decade, the jury was still out on this question.

PRINTERS TO THE PEOPLE

With printers for the masses, print really gets unchained. This is certainly a top 1990s story. The gist of it is back in our Records "Intermission." In 1999 a couple of vendors cracked the $50 barrier, at least for a while. Better yet, there were a number of free printer deals. Consumer marketing swelled through a growing number of channels. Joining the computer retailers were a variety of mass merchants (even Wal-Mart!), mail order, and vendor and distributor e-commerce Web sites.

Ad budgets skyrocketed and the media mix changed. In the 70s, who would have dreamed a printer company would place an ad costing hundreds of thousands of dollars on TV on Super Bowl Sunday? Why advertise printers to a consumer market that is so competitive the vendors are selling many models at or below cost?

Part of the answer is that in this age of solutions, it is not hip to get overly focused on the product. The product is just a vehicle for a service. Telephone companies don't worry about making money selling telephones. The money is in the service. In low-end printers the service is basically the supplies, especially ink.

This is not confined to consumer printers. Tektronix, with its

offer to lend customers Phaser color printers, was obviously looking for supplies revenues. Holt noted that supplies are even more important at the high end than at the low end.

Printer technology over the past twenty years has been in part influenced by strategies to lock out third party suppliers. In the 1980s generic expendables were commonplace. Independent toner companies sold dry toner for all the major xerographic products. When Canon-CX-based laser printers appeared featuring a cartridge that contained the drum as well as toner and developer, it looked like an impossible challenge to the

Well, not quite free. Consumer printers are price-sensitive and need to sell themselves. In many stores, salespeople are scarce, as was the case in this Staples outlet, May, 2000. (DRA photo)

independents. But it wasn't. Some third party vendors have been successful selling laser cartridges as well as re-inking kits for bubble jet printers. As a percentage of supplies, third party sources are

believed to be losing their already minimal share. But they still exist, which is probably good, since they impose at least a bit of pricing discipline on the OEM vendors.

Price Wars. The major recent "race to the bottom" in pricing was spearheaded by HP's Apollo subsidiary and Lexmark. During the summer of 99 these two companies both cut the effective price of their entry-level ink jet printers to $49 after rebate. Just a few months earlier Apollo had introduced its P-1200 at $79, believed to be the lowest priced ever at launch. Lexmark was close, around the same time, with its Z11 at $89. The pressure was on Lexmark which had vowed never to be undersold. Their earlier 1100 Color Jetprinter was found in some stores priced as low as $29 after rebate, apparently a move to clear inventory to make way for newer models. Canon was close behind with its entry level BJC-1000 at $49 after rebate around the same time.[*] The vendor price wars were intensifying. Competition among the merchandisers depressed street prices even further, such as the offers of a free printer if bundled with a PC purchase.

Factory rebates are seen as a flexible way for vendors to accommodate demand cycles. When the rebate offer expires, it constitutes a price hike without looking like one. Another advantage is that redemption rates are surprisingly low – according to Lexmark just 40% to 60%.[†]

INDUSTRIAL DESIGN

Design has many dimensions, among them manufacturability, ergonomics, brand recognition, and of course styling. The fickle and brutally competitive consumer marketplace that unfolded in the 90s imposed new demands on the design of printers.

In the words of one industrial designer, the first mission of design, in the consumer arena, is to get attention. With ten or

Oops! (reprinted from Spectrum newsletter, December, 1999, courtesy of I T Strategies, Inc.)

[*] *Hard Copy Observer*, Lyra Research, August, 1999 issue.
[†] Ibid.

twenty affordable printer models lined up on the store shelf, the one with distinctive design is the one most likely to get eye-share. Some view design as advertising, since the best advertising is buzz. Create a design that gets people talking.

In August, 1999 I T Strategies published a timely "virtual forum" on industrial design. Excerpts follow, reprinted with permission.*

The participants:

- Stephen Schultz, Tektronix 840 Designer Edition product manager;
- Tom Pangburn, Lexmark, industrial designer;
- Pete Mendel, Lexmark Design Manager;
- Will Fetherolf, Senior Industrial Designer, HP/Boise; and
- I T Strategies consultants Marco Boer, Mark Hanley and Patti Williams.

Design sells. In July, 1999 Tektronix announced its Phaser 840 Designer Edition Color Printer aimed at the "creatives," the graphic designers drawn toward Apple's colorful G3's and iMacs. "We took a cue from Apple," Schultz explained. "We want it to look different. We want people to be talking about it. I need people to ask me, What is solid ink? Why is solid ink different from laser?"

Getting attention, Schultz believes, is the first mission of design. If you can't get the attention of the buyer, all the features in the world are lost. In printers, designers (or the managers they answer to) have been cautious about taking chances, and the results have tended to be look-alike products. Did it take courage to commit to a strong color? Strong colors can be a gamble since color evokes an emotional response. In today's cookie-cutter world where many people are feeling ever more need to express their individuality – especially the professional creatives – it is looking like Tektronix has done its homework right and will cash in on its gamble.

Curves and Ergonomics. Lexmark is another company that has taken

design gambles which appear to have paid off and also to have triggered a trend. When the original Optra printers came onto the market with their sensuous curves in late 1994, they made waves. According to one journal (*Hard Copy Observer*), the new generation of Lexmark lasers was "arguably the best-looking laser in the industry," an impressive achievement in contrast to the design of the company's previous laser printers.

Design, Mendel of Lexmark affirms, is more than aesthetics. Another dimension is what he has termed "intuitive ergonomics." This means making sure the product looks like how it prints. Ease of use. "Take the shape of a button, for example. It needs to hint at its intended function. We're also guided by the imperative for rugged, confidence-inspiring design. Our products are strong and reliable. They should look that way too. On the display shelves the shopper will walk up to it, notice how it is obviously strong and durable, and then will look at the price tag and won't believe it costs so little. It's the first impression, as the cliché goes, that sells. You can have the best technology in the world but if its ugly, it's a strike against you. And we also want the brand image to be obvious. That's another first impression we're looking for."

Good design actually costs less," Mendel asserts. "We look at our manufacturing process. In the case of our Model 2030, which was our first IDEA Gold award, Tom looked into all the potential manufacturing problems and designed them out. You need to know the technology and the manufacturing process.

"Then there is relating the shape to the materials. Plastic is a great material since it can be shaped so many ways. With our original 'sculptured' product, the Optra, we took a leap to capitalize on our design goals, to establish clear-cut brand image for Lexmark, and to build a series of printers that look like what they do. We elected to push the envelope with our curves, not only for brand image and having the product look like what it does, we wanted to give it a more humanistic feel. Interacting with this design is more humanis-

* SPECTRUM Newsletter, IT Strategies, Inc., August 1999 Issue

tic, it's like holding a . . . well, a hand, or whatever; it's not like holding a brick. It's a product that is fun and interesting to look at."

Lexmark, at least for now, has not fielded adventuresome colors. The designers agree since the PC world tends to be beige, it makes sense for the printers to be a compatible neutral color. Also, they consider the economics of both manufacture and distribution. In the factory, with multiple colors, there needs to be multiple plastic molding tools, or an added purge step between runs to clean the tools. Distributors would have an added level of model variations to stock. And in the final analysis, Lexmark is responding to what their customers ask for.

Pangburn thinks in terms of shapes. "It's hard to say what the future trends will be. Products will not evolve into jelly beans. I expect we'll be getting sweeping curves, but also crisp edges. Curves are interesting and humanistic. But there is power in crispness." Their ongoing goal, in short, is to create products that are not intimidating, that are fun to use and look at."

Curves have more than emotional or aesthetic justification,

Boer observes. This trend has functional origins as well. "It takes a while for people to learn from bad design. When printers were flat, they were a perfect spot for people to park their coffee cups, so there were accidents. With curves, you can't do that. It's good function. It took a long time for designers to move the power switch from the back, where it's hard to find, to the front. A lot of other ergonomic improvements could still be made. Operator manuals are never around when you need them. Why not a rack on the back of

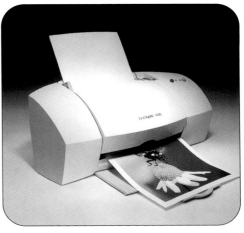

1.

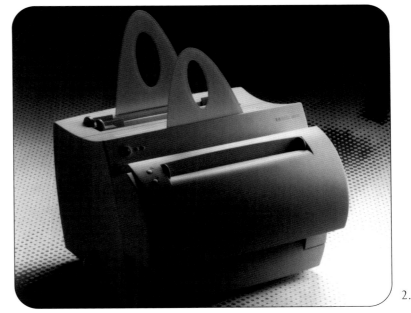

2.

3.

Here are three recent printers which have been lauded for outstanding design.
1. The Lexmark 3200 Color Jetprinter which won an IDSA bronze in 1999.
2. HP's Fetherolf, collaborating with the design house ZIBA, won an IDEA silver in 1999 for the HP 1100 MFP. The modular printer/scanner/copier was cited, in part, for its small size, low cost, concave surfaces, and "whimsical base." The scanner module, on the front, is removable. The IDEA award program is sponsored each year jointly by Business Week magazine and the Industrial Designers Society of America.
3. The Tektronix 840 Designer Edition, 1999.
(Photos courtesy of Lexmark, HP, and Tektronix, respectively)

the printer for manuals and diskettes? If 3-hole punched paper needs to be fed in a specific way, why not put some indicia on the printer to guide the user?"

The Lexmark design group, besides working on current products, experiments with more visionary designs. As in the automobile industry, they invest in more radical, futuristic modeling. This program was until recently confidential. But we're told lately they have gone public with some examples which provide an interesting glimpse into at least some designers' views of what we might be seeing in the longer range future, which in the printer industry will be just a few years.

Future concept designs created by Lexmark. (Courtesy of Lexmark, Inc.)

"Products will not evolve into jelly beans. Curves are interesting and humanistic. But there is power in crispness."
-Tom Pangburn

Commitment from the Top. In this industry HP's Fetherolf sees Apple as clearly the design leader. Their iMac color assortment makes good business sense in that dealers needed to buy five at a time. With the iMac, blue is turning out to be the most popular. Apple capitalizes on design because that function comes down almost from the top, from the level of a VP. In a 1998 release announcing an Apple flat-panel display, Interim CEO Steve Jobs states, "The Studio Display's new industrial design, high quality image, powerful features, and aggressive price demonstrate Apple's return to the fore of product innovation." Steve Jobs lists design first — ahead of function, features, and price!

HP, with their traditional "putty," has been conservative with color. According to Fetherolf, they have researched the psychology of color and tried out various colors with focus groups. This has evoked some positive feedback. But the color issue is complex. "Darker colors are more costly," he observes. "They are more costly in manufacturing and inventory control and distribution. Also, you have to be careful in a global market. Take white. In the USA white is associated with purity and virginity. But in Japan it's the color of death."

In styling Fetherolf's key theme is "truth to materials." "I see plastic, for instance, as a plus. Plastic should look like plastic. You can make it into any shape. But you can go too far. Our design is not as 'soft' as some printers, although we do have compound surfaces radiused in two directions."

Discipline and Balance. The freedom of form that plastic offers needs to be tempered with discipline, which seems to be achieved at HP for both aesthetic and business reasons. One company, however, seems to have gone a bit further with this freedom. The bargain-basement P-1200 by HP subsidiary Apollo features a compound concave/convex profile that can be literally vertigo inducing. Here is a case when styling is perhaps being used purely for product differentiation and to compensate for generic, off-the-shelf ink jet innards.

The IT Strategies consultants argue for balance. "I don't really care if my printer looks curvy and friendly," Mark Hanley confides. "I just want it to print! I'm a lot more interested in what the printer can do for me than whether it looks stylish. I think a lot of people like plain vanilla. The biggest constituency is not afraid of boredom. With a design statement, you are making an emotional appeal, not a functional appeal and this can turn some people off."

Design needs to be seen as an element of customization to a

specific market, Hanley believes, noting there are some markets which are 80% spec and 20% styling and others which are the other way around. Design in itself creates a specific, vertical market. It can also turn off the rest of the market, so it becomes very important to get it right.

The design bottom line, then, looks something like this –

First, be careful. Appreciate its power and its complexity. Consider its relationship to market segmentation and to matching product and market, which means knowing the market at least as well as the product. To echo design guru Charles Eames, it boils down to identifying, defining, and addressing the real *need* of the targeted market. Design may change but good design principles do not.

APPLICATIONS

At the outset, the announced scope of this history was defined as the general-purpose and other page-size printers which we are familiar with in business, home, and as print consumers. Turning now to applications, it is time to at least acknowledge the vast "other world" of digital printing (see box, page 206).

Secondly, looking at the 90s, it needs to be noted that resourceful matching of technology to needs – and especially advances in inks and ink-substrate combinations – has birthed an incredible diversity of new, creative, even unexpected applications. It seems appropriate to bring down the curtain on this decade with a sampler of a few such applications and a gallery of wide format and other photogenic applications.

The following assortment of vignettes illustrates how deeply digital printing has permeated our world and hints at the potential of print unchained for fun, profit, good, and yes, for evil.

3-D Prints and Transparencies. The StereoJet process is a 3-D technology developed by The Rowland Institute of Science, a Cambridge, MA non-profit research center established by E. H. Land, the founder of Polaroid and inventor of one-step photography. A conventional ink jet printer is used to produce high-quality, full color digital stereoscopic prints and transparencies. The process has been licensed to labs in San Francisco, Toronto, London, and Maynard, MA. The system uses dichroic inks and proprietary coated media. The left and right eye images are printed onto opposite surfaces of the sheet with the polarizing axes oriented at 90° to one another and 45° to the edge of the sheet. To experience the result, polarizing 3-D glasses are worn.

Map-making software opens the door to do-it-yourself, professional-quality maps. (Reprinted courtesy of Map Maker Ltd.)

Who needs it? Three dimensional simulation is helpful to all sorts of people from architects to automobile designers to researchers doing molecular modeling. In the film industry 3-D images can replace hand-drawn storyboards to preview scenes. The images aren't cheap. But as an easily portable display or presentation in place of a $5,000 architectural model, it is a bargain.

Do-It-Yourself Maps. Mapmaking is complex and there are a lot of graphic software packages for professional mapmakers. Inexpensive, color digital printers with the right software have opened mapmaking to almost anyone. One source of easily learned mapmaking software that complements general-purpose printers is Map Maker Ltd. in Scotland. They say their application was origi-

nally designed for projects in developing countries where low cost and ease of use was essential. Professionals in fields such as public health, archeology, geology, and agriculture have found this capability essential for their work. It runs on Windows, occupies just 5MBs, and is therefore appropriate for laptops.

Essentially, the software allows users to download and modify existing maps, link maps to data, and provides powerful drawing and editing tools. Printing can be on page-size sheets which are

The Other World of Digital Printing

Many non-document applications have been with us since the beginning. Early applications such as listers for cash registers and industrial data logging, labels and bar code printing are as vital as ever. Now there are vast new frontiers in other areas such as industrial printing and textiles.

In the 90s, mainstream digital printing continued to grow, albeit at a slower pace. The paperless office was a phantom haunting the industry for decades, but paper volume continued to grow. Now trends are emerging that point to – if not a paperless world – at least a less-paper world. It is this reality that leads the consultants at I T Strategies to become almost messianic about other relatively neglected markets.

"Turn around, look out the window at this huge, fertile land that nobody has occupied!" That was the rallying cry of Mark Hanley at an I T Strategies in-house roundtable on industrial printing late in 1999. There is much attention to growth potential in applications now the domain of business offset and commercial printers. But the industrial market is even larger and has tended to be ignored by the major digital printer vendors.

This other world is for other books. For the moment, it will have to suffice to note that it exists and list some major components.

• Industrial Printing (defined as the applications in which the value of the printing is secondary to the value of whatever the printing is on).
Examples:
- labels and packaging;
- product coding (printing on the product itself);
- manufacturing;
- wall coverings, decorative laminates;
- textiles.
• Industrial Processes such as printed circuits, electro-etching, solid modeling.
• Digital Photography.
• Graphic Arts Proofing.
• Wide and Grand Format Printing:
- outdoor advertising;
- point of sale advertising;
- art prints and restoration;
- CAD/CAM, other plotting, mapmaking;
• Data logging and other lister applications;
• Receipt printers in retailing, ATMs, and other terminals.

then taped together. The e-mail address is info@mapmaker.com.

On-Demand Books. Two guys in college in Virginia asked a logical question: "Why not sell the book first then distribute and produce it?" They decided to go into business together, forming Sprout, Inc., based in Atlanta, GA, in 1997. To date they have captured the attention of the press, one angle being that the system could save local bookstores. The idea is for the bookstore to have Sprout equipment that will let customers browse titles from the Sprout database and view content on-screen. If the customer wants a book, in-store equipment can print out and bind the book on a while-you-wait basis.

Is it happening yet? Sprout is no doubt not alone with the concept, but they may be the farthest along. Co-founder Pat Brannan early in 2000 said they were in the prototype stage. What kind of printers? "We're printer agnostics," Brannon noted. He said they are working with HP but also other vendors. One configuration is printing pages on an HP 8100, the cover on an HP 8500 color printer, and then putting the book together with a desktop binding machine. The whole thing, Brannan said, would take a maximum of fifteen minutes.

Hardware and systems for digital book printing are multiplying. Among the vendors offering short run or while-you-wait systems include Xerox, Océ Printing Systems, Xeikon/Nipson, IBM, and Scitex Digital Printing.

On-Line Postage. It's been a long time in coming but as of this writing it looks like it's happening. The system provider (not the USPS) markets the software for purchasing postage and "Information-Based Indicia (IBI)" via the Web. This means almost any general purpose printer can become in effect a postage meter, the first major expansion of postage indicia since postage meters originated in 1920s. The U.S. Postal Service began working on the system in 1995, and since then several vendors have been

jockeying for a position in this potentially huge market. These include two dot-com companies, E-Stamp and Stamps.com, plus the established postage equipment firms Pitney-Bowes and Neopost. During the late 90s several pilot programs were authorized and in November, 1999, Stamps.com advertised the service with a free, one-month introductory offer that included $25 in free postage.

Cephalometric Medical Imaging. Color ink jet and laser printers are being used as part of a wide variety of medical imaging applications. Image capture, manipulation and printing packages are now being used successfully by orthodontists. Two vendors have been American Orthodontics and Orthodontic Processing. Essentially, the system lets the provider and the patient get a look at the problem and probable outcome with varying treatments. The patient's head and jaw profile is scanned into the system and X-ray images overlaid. Then the images can be modified to show how, if all goes well, the profile and bite will look after treatment. As of a few years ago, a variety of color printers were being offered by Orthodontic Processing, among them the Epson Stylus Pro, Tektronix 440 Phaser and the Apple LaserWriter 660.

Body Decorating. Skin. Now there's a demanding substrate: not very porous, variable moisture and temperature, stretchable and full of creases. Palladium Interactive of San Rafael, CA rose to the challenge and in 1998 introduced "Tattoo Time," an ink jet printing application and transfer material developed by Avery. The Palladium CD-ROM contained the application, transfer sheets, and an assortment of clip art images. Original text or imported images could also be printed. There was a boy's version and a girl's version. According to the adventuresome editor of *Color Business Report*, applying the tattoo was easy, the image was not water resistant (and was not specified to be) and the transfer adhesive was more

water resistant than he would have liked.[*] As of late 1998 the kits were being marketed at around $20.00 at CompUSA, Wal-Mart, and Kmart.

Pastry Decorating. It can be your logo, your picture, or text on cakes, cookies, or other edibles. Sweet Art Company of Olathe, KS has been marketing an ink jet system for pastry decorating for a couple of years. Tom Hall at the company explained the technology. The key patent, he said, is how to remanufacture the HP ink jet cartridge to handle USDA grade food coloring. Airbrush systems have been around for a long time for this application, Hall said, but his inkjet system is a lot less expensive. Bakers using the system say an ink jet printed cake fetches a premium of from $6 to $15 per cake with an 8 x 10 color image printed at an ink (or rather, dye) cartridge cost of $.50 to $1.70.

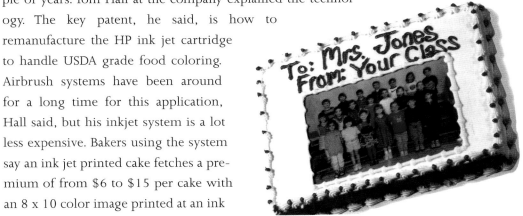

An edible birthday greeting. (Courtesy of Sweet Art Co.)

The high-end systems, which can cost over $10,000, print directly on the pastry and include a computer, scanner, monitor, software and printer. More recently a sub-$1,000 system was introduced that prints color images on an "edible icing sheet" rather than directly on the cake.

Sweet Art supplied a selection of rave reviews from customers. My favorite: "I've been in the baking business fifty years. This is the best new product I've seen in at least thirty years." This is attributed to Dan Jones, Executive Baker, the Pentagon.

Mass Customized Printed Apparel. Every body has unique metrics and

[*] *Color Business Report,* November, 1998 issue.

every mind has individual aesthetic preferences. Imagine this: For a new outfit, you take off some layers and go into a booth where a two-dimensional grating pattern is projected onto your body and the contours captured by a sensing camera. The scan data is processed into a 3-D virtual image which becomes input to an agile manufacturing environment where clothing is tailored for a perfect fit and digitally printed to your spec. As of the end of 1999, pilot systems had already been developed and placed by [TC]², the apparel industry R&D and education center in North Carolina. This system, they say, "leverages information technology to integrate the production efficiencies of mass production with the individuality of the craft era."

Counterfeiting. Every leap forward in technology has an up side and a down side. This brings us to the evil: counterfeiting, which is a form of theft. First it was color copiers, and more recently inexpensive ink jet printers with their amazing print quality that have become an affordable tool for counterfeiters of currency, credit cards, and other money-linked documents.

"Sure, I can do it, almost anyone can do it. I played around with my scanner and ink jet and got pretty good $20 bills. It was a blast! I'd never think of trying to pass them, but I do wonder if they would pass."

That was the response of Compu-Doc, my computer consultant, a while ago when I mentioned counterfeiting. Over the past decade a lot of adventuresome techies have gone that next step and tried to pass their home-printed currency. According to the Boston office of the U.S. Treasury Secret Service, tracing counterfeit bills back to the source more and more often leads toward teenagers. "The youngest ones we've dealt with are 13 or 14," noted special agent Timothy P. O'Connor a while back. Some work at home and others, surprisingly, produce their phony bills on school computers.

In the past, most counterfeit was produced by skilled professionals from laboriously crafted printing plates. As with normal commercial printing, there was economy in long runs. Counterfeit seizures were less common than today, but the batches much larger. Now it is getting more decentralized. "In the past," O'Connor said, "the passers of bills were like mosquitoes. Now the printers of the bills are like mosquitoes."*

A source at the U.S. Secret Service in Washington, DC was clear about the new wave of counterfeiting ushered in by digital printers and copiers. "Five years ago only about one percent of the volume was computer-printed counterfeit," he noted. "Today it is 46%. The number of cases has gone up, but the size of our seizures is much lower. Even the old pros, the drug and organized crime gangs, have switched from printing plates to the computer and inkjet printers. The quality has gone down and so has the passing rate. The percentage of counterfeit being circulated is still pretty low. We have about $500 billion of currency in circulation at any one time, and we estimate last year only about $139 million of this was counterfeit."

The new U.S. bills in denominations of $20 and up have a number of features intended to foil low-cost copiers and printers. Duplicate serial numbers is an obvious clue. The words, "United States of America" are slightly raised. There is an embedded line of fine print ("USA Twenty") that is virtually invisible except by holding the bill up to the light. Digital printers lose fine detail. Counterfeit bills generally lack this feature (i.e. "eyes are dull, lifeless"). And there are no doubt other, more mysterious safeguards.

Yet another print quality challenge for the digital printer industry.

* "Home Computers a New Tool for Teenage Counterfeiters," *The Boston Globe,* March 24, 1999

AN APPLICATIONS GALLERY

Medical Imaging. Art, computer graphics, and digital printing combine for medical imaging used in research, teaching, diagnostics, and litigation. This knee is an example, from Medical Renderings & Images, a service that creates high quality case-specific exhibits for use in trial and mediation. Many of these illustrations are hand drawn, then scanned into a computer. The artist uses Adobe Photoshop, and Adobe Illustrator. The images are then written to CD-ROM and printed out digitally (generally 30" x 40") and mounted for demonstration at trial. (Images courtesy of Medical Renderings & Images, www.medicalrenderings.com).

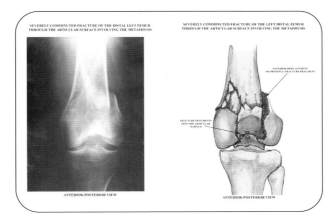

Outdoor Advertising. At intermission time, we saw what might have been the world's largest billboard, the Shanghai Coke spread that was digitally printed on strips of Ultraflex. This Anne Klein billboard, is another example (Courtesy of Ultraflex, Inc., Rockaway, NJ).

Avery Dennison supplied the material for this spectacular, psychedelic bus wrap. Bus wrap is a system, in this case consisting of the following Avery products: DOL 4100 Perforated Window-Film Overlaminate; ETM 4001 Perforated Window Film; ETM 4201 Bus Wrap Vinyl Film; and DOL 1000 Glass Clear Cast Vinyl. (Courtesy of Avery Dennison Graphics Materials Division, Painesville, OH.).

Vehicle wrap makes any truck or car an eye-catching billboard. (Courtesy of Castle Graphics Ltd., Concord, MA)

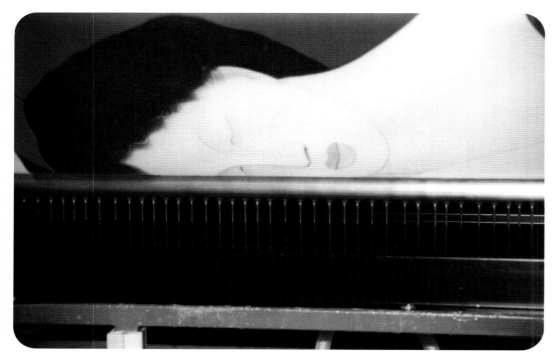

Indoor Advertising.
New media expand the possibilities for indoor advertising. This 3M FloorMinder is digitally printed on marking film and protected with an overlaminate, designed for applying graphics to high impact environments including supermarket floors and counter tops. (Courtesy of Castle Graphics Ltd.)

Wet Laminating. Weather and UV protection for outdoor digital prints are generally achieved by either hard or liquid laminating. In hard laminating a transparent laminating material with an adhesive coating is bonded to the print. The cost per square foot can range from 9 cents to $1.00 depending on the degree of protection needed. Wet laminating is the future, according to David King at Castle Graphics. It lasts longer (up to five years outdoors), can cost the same or be cheaper for the same protection, and can't peel or chip.

In wet laminating the digital image is brought into contact with the bar that transfers a water-based Valspar lamination fluid, the excess is squeegeed off, and the coating heat set. In this photo, the print is shown behind the bar with the excess lamination fluid artfully dribbling down to a trough for recycling. (Courtesy of Castle Graphics Ltd.).

Textiles. Textile printing is seen as a major opportunity for digital technology. This is the new ColorSpan FabriJet XII DisplayMaker in action. The 72-inch wide ink jet machine prints full color on non-backed fabric at up to 33 square yards per hour. Applications are design sampling and short-run fabric printing for the fashion and home furnishings industries. (Courtesy of MacDermid ColorSpan, Inc., Eden Prairie, MN.)

Art Prints and Photo Restoration.
Jonathan Singer Editions of Boston, MA claims to bring together "the sensibilities of traditional fine art printmaking with the world of digital images to produce prints of unique beauty and quality." By fine art standards, this method of printmaking is also economical. By commercial digital printing standards, it looks like one of the highest mark-up specialties. Singer's published printing cost in 1998 was $.27 per square inch (for a 2 x 3 foot print, that comes to $233). ColorSpan offers a wide format printer designed specifically for fine art reproduction, the Giclée PrintMakerFA™ shown here producing the firm's popular elephant image. Below are some samples of media possibilities that can include canvas, watercolor papers, rice paper, leather, backlit film, wood veneer and other

non-absorbent materials. (Courtesy of MacDermid ColorSpan, Inc., Eden Prairie, MN.)

These examples of photo restoration and colorizing show what can be done with graphics software and a high resolution digital printer. According to Mark Whitney, a graphic artist at Anderson Photo, it's no big deal. He scans in the photo and patches it up using Photoshop. With the rubber stamp tool or just by cutting and pasting he grabs a chunk of image from elsewhere on the photo that matches the missing area and copies it where needed, blends the edges, and that's it. (Courtesy of Anderson Photo, Concord, MA).

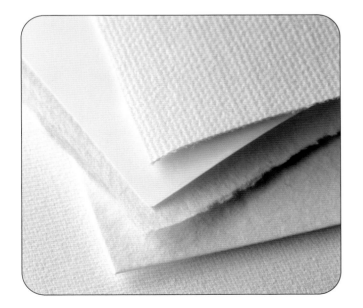

Digital Restoration (Basic)

Digital Imaging (Colorizing)

PART THREE

Coda

The digital world is upon us, reaching more and more people in an ever-smaller world, creating unimagined opportunities for print, and posing new challenges to a reconfigured digital printing industry. (Collage by PhotoDisc)

CHAPTER 8: CONCLUSIONS, THE NEXT DECADE, & BEYOND

Drawing conclusions based on the oral histories in the various decades can be tricky. Hindsight is always 20-20, and the context changes. What is right for the 70s is likely to be wrong for the 00s.

The interviews in each decade reflect a definite progression, from the entrepreneurs and inventors of the 60s and 70s, to the managers and team-builders of the 80s and 90s. Under Erwin Tomash, Dataproducts placed the company above the product, at least during its formative years. Centronics and Printronix offered interesting contrasts in the 70s. The inventor-entrepreneur Robert Howard flew high when his products and market demand were in synch, but was unable to respond to new competitive demands and became perhaps overly focused on manufacturing, which was bucking a sea change. Hackborn's HP printer business was a successful demonstra-tion of a large company able to reinvent itself and apply its trusted principles to two quite different business models to win domination of large portions of the ink jet and page printing market. Tactics changed. The principles did not. Chip Holt of Xerox comes through as the ultimate technology manager and team-builder, appropriate to the 90s.

The hypotheses listed in the Introduction are generally affirmed by this history. These and other currents have been brought into relief as "seasons" which need to be accurately read for a company to prosper. They are a context within which all players need to operate. The seasons of the future will differ. But to look ahead we have to know where we currently stand and how we got here.

The broad themes that flow through this history fall generally into four movements.

Acceleration. This might be considered in terms of technology acceptance, time to market (from invention to commercialization), and product life.

The conventional view is that everything is accelerating. This is certainly true of technology acceptance (see box). But it does not seem true for time to market in terms of commercializing an invention. It took Carlson about ten years. He and Kornei demonstrated the concept in 1938, and Haloid launched a marginally successful commercial implementation in 1948. Ten years. It was another eleven years before the incredibly successful 914 appeared. But in Chapter 7 we saw how twenty years later it took Xerox around ten years to create the DocuTech. Ink jet printing, which became commercially viable somewhat later than electrophotography, underwent a considerably longer gestation period.

Product life cycle is also commonly seen as accelerating, but this, in part, depends on how "product" is defined. The Teletype Model 33, IBM 1403 family, and Digital Equipment DECwriters are sometimes cited as examples of the longevity of the products of yore with their lifetimes of ten years or more. In contrast, Dick Hackborn talks about obsoleting his own products with a continuous roll-out of LaserJet models.

In May, 1999, HP celebrated the 15th anniversary of the birth of the LaserJet. During these fifteen years, by their count, 38 models bearing the name "LaserJet" were introduced – well over two per year! Some of these were truly new products. Others were variations on a theme. None was truly a new "invention." This roll-out of products (or at least models) is certainly not evidence that product life cycles are accelerating. It can rather be seen as an example of how product strategy is now a function of marketing as much as – if not more than – technology.

What is really accelerating is the speed at which the market window flies by. Centronics, for instance, caught one window perfectly, but missed those windows which followed.

Technology Acceptance

According to a report on the "digital economy" by the U.S. Department of Commerce, it took forty years for electric power to reach 80% of U.S. households. The Internet is cited as a remarkable comparison. "Radio was in existence 38 years before 50 million people tuned in. TV took 13 years to reach that benchmark. Sixteen years after the first PC kit came out, 50 million people were using one. Once it was opened to the general public, the Internet crossed that line in four years." [*]

Maturity, Concentration, and the Ascendance of Marketing. Over the fifty years, we have seen the number of competitors grow from just a dozen or so to almost a hundred, and then slowly shrink. More significant than the number of companies is how market share has changed. In 1980 IBM was the largest vendor, but its share in terms of units was only 28%. More than a dozen companies held significant chunks of the remaining 72%. Today HP and Xerox dominate, followed, well behind, by just four or five vendors with significant shares.

Technology has also matured and concentrated, with ink jet and electrophotography now dominating.

In a mature industry, the rules change. Competition becomes more intense. Product development takes a back seat to marketing. This, in turn, favors the entrenched, large companies that have the resources to build the channels and buy the advertising.

It follows that the odds are stacked against the small time start-up, and certainly the lone inventor or entrepreneur. As hypothesized in the Introduction, it is now teams rather than individuals, people and organizational skills rather than technology, engineering, or manufacturing skills.

[*] "The Emerging Digital Economy," Chapter 1, U. S Department of Commerce, April, 1998, http:// www. ecommerce.gov/danc1.htm.

Changing Products and Product Focus. Products have obviously undergone radical changes over the half century. Significant at this point is the way in which product lines are blurring. There is convergence among digital printers, copiers, digital presses and plotters. It is all going digital and the old product definitions don't really work anymore.

The really significant sea change, however, is that products themselves are becoming secondary to know-how, services, and solutions. The first step in this eclipse of the product was ever-growing preference for outsourcing of the marking engine. The engine has become a component that finds its way into printing systems through new channels and sometimes additional layers of distribution. And once the central product becomes a component, this means the industry loses control of its destiny. What was formally the "printer industry" continues to shrink. Or, looking forward rather than backward, it is becoming part of something much bigger.

What is this "something bigger?" At this point, it's not entirely clear, but it surely has something to do with the ascendance of the Internet and control of the great stew of information and images flowing through it. More on this below.

WHAT IT TAKES TO WIN (OR AT LEAST NOT LOSE)

Deep pockets, agility, and moving beyond a focus on inventing and manufacturing or even on products appear to be keys to success in the new millennium. That is the big picture. In addition, a number of specific lessons can also be drawn from the experience of companies covered in this history. Toward the end of 1999 I ran some of the oral histories by Boer and Hanley of IT Strategies and asked if they saw lessons that might be drawn from them. Being consultants, they had a lot of opinions, a few of which follow.

Tunnel vision. Companies need to guard against complacency or tunnel vision, against being lulled by success. Dataproducts, for example, was riding an exciting growth curve. But they needed to know when to jump off that curve and begin building a new one. Companies in fast-changing technology businesses need to reinvent themselves to maintain growth and profitability.

"They did things by the seat of their pants for a while," Boer observed. "They figured it out and then became complacent, so they didn't see the next wave coming. And that's a lesson I think we're seeing from many of these printer companies time and time again, such as Citizen and Star Micronics. They aren't around anymore."

Is the problem finding the resources to make the leap to another growth curve? "No, it's not necessarily the resources," Hanley asserted. "It's the thinking! The people who run the corporation can get so deeply involved in the day-to-day issues that they don't think outside of that. They need to look at what their next step needs to be."

Harris Corporation was cited as an outstanding example of a company which has succeeded in staying ahead of the wave (see box on page 219).

Restrictive Mission Statement. A broad mission statement helps. "Rather than build a company around a product," Boer suggests, "build it around a set of values or rules. It needs to be broader than any single product or service, such as addressing a mass market or how people communicate — a broader mission so you can jump around a bit more."

Cover-Up. Hanley talked about denial. "Not acknowledging a crisis when you are in it is a common pitfall for the public corporation. If the boss chooses to lie to support the stock, the company family is going to believe it. When you have a strong corporate cul-

ture, it's hard to play by the rules. It's been called the windbag effect." Reading a company annual report, Hanley believes, can be a clue. Hanley's Law: the more robust the verbiage, the greater the problems.

Ingrown Management. Companies are especially vulnerable to Hanley's law when management becomes too ingrown. Loyalty and promoting from within has its up side. But if you look at a company and see that key managers have not changed very much over the years, this could be symptom of inbreeding, of worrying too much about offending some one. Companies need fresh air.

Reactive Acquisition Strategy. At some point it may make sense to be acquired. But it can't be a fire sale. With good planning, by being proactive, an acquisition can work for all parties. Dataproducts, for example, did not seem to see they were in a crisis situation in time. When you are in a crisis it is hard to see it.

Any printer company with revenues under $500 million or so needs to envision a point at which acquisition would make sense, Boer suggested. "You can imagine a company such as HP or Xerox or even Lexmark, with its $3 billion in sales, taking over Tektronix," he added "Tektronix is a $750 million company, but their stock price is low right now. Maybe that would not be a bad thing. A company such as Seiko Instruments just struggled and struggled along with its printer business and in time just faded away." (Note: this was before Xerox's acquisition of the Tektronix printer business was announced.)

Conversely, there's the divestment option. Even a company as successful in printers as HP might leave the industry. "I know, that seems absurd right now," Boer admitted. "But if you fast-forward the clock and look at where the push is within these organizations, it is all e-commerce and the Internet, right? Printers are the big cash cow now, but on the laser side they are going to collide with Canon. As HP builds its color laser business it will be going up against color copiers. Canon is the color copier leader, so it means HP will be competing against their major supplier, which is not a good thing. So one option would be to buy out the Canon copier business, which is very unlikely. Or they could develop their own engine. Or fade out of the business, which would be easy for them to do very profitably over many years."

This shifts the narrative from lessons to predictions. It is time to broaden the dialogue.

THE NEXT DECADE AND BEYOND: A VIRTUAL FORUM

The future will certainly not be more of the same. In fact, with the advent of the Internet some see the turn of the millennium as a watershed, a sharper than normal breakpoint in history. Even in the normal course of events the future tends to hold surprises, pleasant and otherwise.

Group consensus should be a more likely predictor than the visions of any one person. Years ago a planning technique called The Delphi Process made use of several rounds of questions circulated by old-fashioned mail. Cycles of questions could circulate forever since people always have second thoughts and reactions to the thoughts of others. Not wanting to take forever, this forum abbreviated the process.

Most of the contributions below are the initial responses to questions emailed in May, 2000. People were asked to respond only to the questions that turned them on, whether one question or all ten. In addition to the original panel, some views of several other industry leaders were inserted into the dialogue because they fit in so nicely. Not unexpectedly, participants stressed the significance of the Net in transcending limitations in time and geography. So the "virtual" nature of this forum seems in keeping with the times.

The Participants:

- Benny Landa, Founder and CEO, Indigo N.V., Maastricht, The Netherlands
- Andreas Luebbers, Marketing Director, Printing and Publishing Business Group, Océ Printing Systems GmbH, Poing, Germany
- Frank J. Romano, Melbert B. Cary Jr. Distinguished Professor of Graphic Arts, Rochester Institute of Technology, School of Printing Management and Sciences, Rochester, NY USA
- Mike Willis, Managing Director, Pivotal Resources, Cambridge, England
- Mike Zeis, Editor and Publisher, Color Business Report, Blackstone Research Associates, Uxbridge, MA USA
- IT Strategies, Inc., Hanover, MA USA: Marco Boer, Consulting Partner; Mark Hanley, Managing Director; and Patti Williams, Consulting Partner.

Please note:

As a virtual forum, the main, initial responses of a given participant to the questions did not benefit from the responses of the others. That is why you may notice some redundancy and discontinuities in transition and tone.

Reviewing this dialogue, I found it interesting to consider how the interchanges might have differed had this been a face-to-face forum. As with everything in life, there's an up side and a down side. Without the Internet, this forum would not have happened. The virtual format also has merit in that participants can speak in their own time and from their own agendas, colored by their individual personalities. In face-to-face discussion, what is the contribution of spontaneity and interpersonal energy? To what extent does it obfuscate, and to what extent does it fuel honesty and creativity? As the world becomes more virtual, the nature of discourse will surely change.

Harris Corporation –
A Case Example of "Reinventing" the Company

Sometimes a company can evolve successfully from one core business to one or more new businesses. Harris Corporation is an impressive example. Their journey began with inventors working in the back room of a jewelry store in Niles, Ohio. Today's Harris is a multi-billion dollar electronics company based in Florida. One lesson is that financial resources can be traded for expertise. Another is timing. The Harris story covers over one hundred years.

A Harris Chronology:

1895:	Harris Automatic Press Company established
1906:	First commercially successful offset litho press shipped
1926:	Merger with Seybold Machine Company and Premier-Potter Company to become Harris-Seybold-Potter
1931:	World's first four-color offset press shipped
1944:	Harris-developed computerized bombsight used for precision bombing in World War II
1955:	Harris-Seybold becomes a public company
1957:	Merged with Intertype and began transition of type-setting from photographic to digital
1957:	Acquired Gates Radio Company, a pioneer in radio broadcast equipment
1967:	Merged with Radiation, Inc. of Melbourne, Florida, described as a leader in information handling, aero-space, and digital electronics
1978:	Harris Corporation moves headquarters from Cleveland to Brevard County, FL
1983:	Harris divests all printing equipment operations to become a pure electronics company; uses proceeds to buy Lanier Business Products.
1988:	Acquisition of GE Solid State which includes the semiconductor operations of GE, RCA and Intersil

As of 1995, its 100th anniversary, Harris claimed 27,000 employees and sales and earnings at an all time high. The revenue trajectory is impressive. It took them 70 years to reach $100 million in sales. In the next 30 years annual revenues had reached $3.4 billion.

Participants reviewed and approved the finished forum, but were encouraged to keep their inevitable second thoughts to a minimum. Also, in a fast-moving world, opinions reflect a given point in time, in this case, the first half of 2000. Since the life of this book will hopefully be years, this needs to be viewed somewhat as a time capsule. Were the questions to be asked in 2001, many responses might well be quite different.

Technologies

To start, let's think about technology. I'm wondering if you see any alternatives to ink jet and electrophotography that hold promise of displacing these dominant digital hard copy technologies in the next decade?

WILLIS: No, mostly it will be evolution, not revolution. Most of the technology shake-out was back in the 1980s. The established market players already have technology and are unlikely to jump to something radically new, and it is questionable how far new players or start-ups can get in what is now mainly a mature market.

Willis

WILLIAMS: Printing technology in general is divided into analog and digital. Of today's digital printing technologies, ink jet and electrophotography will take their place alongside analog printing technologies – offset, screen, flexo, gravure and letterpress. Many R&D dollars have been invested – and continue to be invested – in ink jet and electrophotography. These technologies are not mature but they have proved themselves over years of use. Alternative digital technologies are certainly on the horizon; however they will also require time to "prove" themselves.

Between ink jet and electrophotography, ink jet is special because it is a non-contact technology which gives it great flexibility in terms of substrates and shapes that can be imaged. In analog

Williams

printing, screen printing has been known for it's flexibility in printing on a wide variety of substrates. Ink jet printing is the digital equivalent to screen.

I would take issue, by the way, with the words "hard copy" in the question. Those modifiers would have been true in the days when digital printing was primarily printing documents or "copy." Today digital printing is used for a wide variety of output from signs to fabric to carpeting.

HANLEY: Ink jet and electrophotography look like the only foreseeable technologies at this time. Electronic paper will have limited specialist applications. Nothing foreseeable will replace the usability paradigm of paper.

BOER: There are about twenty other types of novel print technologies being nurtured under development, many derivatives of ink jet technology. They show great promise and will in most instances continue to show great promise for years to come. As a general rule, perhaps one of the twenty will succeed to become as popular as electrophotographic and ink jet technology. However, in the near-term (five years) I foresee no other technology replacing either ink jet or electrophotography.

Boer

ROMANO: Electrocoagualation, phase-change solid ink, and technologies that deposit a pigmented substance directly on a substrate are potential competitors. But don't hold your breath — toner and inkjet will dominate hard copy reproduction for the next 20 years.

It might be interesting to get less abstract. Suppose God said to you, "You have to invest all your life savings in just ONE digital printing technology and hold that investment for ten years." Which technology would you pick and why?

BOER: I'd invest in piezo ink jet technology from a well-established Japanese manufacturer with a proven track record in piezo head manufacturing — perhaps Epson, Brother, or Konica. Piezo ink jet technology shows the most promise in jetting the widest range of inks needed for industrial applications, and for high-volume applications is likely to be more economical than thermal ink jet.

WILLIS: Well, life savings don't amount to much with 3 children either at college or getting close, but what's left would be invested in ink jet. I believe that ink jet still has a long way to go in office environments, but there's more than the office, there are the new industrial applications like packaging, textiles, laminates, flooring and so on.

Romano

ROMANO: We like the idea of an ink made from ozone that will deposit it on paper instead of in the atmosphere. We would invest in anything that is consumed and needs to be bought again — like childrens' shoes or ink jet ink.

ZEIS: I see this choice related to the displacement of offset with digital. As digital technologies become more facile, they are taking on more and more work that used to be offset (analog). The print volumes are huge, so the supplies opportunity is huge. Presstek and Indigo and Heidelberg and Scitex Digital Imaging are among the present participants whose products are digital but may be sensible additions to a commercial printer's equipment list. So which technology is that? If I had to pick between Scitex (ink jet) and Indigo (electrophotography), I'd go with ink jet, because of its innate simplicity (squirt ink, to oversimplify) and because colorant is used or consumed only on those portions of a page that are to be imaged. I see the opportunity in the next ten years in ink jet technology that will displace commercial offset.

Zeis

HANLEY: Rather than get focused on technology I'd rather think about companies. Technology is just a vehicle. The company is the payload. Xerox, I predict, will be able to re-invent itself as a color print systems integrator at very high volumes. But if I had to invest all my savings in one company, it might be HP. They are smarter about the nature of business communications and the role of the electronic image than any other vendor. A company like HP — in fact maybe primarily HP — is the vendor best positioned to make the transition.

WILLIAMS: I'd find an application and then see what technology is best suited for it. If I had to make this investment tomorrow it would be in wide format thermal ink jet systems found in what most print providers today are doing. I think there is huge untapped demand in segments served by this group of printers for digitally printed products if a way could be found to more easily bring together the potential buyers with the printers. If a way could be found to easily create, design, and produce digital products it would open the market up to the next lower tier of less computer-literate buyers. For example, I recently gave a talk on digital printing to the Na-

Hanley

tional Dog Groomer's Association. Not a very computer-literate group, but they have access to million dollar images of pets. Consumers today spend incredible amounts of money on their pets. Seniors in high school are having their senior pictures taken with their family pet! What would these consumers spend to have a picture of Fido on a tie (What a great Fathers Day gift!!!) or a scarf (What a great Christmas gift!). These products would sell. The problem in the market today is that there is no system in place to facilitate the process. If such a system were in place, the rewards would be huge.

Application Areas

Technology for sure interrelates with what I'd call application areas or markets. What application areas will present the best opportunities for profitable growth over the next ten years? Commercial printing with traditional technologies is still a lot bigger than digital printing. Will that always be so?

LANDA: Offset printing didn't by chance become the dominant technology worldwide. It's a brilliant process, and we at Indigo are progressively, one step at a time, borrowing from this one hundred-year-old technology. The first thing we borrowed was the use of printing ink, the second was the use of an offset blanket and the third was a refined inking roller system.

But we are only just beginning and you can expect to see other benefits that have evolved in offset to also be embraced by Indigo and other manufacturers as digital printing progressively moves into a dominant position. Remember that offset printing started off as a niche process for printing posters and sheet music at a time when letterpress dominated these and every product category from books to newspapers.

Landa

Now the battle will be between digital offset color printing [Indigo technology] and mechanical offset color printing. And mechanical offset color printing is fighting back, just as all mechanical technologies have resisted digitization. It was inevitable. The way they are doing it is with hybrid products and transition technologies. Take for example the typewriter — we

Luebbers

had the mechanical typewriter for a century. But then, before it became an all-digital process with computers and laser printers, there was a transition period with the electric typewriter. For thirty years a hybrid technology bridged the gap.

Phototypesetters bridged the gap from hot metal to the digital creation of type. Offset printing is fighting digitization with hybrid machines — mechanical printing presses fed by CTP [computer-to-plate] systems and on-board digital plate-makers. But it is inevitable that everything that can become digital will become digital. While I'm not dismissing the role of these products, speaking strategically it's a period with a start and an end. And in this digital era I believe these transition technologies are foredoomed to a relatively short life.

HANLEY: If one starting point is traditional print, we accept that it is significant in its own right; it has created it's own culture. It won't go away. But, as image communication becomes fluid, print needs to become digital. The traditional print people have a strong card that goes beyond their technology: they know the commercial print user and this could serve them well.

Digital print people understand front end image-capture. But they still generally need to absorb what print means to business and industry.

Traditional print people, on the other hand, understand this —

but they are limited with the current technology. There is a lot they can't do such as short runs and variable printing.

We probably should talk about them as "converters." Printing is not the final step. It's not just printing. It's also cutting and folding and everything else to get the document into the form the target audience needs. They understand the ultimate user but they need to expand their sights. They need to become intermediaries. Conversion intermediaries. They have the know-how to make the product. But, if they don't make the transition to intermediaries, you might as well write them off.

In the long run we're talking image services. Offset is just one and one which does the job now across a huge spectrum of applications. But they must be ready to make a transition and many hopefully will in due time.

Most print services will be accessed on a pay-as-you-go basis by users at their desks, for fulfillment mostly out-of-house, although some big volumes will be coming back to departmental printing locations.

Distribution will be nearly entirely through business communications system houses except for very low end stuff. Business communications systems houses will be computer services vendors as well as printed media service providers.

Print does not need to shrink dramatically. It does need to become more involved in the services related to it. Some large printers are already making this transition. There's Donnelley and Quebecor. The business plans of such companies include data management, remote communications, secure communications – not necessarily presses. Companies need this broader mandate.

LUEBBERS: I agree. The printing and publishing industry needs new definitions to meet the challenges, extended business models and new marketing strategies. It also needs to adapt its organizations to complete these changes. Thus, there are opportunities for new providers who combine competence in printing technology, information technology, database management, electronic media, communication and networks – companies who can chart a course for their customers through the changes to the industry. Océ will be one of them.

ROMANO: Regarding future markets, I can be more specific. How about personalized direct marketing and one-book-at-a-time. Look for M&Ms with ads on them.

WILLIS: The big growth area will be industrial printing markets, mainly on to non-paper substrates. Regardless of the paper versus electronic debate for the office, we will still wear clothes (or at least I intend to!), buy groceries, lay carpet in our homes and hang drapes or blinds in our windows. We are talking here about printing for decoration, not the communication of information. The drawbacks of existing industrial printing technologies such as flexo and screen printing will be overcome with ink jet technology, particularly for short runs, specialties and new products and services.

To get there we need further development of piezo printhead technologies and a lot more development of the ink jet inks. This is going to require development efforts throughout the value chain, from the raw material suppliers right through to the equipment manufacturers. In the last year or so we have seen a tremendous surge in activity in this area, with the development of new inks and new machines for production printing. The obstacles seem to be mainly time and money now.

BOER: Graphic arts and Industrial printing offer the largest opportunity for digital printing. Users of print in those markets spend over 10 times the amount of money on print than they do for office and home printed documents. Near-term, decorative printing of home furnishing is most interesting, followed by packaging and label printing for consumer goods, and ultimately textile printing will be the largest market yet for digital print.

WILLIAMS: I take issue with the word "traditional." We in the industry must stop thinking of digital printing as a non-traditional printing technology. The world of print is now divided into analog printing technologies and digital printing technologies. Technologies will be chosen for the specific advantages they bring to a specific job. For example, a long run of wallpaper might be done using gravure printing. Custom runs of wallpaper might be done using ink jet printing. A company that has the capability to do both analog and digital printing will be the company in the future that will be able to offer a range of products and services that will provide added value to the customer. Therefore I predict that over the next ten years commercial printing applications will be satisfied with a mixture of analog and digital printing technologies.

HANLEY: Right now digital printer people find it is easier to focus on documents. So many people mistakenly think that's all print is about. Yes, I agree there is a much bigger market out there and that's the so-called "industrial" market.

Document printing is important. But more is done than is needed. And it is vulnerable to the impact of the Net. On the other hand, industrial printing is essential. You can't buy a can of Coke unless it is printed. It's driven by consumer economics. It is one of the highest value markets and one of the fastest changing.

And more importantly, it is a market where the limitations of the print technology they are now using are most strongly felt. The economic drivers are far greater in industrial for "print rationalized" digital printing. The industrial print people are well educated regarding the demand problem. They know the limitations of their current technologies. But conversely, they have little education about digital printing. Many are not even aware there is such a thing. But from the standpoint of the digital printer industry, here is guaranteed demand!

The digital print vendors don't have to convince that market of the need. The demand is there. Right there you have a strong argument for industrial digital print. It is a smarter place to go if you want to grow in digital print. I can't say it will happen for sure. It may happen. But technology tends to follow the dollars. It is a historical fact: Over the last few years, there has been almost no work going on adapting "print rationalization" to the needs of the industrial market.

The Internet

What about the Internet? Will it displace digital print, or print in general, over the next decade?

BOER: The Internet and digital print will complement each other for the most part in the coming decade, but in another twenty years video telecommunication will be so good that hard copy will become less critical as a daily tool to function in the modern world.

WILLIS: It will considerably reduce the amount of paper in offices, but I think the use of paper will increase in the home. Print in general? I still think newspapers, magazines and books will be paper-based for a decade or two yet, at least. They will be supplemented by electronic media, just like radio and television didn't kill off print either.

ZEIS: The Internet is already displacing print, of course. Over the next decade, the Internet can be expected to displace a lot more. Replacement of print is unlikely in the next decade, though.

Many printed documents will still exist in two forms. I have a hard copy of the MBTA's Franklin Line Commuter Rail Schedule in my desk drawer, dated May 12, 1997. I won't ever need to pick up another one. If I need to know when the trains run, I will find out on the Web. But can the MBTA stop printing schedules in the next decade? Not likely. Within 25 years? A much better chance. (I

can also learn about schedules by phone, which is a system much closer to reaching the full population. Virtually universal phone access to schedules has not stopped offset runs.)

Where the connected universe and the target market are nearly the same, the printed version is threatened by the Web. I was told that ham radio operators used to be able to buy a printed directory containing call letters and names of other ham radio operators. Not any more. The information is said to be available for nothing on the Web. That's displacement

ROMANO: Simple answer: no. But it certainly will affect the volume of print on paper. Answer the question: if the Internet never existed, would there be more print? The answer is yes. But the needs for all forms of communication will grow.

The world of information dissemination is changing. More and more of the information and entertainment we encounter is in electronic form. By 2020, we envision:

	1900	1950	2000	2020
Paper	61%	57%	49%	25%
Electronic	—	18%	36%	65%
Other	39%	25%	15%	10%

But the pie will be much bigger so 25% of the total will still be substantial.

WILLIAMS: The Internet is both an enabler and a replacement. The World of Print is divided broadly into documents, such as catalogs, product specifications, manuals; and industrial applications such as wall coverings, textiles, floor coverings, CD ROMs, packaging, etc.

In the document world, where the product is print (you don't buy a brochure with no print on it), the Internet can be a replacement by providing virtual documents available on line. The document may ultimately be printed, but in the office rather than at the commercial printing establishment that printed it in the past.

In the Industrial world, print is subservient to the final product. For example, the printed design on a scarf cannot be replaced by the Internet. In this scenario, the Internet will enable digital printing by enabling the movement of the digital image, from creation, to design, to proofing, to final product.

LANDA: Everyone is concerned about the Internet's impact on the printing industry. I believe the Internet will help promote, not hinder the growth of commercial printing. Clearly even conventional printing is going to be made easier and more efficient to manage through these powerful Internet tools. One has only to look at the new online services linking buyers and printers such as Noosh.

But the Internet won't replace printing. In fact, by enabling immensely powerful database access to vast populations and already to some two trillion documents, the Internet can facilitate huge growth in personalized, targeted printing – especially direct mail. This is the most powerful customer relationship management tool of all. So, yes, one day our children's children may live in a paperless society. But in the meantime, for all practical purposes, all printing will be digital, and that will be in our lifetimes. In fact, for most printing, in less than one decade.

You need to look at both the Internet era and the digital printing part of the Internet as an enormous revolution – the "gold rush." In the end, who really profited from the gold rush? It was not the gold diggers. It was Levi's jeans, the railroads, those who provided the tools, the capabilities for this madness to be supported.

So in the upheaval that we're facing in the Internet era, the future of publishing is targeted advertising, targeted articles, customization. Those companies that have the infrastructure to enable publishers to produce these types of publications will be the winners. I don't think it's necessarily the large companies. It's the smaller companies who are creating these trends – the entrepreneurs. They're going to be the big winners.

HANLEY: Yes, print on paper, taking advantage of the power of digital technology will remain important. But it is only a piece of the picture. Printing is not what it is really about! It used to be the only way to communicate on a large scale, but printing is not king anymore. Digital print vendors will become servant to the larger industry.

So, what is this larger industry?

It's about imaging before printing. Imaging is much bigger, it can be virtual or hard copy. I think what we have to get our minds around is the new, bigger reality, which we might call "image communications technology." Or maybe just the Internet. Printing is just a subset.

In this new reality the vendor that succeeds will have to be a composite vendor, offering computing service when you need it for support, and, finally, content which may or may not be print. The vendor of only printers will be obsolete.

I see the Internet as a monolithic communications technology which is growing exponentially while in comparison, the printer industry remains relatively flat. Businesses not connecting with the Net are suffering from an illusion.

Distribution needs to change. The places where technology will be sold is within higher level communications services. These service vendors will sell hardware as a component. Here's how I see the levels of distribution:

First, the old fashioned way which is "I make it, you buy it with my brand on it."

The newer level is to be an integrator: Take the print engine and integrate it into general or special-purpose print systems.

The newest: the printer product is embedded within the networked image communication system. It is nothing more than a component. I don't see this as diminishing its importance; it is just making it part of something bigger. These system component suppliers won't call themselves printer vendors; if they're smart, they'll call themselves information technology suppliers.

The old printer industry is dying, it's being co-opted against its will. HP is the only vendor from that old industry that I can think of that accepts this philosophy, that accepts the reality that the game is now the much bigger world of communications.

LUEBBERS: Yes, at present we still tend to view the printing industry in isolation, but ultimately there will be no demarcation lines between printing, publishing, and the Internet. Interactively-selected content, tailored to meet the needs of the customer, must be made available rapidly, anywhere in the world. Driven by the need to offer services on a global scale, international co-operations will be more significant than ever.

The core demand is for integration of digital workflows, customer databases, distribute-and-print concepts, single-source publishing, and electronic media. The technological features of the printing lines do not matter as much, even though these development cycles are also accelerating.

Digital printing is one of the answers to the challenges of the future, because there is no need to switch from one medium to another as the job passes down the workflow. Hence, digital printing will become more prevelant. We can expect to see new applications; digital printing will also play a role in volume printing, and special applications will become commercially available. Software will become more important, evolving into an integral element of the print solution. The integration of multimedia will create new output possibilities alongside print. This is where dynamic imaging technologies will come into their own.

Hard Copy vs. Display

Thinking about the impact of the Internet era on digital print, it might be interesting to get more concrete. Suppose for the rest of your life you had to be confined to EITHER hard copy or display, but not both? Which would you choose, and why?

ROMANO: Hard copy, but then I am old. Ask my grandchildren — they will pick the display, I would bet.

WILLIS: For me, it's got to be hard copy.

WILLIAMS: If I had to choose one or the other, it would be hard copy. Why? Because it's what I know and am familiar with. I grew up in a world where until just recently, there was really no alternative to paper. I'll bet you would get different responses to this question if it were posed to graduating college seniors.

BOER: This is not a fair question, we need both. I would choose hard copy for leisure reading, and electronic display for all my other informational/transaction/reference needs. As we are asked to absorb more and more information, electronic display (especially graphical images) will be the only way for us to absorb the data. Hard copy will be important as a learning tool, just as Latin remains invaluable as a foundation for learning languages.

ZEIS: I don't accept the premise. There is no reason to have to pick "EITHER." (That being said, I'd choose the display, because information from almost any source can be presented on a display, and because I can alter the appearance of information to suit my preferences. Further, print can be limiting, at least as a publishing medium. A writer writes, an editor edits, and desiger designs pages and lays out the material. Then a printer prints, usually many copies if the material is published. I don't want a novel or text book that has not benefited from what editors and designers contribute. But, considering the tough choice you are forcing me to make, maybe I don't need what the printer contributes.)

HANLEY: I would take hard copy because it imposes limits on information and choice on the final user. In our age the big goal is to filter information. In the future, this will become even more important.

But we probably need to distinguish between private life and work life. In private life the media mix will be wireless screen diary with phone attachment and a terminal with mini-printer for business and lots more books than ever on paper for pleasure. In business life, all decision documents will remain in paper and everything else instantly available everywhere wirelessly.

Thinking about hard versus soft copy, remember, hard copy serves the need to filter information. Business is getting inefficient due to information overload. Everyone is trying to do too much. Look, the human attention bandwidth is finite. Our information capacity has a limit and I think we have reached that limit already.

There are more and more vehicles for information, but there is a hierarchy and I think print will be at the top. Print is the last step in the information chain, and — after creation of the information — probably the most expensive to produce and distribute. That gives it a certain credibility. So I'd put my bet on it. You can't do without print.

Paper Prospects

That brings us to "attention scarcity." Don't human beings have a limited amount of attention to spread among all the media calling for attention? Around what year to you think demand for printing papers will peak out?

WILLIS: Sorry, what was that? Oh yes, attention span. Isn't it interesting that the amount of junk mail we get doesn't seem to have fallen even though email and the Web are now so prevalent. Of course there is plenty of junk email, but the existing junk mailers must consider email junk mail to be worthless. I think this indicates that we are going to be much more comfortable with

Printed products	Tons Paper Millions 1998	Percent	Tons Paper Millions 2020	Percent
Periodicals	2.40	4.29	1.92	4.79
Newspapers	9.80	16.50	6.80	9.87
Books	1.70	3.04	1.40	3.03
Catalogs	1.90	3.39	1.51	2.19
Direct Marketing	2.20	3.93	2.40	3.48
Directories	1.10	1.96	0.90	1.31
Financial and Legal	2.90	5.18	2.10	3.05
Packaging	20.90	33.02	36.90	40.54
Technical Doc	2.60	4.64	1.40	2.03
Advertising	4.60	8.21	8.8	11.26
Stationery	1.90	3.39	1.25	1.81
Internal-Forms	1.80	3.21	1.1	1.60
Miscellaneous	2.20	3.93	2.44	3.54
Cut Sheet/ Blank Paper	3.80	5.30	7.93	11.51
	56.00	100.00%	68.92	100.00%

Source: RIT

to peak by 2010. The Internet can replace many of our transaction printing needs, the stuff that fills our mailbox at home everyday. But printing of what we call consumer goods will never go away. You can't sell the cereal without print on the box, nor the sofa without print.

ROMANO: No, demand for paper will not peak out. Paper is a push medium, which means I can direct it to YOU. The Internet is a pull medium, which means I have to find what I want. You need both. Remember, they said that TV would kill movies.

Over the next 20 years print will change. The volume of printing will grow but it will shift to on-demand technologies. But each category of printed product will have different dynamics. I'll attach our projection (see box).

WILLIAMS: I think demand has already peaked. I say that using the term "demand" very broadly. Only a few years ago, paper was the only choice as a repository for data and information in applications such as encyclopedias, catalogs, bills. The paper model has been in place for thousands of years. There was no choice other than paper. That is no longer true. Paper is only one choice among many including CDs, video, and Web documents. Paper has gone from "the only" to "one of many" and not always the most efficient one. While volumes may continue to grow, the demand model has shifted. Today, it's not about the paper, its about the image.

Print Volume

By one estimate, the global market for digital printing equipment and supplies is now close to $100 billion. What might it be in five years? Ten years? What about the market for print?

printed media for a lot longer than a lot of pundits are saying. Printed paper is tactile, some mailing pieces even retain the odours of the printing process because they are stuffed in envelopes on-line. But one day it's going to go, and I have visions of taking my grandchildren around museums pointing out displays of junk mail and reminiscing about the lost art of tossing paper into the waste basket.

BOER: After continuing to grow for the next few years spurred on by Internet printing, etc, I expect demand for document printing

ROMANO: Print will grow at 2% for the next decade.

BOER: The digital print market has been growing between 6-10% in revenues annually over the last five years. Unless there is a rapid spurt to digital industrial printing which is unlikely, the size of the market is likely to be no more than $130 billion by 2005. After that time, digital industrial printing will have become a necessity to sustain the interest in digital printing. It is not unreasonable to predict that industrial digital printing will add an additional $50-$100 billion to the $130 billion, making digital print an over $200 billion market by 2010.

HANLEY: Value of the digital print engine market including consumables is likely to be up 60% over 5 years. Value over ten years will derive best from industrial applications, followed by offset document applications. More importantly, consider this: the value of digital print services to users will exceed the value of total hardware and consumables sales by several factors over ten years.

WILLIAMS: Forecasts should take into consideration that digital printing represents a shift that will generate value. Remember, with digital printing it's not about simply replacing one printing technology with another. That is, it's not just printing a package with ink jet technology rather than flexographic. Rather, digital printing allows the creation of new products and services not possible with analog printing. Instead of printing the package and inventorying it for future use, digital printing can print the package when it is needed, in the quantity it is needed, and also can customize the package for the particular store, season or sale. There is a value associated with the systems required for this type of printing. As we move into industrial markets, smart vendors will learn how to tap into the value created by digital printing.

Looking Further Ahead

Any thoughts on farther-out developments that would impact hard copy?

WILLIAMS: Again I will take issue with the term "hard copy." Digital printing has expanded and is expanding into applications far beyond the world of hard copy and into industrial printing and graphic arts applications. The real value will come when we can link the two segments together and have the capability to do integrated sales and marketing campaigns where the same image can be easily and quickly used for the product as well as the advertising collateral.

BOER: Further out developments impacting hardcopy will no doubt be generated from other forms of electronic display, whether it is on electronic paper or LCD screens. One large enabler is likely to be wireless technology, which will manage data and graphics in the background without involving the user. Think of packaging with wireless transmitters tracking the product from the manufacturing plant to the cupboard in your home. Automatic replenishing, billing, payment, etc. will eliminate the paper trail.

ROMANO: Mind-to-mind communication.

Reality vs. Virtual Reality

One further-out prediction is that people will be increasingly "enslaved" by the pursuit of pleasure and that the line between virtual reality and real reality will dissolve. Likely?

WILLIS: I can't wait!

ZEIS: I would expect that people will become more adept at tailoring their own entertainment and recreation. Technology will

become an ally in such pursuits. An antique dealer used to roam around flea markets to stock his stores. Flea markets are still there, but the antique dealer complains that the good discoveries are not – they are on eBay. So the dealer down-scales his retail operation, and hangs up a digital shingle. The new order replaces the old order. If the dealer likes the musty smell of an antique chest of drawers, he will still take delivery of the items he buys and sells. But technology will enable him to become an agent, if that suits him.

The point is that technology can broaden some of our horizons, and change the way we work and play. When it comes to recreation, we want to be efficient about facilitating recreation, but being efficient about the recreation itself seems silly to me. I fish for trout. If technology enables me to catch a trout with every cast, I would be more successful at fishing but less successful at recreation.

WILLIS: Once a successful e-commerce retail site is established can't it just be driven by robots and artificial intelligence? Why will we need paper? Why will we bother to work? How will we possibly use all that leisure time? Where have I heard all this before?

WILLIAMS: The world of virtual reality is truly the world of the Individual. One's own world. What if virtual reality is used to create one's living room or wardrobe? How would this be replicated in the real world? Through digital printing! The designs are already digital, quantities would be small, digital printing would be the way to do it.

While the processes to allow this to happen are not yet in place, I feel very confident that in not-so-many years, we will create virtual worlds that we will be able to duplicate in the real world.

BOER: In order for virtual reality to dissolve the line between what is real and what is artificial, three of our other senses will need to be electronically controlled: smell, touch, and taste. We are getting closer to managing the senses, enabling more efficient medical help to extend and ease our lives. However, I am too cynical to believe that eternal happiness will ever be found through electronic means. Virtual reality if anything will speed up our daily lives, making true solitude found in the remaining remote corners of the earth even more valuable. For the world of print, it may mean that hard copy will offer solitude and happiness to those who cannot reach the quiet corners of the earth.

Et Cetera

Anyone have other comments or predictions?

WILLIS: I predict this book will not be available electronically in the next decade. It is designed to be read casually, not sitting at a screen, and maybe to be dipped into rather than read sequentially and therefore is a good example of what works best with paper media. Publications for electronic media will be different, taking advantage of the capabilities of that media and avoiding the limitations, and we will interact with them differently.

ROMANO: Just draw a trend line from every new technology showing it going up and every old technology going down. More memory, more storage, faster scanners, more quality, more color, more speed, faster communications, more bandwidth, computer-to-anything and everything, delivered on-demand, immediate, even yesterday.

WILLIAMS: Images – rather than type – will dominate the world. And digital printing will make them easier to produce.

It will be a world of "Show" rather than "Tell:" Of realistic images everywhere: color maps in paperback books where before was only text; pictures of hamburgers on trucks where before it said

McDonalds; pictures of products on boxes instead of lettering. Sometimes this image of the future saddens me. What/how will people read in the future if the emphasis shifts from text to images? And the flexibility to put images everywhere can become intrusive. Recently, lunching at a restaurant I thought about advertising. I saw all of the places where advertising will be in the future: on the table cloth, the salt and pepper shakers, the cream pitcher and on and on. I see us reaching a point of sensory overload where we will begin to shut out the images that will be all around us.

Power to the individual: Digital printing will allow new cottage industries to develop. No longer will a screen printing press be required to print fabric or an offset press required to duplicate artwork: designers can print out short runs of fabric to create custom clothing; artists can print out artwork on their digital printers. Families can print out family portraits and create art work as gifts for relatives and friends — or even for themselves. I have created digital prints that hang on my walls at home.

Changing relationships between manufacturers, retailers and buyers. It used to be that the manufacturer controlled the process because he had the equipment. During the 80s as retailers got bigger and bigger, the power shifted to them because they had the customers. In the future, the power will shift to the buyer who will have the power via the Internet to buy what he wants, when he wants, in the amount he wants.

BOER: I have a rhetorical question — will reading ever become an obsolete skill or will it thrive economically in the modern world? Arguably graphic communication will empower billions to express themselves in ways they were never able. Many of the great inventors and contributors never graduated from college, and dyslexia would become a historical anomaly. As people we've been communicating with graphics since the cavemen. Will a digital world allow the masses to express and contribute to the eternal pursuit of happiness?

REALITY, SUPPLY, DEMAND

The two most provocative themes that emerge from all this — at least for me — both touch on human psychology. Beneath all the technology, beneath all the issues around business strategies and products and industry structure, lies human consciousness. The two themes are closely related. The first is "reality." How is reality defined and how does it relate to where we choose to invest our attention? The second is the immutable law of supply and demand.

Reality. That word prompted provocative forum responses. Exploring the interaction between technology and consciousness could well be grist for another book. For now, a few observations. We are divided between the material demands of life ("hunger") and the desire to escape. More and more choices are presented to us to satisfy both. In the case of both escape and hunger, the line between reality and virtual reality will progressively blur.

Neither the escape nor hunger impulse is new. It is a continuation of evolution. Since the beginning of time humans have felt the need to move beyond their normal reality. For some this might manifest as a search for spiritual experience. For many others, it assumes more mundane forms of secular escape. Early media were the village storyteller and medicine man or woman.

As technology and communication advanced, the vehicles for escape multiplied and became ever more powerful. Live theatre and storytelling and traveling minstrels and circuses gave way to film and radio and TV and video games. Now we hear about telephone sex and Internet porn. Movies are likely to evolve toward feelies. This does not necessarily mean wiring up the sensory organs. More likely, it means going directly to the brain.

Inventor/entrepreneur Ray Kurzweil, in a presentation at Seybold/Boston, March, 1999, predicted within the next decade stimuli and information would be piped directly into the human

brain. "Thirty years from now," he continued, "there will be no difference between real reality and virtual reality."*

Far-fetched? Perhaps not. Associated press in January, 2000 reported the first artificial eye, a small camera wired directly into the brain of a 62-year old man who had been blind for 26 years. The results were far from normal sight, but this represents a beginning. Nano-technology is coming.

On the escape side, then, we may ask, What role will printed words and images play in satisfying this need? Here I wonder about the addiction paradigm. Up to a certain point a larger and larger dose is needed for the desired response. Words on paper are a relatively low level stimulus. Printed words, despite their convenience, are passive. The dense, page-long paragraphs of, say, a James Joyce novel are intolerable to today's video generation. People (well, at least some people) need progressively more aggressive stimuli.

True, book sales, including fiction, continue strong for now. But it has been noted that throughout commerce and society progressively fewer of the printed words produced are actually consumed. Yes, so far the media feed off one another. But in the long run, it seems inevitable that print volume will peak out.

On the hunger side, we need to engage with the physical world and issues around physical comfort and security. This means earning a living and dealing with the economic side of life. It means satisfying acquisition impulses (which can be perverted into escape). It means business. As more and more business as well as social activity occurs on line, our print habit will in time surely fade away. The gears of industry and commerce will be given more and more intelligence so units that need to interact will often do so directly, without human intervention or monitoring. And when humans are in the loop, the information they

will need to act upon will be presented less and less in hardcopy form.

Today documents still make most of us feel secure, even as they bury us. But more and more people opt for direct debits rather than writing checks. And although we buy and sell stocks representing large sums of money, most of us never actually see our stock certificates. There may be certificates somewhere, but we could care less.

Attention-Scarcity. The second megatrend has to do with that old economic foundation, supply and demand. We pay increasing attention to the carrying capacity of the earth and our atmosphere. But it doesn't seem as though there has been much thought about another finite commodity, one that is abstract, but real. That commodity is attention.

Even though media tend to feed off one another and multiply, it is easy to forget the other side of the relationship: the mind of the consumer. The media can multiply forever, but it is wasted without being partnered with the attention of the targeted audience. More and more messages clamor for a limited pool of human attention. Demand for attention keeps rising, but the per capita supply is fixed. This means it gets ever more valuable. Economists talk about "scarcity value." In terms of marginal utility, there can be some substitution in our "market basket" of attention. We can trade family chatter at dinnertime for TV, or sleep for a late night movie. But there are limits.

As the fixed supply of attention becomes increasingly in demand, the media seeking to access it work harder. In a crowded, noisy room to get attention you raise your voice. We see this raised voice in the ever increasing volume of catalogues and telemarketing. We see it in more and more messages plastered over any available spot — from grocery store "floor minders" to bus wrap. A recent start-up proposes car wrap: they will give you a car if you commit to using it as a mobile billboard. We see it in more

---

* Seybold Seminars are sponsored by ZD Events, Inc. Foster City CA. Ray Kurzweil is currently with Kurzweil Educational Group, Waltham, MA. He has pioneered omnifont OCR, text-to-speech synthesization, and authored several visionary books.

focused messages, the increasing use of one-on-one marketing which our industry hails as an opportunity for more and more on-demand, personalized digitally printed messages.

So, at this point in history, the increased demand upon our limited supply of attention is fueling the market for digital printing. More and more will be spent in ever more creative, aggressive, and targeted attempts to capture the limited supply.

How far can this go? For the next decade, it would seem this supply-demand paradigm will continue to feed digital printing. However, beyond this, a reaction seems inevitable. Our more or less fixed attention capacity will be increasingly taxed by ever more frantic efforts to access it. One reaction may be retreating to the expanding array of escape vehicles. Some of this escape may be print. But much will be more intense non-print. Another reaction may be to rebel, pull the plugs (or turn off the switches if, indeed, in an age of micro-miniaturization and genetic engineering we can find the switches) and retreat to a low-tech lifestyle.

At this point predictions no doubt are skewed because we are in the midst of Web fever. There may be a Web-reaction as well as an "attention recovery" reaction. Interconnectedness has downsides which include privacy violations, lack of information "filtering," and vulnerability to viruses. As noted by Romano, the Internet is a "pull" medium. Going to a Web site is like asking for a catalog. Print is a push medium and is needed for balance. How this will play out in quantitative demand for print remains to be seen.

But one message of this history seems clear: in time, whatever print there is will all be digital, and where there is print, more and more will be recreational or, better yet, artful. More of our limited attention quotient will be devoted to the peace of beauty and harmony. In this vein, Boer's vision is worth repeating:

"Virtual reality if anything will speed up our daily lives, making true solitude found in the remaining remote corners of the earth even more valuable. For the world of print, it may mean that hard copy will offer solitude and happiness to those who cannot reach the quiet corners of the earth."

In the long run this may prove to be the most enduring and treasured gift of digital printing.

The End.

GLOSSARY

Terms and Acronyms, Defined as Used in this Book

Emphasis is on terms that have special, printer-relevant meanings not found in a normal dictionary. Readers are referred to graphic arts glossaries for terms that are associated primarily with traditional (non-digital) printing and the graphic arts. An excellent graphic arts glossary is contained in *Pocket Pal* published by International Paper. A graphic arts-oriented digital printer glossary is in the Delmar *Pocket Guide to Digital Printing* (see Bibliography). Trident International (1114 Federal Road, Brookfield, CT 06804) has published *Inkjet Glossary*, a compact, focused glossary that also includes technology diagrams. The *Lyra 1996 Guide to the Printer Industry* (see Bibliography) contains a glossary. As of this writing Lyra also has a glossary said to list over 1,100 terms posted on its Web site, www . lyra.com.

-up. Two-up, three-up, etc. describes printing multiple pages side-by-side on a single sheet or web. A common print width target is 17-inches since it allows printing two standard 8.5 x 11-inch pages 2-up.

addressability. The number of locations per unit of area that a digital printer can place a mark. cf "resolution."

artifact. Unwelcome or unintended software or printer generated image aberrations.

ASCII. American Standard Code for Information Interchange. A widely-used, (probably the most widely used), 7-bit code (for 128 combinations) that supports upper and lower case characters, numeric digits, punctuation, and control codes.

band printer. An on-the-fly, impact printer using an etched steel loop type element.

Baudot code. A 5-bit code invented by Emile Baudot in 1870 that accommodates 32 characters with alternative sets accessed by escape code.

bit. A single, binary (i.e. on or off) unit of digital information. Contraction of "binary digit."

bitmap. An electronic representation of an image, page or font made up of an array of bits. The bits form the image itself rather than being treated as character codes.

bleed. Printing that runs off the edge of the paper, normally not possible for either presses or digital printers without printing beyond the edges of the final document and subsequently trimming off the "bleed."

bubble-jet. A term used primarily by Canon for thermal ink jet.

byte. An 8-bit chunk of digital information, normally representing one alphabetic, numeric, or control character.

carrier. Any substance that suspends colorant in toner or ink jet ink. In dry toner systems, a material that helps distribute the toner. Also called "developer."

cartridge. Any throwaway component. A toner or ink supply. The cartridge in a thermal ink jet printer typically also includes the print head as well. In a low-end electrophotograhic printer the cartridge may contain not only the toner supply, but also the drum and developer system.

Centronics Interface. The de facto standard parallel printer interface used for PC printers and other low-end printers.

chain printer. An on-the-fly impact line printer in which character fonts are mounted on a chain loop.

character printer. An impact serial printer which prints characters from raised type (i.e. not a matrix printer or line printer).

character set. The assortment of characters that can be printed by a digital printer, or that is included in a coding system (e.g. the ASCII Character Set).

continuous ink jet. A type of ink jet printer in which droplets are ejected in a continuous stream; those droplets not needed to create the image are deflected into a gutter and recycled. Also, "CIS." In binary CIS, each droplet is allowed either to reach the printing surface or is deflected into the gutter. In multi-level deflection CIS, the droplets are deflected under a varying charge to position dots in varying vertical positions as the print head moves across the target surface (or as the surface moves under the print head).

continuous tone. Printing in which a more-or-less unbroken progression is achieved in tones from light to dark, as in conventional photography. Most printing processes simulate continuous tone by breaking the image into tiny dots or pixels of varying size, spacing and shapes (in traditional press printing, a "halftone.").

controller. The firmware or software and/or electronics that controls the printer, telling it when and where to generate an image. Also,

the processor — often a freestanding component — which runs the printer software including its physical functions and datastream processing.

cpi. Characters per inch, a measurement of print density, generally for full character printers (cf. dpi for matrix printers).

cps. Characters per second, a measure of print speed for character (serial) printers.

CTP. Computer-to-plate. See direct-to-plate.

daisywheel printer. An impact, serial, full character printer in which the type element resembles a daisy: i.e. flexible fingers radiating out from a central hub.

developer. See "carrier."

dielectric. Describes any material which will hold an electrical charge, at least for a small period of time. An insulator. A characteristic of the photoconductor in a xerographic printer or the paper in a direct electrostatic printer, i.e. the dielectric property is what maintains the latent image through the toning cycle.

digital. Refers to counting or numerical representation (as opposed to analog systems) to quantify the magnitude of a continuous variable. Describes a machine or system that processes, stores, or transmits data, image information, or other variables by counting discrete units.

digital printer. See "printer."

digital press. In a broad sense, any high-speed, non-impact printing system. More specifically, may refer to a high-speed digital printer which emulates the traditional printing press, including the ability to print process color.

diode laser. A solid state laser used in many lower-cost laser printers rather than a gas laser.

direct-to-plate. Imaging a printing plate directly from digital data as part of the pre-press process.

direct-to-press. Imaging a printing plate or image cylinder on-press directly from the digital image file.

distributor. The company that buys products from the manufacturer and resells them to retailers or dealers.

dot matrix printer. See "matrix printer."

dpi. Dots per inch. A measurement of the density at which dots can be printed by a matrix printer or non-impact printer. Also, "resolution."

drop-on-demand. An ink jet printer which ejects droplets asynchronously, as needed, to form the specified image (as opposed to "continuous ink jet").

duplex. The ability to print on both sides of the page, normally in a single pass. In current common usage, this is sometimes also applied to a pair of printers working in-line to print both sides of the continuous form. The form is turned over between the two printers ("tandem" or "twin " systems).

dye sublimation printing. A form of thermal printing in which heat is used to vaporize and transfer dyes from a ribbon or donor sheet to the medium, especially well-suited for high quality, more-or-less continuous tone color imaging. Also, "Dye Diffusion."

electron beam imaging. The technology used by Delphax/Xerox for high speed non-impact printing, earlier termed "ion printing" or "ionography."

electrophotographic printer. A non-impact printer in which light is used to create a latent image that is then developed with dry or liquid toner. Also, "xerographic printer," or "indirect electrostatic printer." Sometimes referred to by the specific light source, i.e. "LED" or "laser."

electrosensitive printer. A non-impact printer which creates an image by selectively removing a light-colored coating on the paper to reveal a dark substrate. Also, sometimes more graphically termed "burn-off" printing.

electrostatic printer. Any non-impact printer which uses static electricity to form a latent image which is then developed with a liquid or dry toner subsystem. Most often refers to a "direct electrostatic" printer that forms an image without a photoconductive intermediary, normally using dielectric coated paper and sometimes classified under the somewhat vague term, "electrographic."

EPG. Electrophotography or electrophotographic.

facsimile. Long distance copying, i.e. a non-digital process of scanning and transmitting images over telephone lines. Also, "Fax."

fanfold paper. Continuous zig-zag folded, pin-feed paper traditionally used in most high speed impact and many non-impact printers. The trade name "Fanfold" was used by forms manufacturer UARCO for its specific forms technology, but has since come into generic usage. Also, "flat pack" or "Z-fold" paper."

font. A collection of text characters designed as a set. A typeface may be seen as a collection of fonts. Digital fonts may be scalable or bitmapped.

form. Any material (normally paper) which has been converted in some way, and perhaps preprinted so that it may be processed (i.e.

printed with variable information or images) by a printer or by hand. Also, "business form." Can be continuous flat-pack, roll, or cut sheet; single or multiple-part; pre-printed stock or custom.

Full-form character printer. A non-matrix impact printer that prints a set of continuous-stroke character images (i.e. not made up of dots) from an interchangeable or integral type element.

fusing. The process by which toner is affixed to the media in a non-impact page printer. Methods include heat and pressure (most common), radiant heat, flash, and cold pressure.

GDI. *Graphics Device Interface.* The Windows operating system for representing images on a computer screen or formatting them for printing.

I/O. Input/Output. Refers to the receipt and transmission of a data stream or files as related to the physical processor, software, or data itself.

image area. The area of the page that can be imaged by a given printer, normally defined by a non-print margin on all sides.

impulse ink jet. See Drop-On-Demand.

in-line. Describes the operation of two or more physically independent units that function as a single system, such as a cutter that works synchronously with a high speed printer to sheet the printed, continuous form without human intervention.

interface. The facility through which data and commands enter or exit a system. Also, "port." Includes standardized mechanical specifications (plug/socket specs), serial/parallel (if parallel, also the number of circuits), data rates, unidirectional/bidirectional, etc. Printer interfaces include the Centronics parallel interface and the newer IEEE P1284 high-speed bidirectional parallel interface. More generally today, "interface" refers to communications connectors (for scanners and network connections as well as printers) such as USB, Ethernet, Firewire/IEEE 1394.

Interpress. A PDL developed by Xerox, described as an "ancestor" of PostScript.

ion deposition. A direct electrostatic technology developed and patented by Dennison Manufacturing and Delphax Systems.

jaggies. An image defect characterized by irregular strokes or edges, an indication of low resolution. Most pronounced on lines that are near-horizontal or near-vertical. Also, "stair-stepping." Font jaggies were effectively eliminated with the introduction of outline fonts.

JPEG. Joint Photographic Experts Group. A standard graphic file compression system, commonly used for saving photographic images.

justified. A block of printed type characterized by a straight margin. Type may be left-justified or right-justified. Justified alone normally means both right- and left-justified.

laser printer. An indirect electrophotographic printer which uses a deflected laser beam to write the latent image on the photoconductor.

LCD. Liquid crystal display.

LCS. Liquid Crystal Shutter. A solid-state, programmable light source used in some electrophotographic printers, mostly in the 1980s.

LED. Light Emitting Diode. A type of semiconductor that glows under the application of a voltage. LED arrays are used for panel displays and as a programmable light source in some elec-trophotographic printers. This alternative to the laser is commonly thought to be more reliable due to fewer moving parts. Today's main users include Okidata, Xeikon and Océ.

line printer. An impact printer that prints lines of formed characters or dots presented a line at a time (although the whole line is normally not printed simultaneously).

lpi. Lines per inch. Vertical density of lines of print, most often used with impact line printers, but also for any text format.

lpm. Lines per minute. A measure of speed for line printers.

LQ. Letter Quality. A somewhat vague print quality descriptor. Generally, the quality achieved by a typewriter with a film ribbon. Daisywheel printers were considered to produce letter quality and matrix printers, more often, "NLQ" (Near Letter Quality). Non-impact printers capable of 300 dpi and above are also usually considered letter-quality.

machine language. Binary-coded instructions unique to a given computer or peripheral; the assembly program translates data and control information in the application into machine language appropriate to the selected peripheral.

magnetography. A non-impact printing technology making use of a magnetic drum and writing head. Nipson (a Xeikon company) is the main user.

marking engine. See print engine

matrix printer. Normally refers to an impact printer that builds the printed characters or image with programmable dots. Also "dot matrix printer."

media. (1) Materials upon which printing is performed. Can be paper, textiles, laminates, films, etc. Also, "Substrate." Strictly speaking, the singular, "medium" should be used, but the term "media" now seems to be commonly used as either singular or plural. (2) The various vehicles for communication including audio, soft, and hard copy. (3) Most broadly, the press (newspapers and television news).

MICR. Magnetic Ink Character Recognition. MICR printing is a character reading system used primarily in the banking industry for checks and financial documents.

microencapsulation. A process by which liquids can be enclosed in polymer capsules, used in carbonless paper systems and Mead's CyColor non-impact printing technology.

monospaced font. A font in which each character is allotted the same horizontal space, whether it be an "m" or a narrow "i." cf. "proportional spacing."

MFP. Multifunction Peripheral. A device that combines several functions, such as printing, scanning, copying and facsimile.

MTBF. Mean Time Between Failure. An index of reliability, usually specified in hours of operation.

NLQ. Near Letter Quality. A relatively non-specific print quality descriptor, somewhere between letter quality and draft quality.

non-impact printer. Any printer that does not require impact to create the image. Most non-impact technologies require contact, except for ink jet.

OEM. A curious term, literally the acronym for Original Equipment Manufacturer. Current usage probably evolved from dropping "distributor" from the term "OEM distributor." As a noun, OEM has come to mean the company that places its name on a product and sells and services it, that is likely not to be the actual manufacturer. It is also used as a verb: Canon "OEMs" its laser print engine to Hewlett-Packard.

on-demand printing. Printing documents, booklets, etc. when they are needed rather than producing a long run for inventory, a capability unique to digital printing. The capability may or may not also include variable printing.

on-the-fly printer. An impact printer which achieves relatively high speed by keeping the type element in constant motion.

off-line. A printer that is not wired directly to a host computer, i.e. not "on-line."

OPC. Organic photoconductor, a light-sensitive, carbon based material used as the active coating for the drum or belt of an electrophotographic printer. Sometimes refers to the entire photoreceptor subsystem.

outline font. A system for digitally representing a type font by mathematical description. Unlike a bitmapped font that requires storing the characters in every size, an outline font is scalable, i.e. just one master representation of each character is needed with the facility to print the font in any size. Invented by Adobe Systems, current offerings include Adobe Type 1, Agfa Intellifont, and Microsoft TrueType. Also, "scalable font."

page coverage. Ink density on a page, which needs to be specified as part of cost-per-page. Also, a component of a speed specification for printers whose speed varies depending on the ink density of the page in question (i.e. not page printers). Five percent is commonly used.

page printer. A digital printer that composes and prints an image more or less a page at a time (cf. serial printer and line printer). The whole page needs to be formatted before printing can begin. All laser printers are page printers.

PC. Personal Computer. Originally IBM's name for its groundbreaking single-user, desktop computer, now a generic term for all such systems.

PC. Sometimes used as an abbreviation for "photoconductor." See also "OPC."

PCL. Printer Command Language. A PDL-type software system developed by Hewlett-Packard that is used with many its printers and also by many other vendors.

PCM. Plug Compatible Manufacturer. A company that builds printers or other components which can be plugged into the product of another manufacturer. Plug compatible printers are most commonly those designed to work with IBM computers.

PDF. Portable Document Format. An Adobe Systems technology which has become a standard that allows users to view or print a file without the file's origination software.

PDL. Page Description Language. Printer driver software on the host system that converts documents or images from "native format" into a format that can be used by the printer software to recompose the image to bitmap form to drive the print engine. Adobe PostScript is one of the most widely-used PDLs. Printers now commonly offer both Adobe PostScript (or a PostScript clone) and PCL.

pel. Picture element, or pixel, a term used mostly within the IBM world.

photoconductor. The light-sensitive material which receives the latent image in an electrophotographic printer. Photoconductors are either organic or inorganic. Organic photoconductors are plastic-based materials; inor-

ganic are commonly selenium, selenium alloys, tellurium, or arsenic compounds.

piezoelectric. Refers to the property of certain crystalline structures to change shape or twitch upon the application of an electrical field. Piezo transducers are the major alternative to thermal actuation in drop-on-demand ink jet printers.

pinfeed paper. Continuous forms with "line holes" on half-inch centers running along each edge, designed to engage the feed pins of the printer's tractors or sprocket feed mechanism.

pitch. The density at which characters, dots, or picture elements can be placed across the printed page. Used primarily with typewriters and other full-form character printers.

pixel. Picture element, a basic building block making up the digital image. A pixel may be a single printed spot, or, more likely, a cluster of spots or dots.

platen. That component of a printer or copier that supports the paper. In a copier the platen is a glass or other transparent plate which supports the paper while it is being scanned. In typewriters and many accounting machines the platen is a round rubber or metal cylinder. In some non-impact printers a "pre-heat" platen is part of the fusing station.

plotter. A printer, usually wide format, designed and/or sold for engineering applications. Plotters originally were defined by a specific technology, i.e. using liquid ink in one or more marking pens moved under digital control (pen plotters).

PostScript. A page description language developed by Adobe Systems which is one of the more popular and which has become a de facto standard for desktop publishing.

ppm. Pages per minute. A speed specification for page printers.

press. A printing press, i.e. the machine used in traditional printing, normally utilizing fixed plates, liquid ink, and contact or pressure. Can be further defined by paper handling (roll to roll, sheet, etc.) and by printing process (letterpress, offset, litho, gravure, etc.). cf "digital press."

press printing. This is a suggested term to describe traditional, non-digital, analog printing, i.e. any printing process using fixed plates. Before digital printing, all printing was press printing so no modifier such as "press" or "digital" was needed. Now the printing world is divided into digital and non-digital. Unfortunately, none of the terms used to describe non-digital printing is particularly satisfying. Complicating terminology are the new hybrid, direct-to-press systems. See "direct-to-press."

printer. As used in this book, a printing device that receives and converts digital input to a visible image applied directly to the final surface. The content and format is normally subject to at least some degree of computer or programmable control.

print engine. The basic electromechanical mechanism that performs the physical imaging and paper handling of the printing system. Also "marking engine."

printer driver. The software on the host computer that converts the page image in the computer into a page description that the selected printer can understand.

process color. The creation of a great variety of colors by sequentially printing the subtractive colors (cyan, magenta, and yellow), plus black (CMYK). The perceived color is actually blended by the vision of the human viewer.

Colors that cannot be reproduced in this way are printed as solid color from pre-blended inks.

proportional spacing. Refers to a type font in which the horizontal space for given characters is proportional to the width of the character. May also refer to a printer that can print proportionally spaced characters. cf "monospaced font."

rasterization. The conversion of high level digital image representations to low level bitmaps that can be handled by a monitor or printer. "Raster" is sometimes used to refer to a two-dimensional array of pixels. The term is derived from the Latin / Greek root for a toothed hoe or rake.

resolution. A measure of image quality which depends on printable dots per inch (dpi) plus a number of other factors such as dot size and consistency, image density, and other variables. Can refer more broadly to the capability of a printer to render tone, color and detail. Resolution is sometimes used interchangeably with addressability, although purists are likely to object. To confuse things a bit, in CTP parlance, resolution seems to be the physical size of the "spot," a group of which make up a halftone dot (i.e. a different "dot" from the non-impact printer "dot").

RIP. Raster Image Processor. The hardware or software that converts an image to the machine language commands and bitmapped data that drive the printer. Also, "controller." The function may be performed by the host computer, by a separate control unit, or by the printer.

RIP. Raster Image Processor. The software and/or hardware that converts print files to a bitmap for printing.

scalable font. See "outline font."

serial interface. An interface with a single data circuit, usually with additional circuits for control data. RS-232C is a common serial interface for printers. cf. "parallel interface."

serial printer. A printer that prints characters sequentially rather than in parallel with a print head that moves back and forth. Serial printers may be unidirectional or bidirectional.

slew. High speed vertical paper feeding over non-print areas.

SIPS. Square inches per second as a measure of print speed.

solid ink jet. Ink jet printing in which ink is liquid only when heated. The liquified ink is jetted on demand and "freezes" almost instantly upon striking the intended substrate. Termed "phase change" ink jet by Tektronix, a leading proponent of the technology.

spot color. In digital printing, the capability to print a second, solid color in addition to black. A color, not generally black and or one of the three process colors (CMY), usually specified by its PMS number (Pantone Matchine System)

substrate. See "media."

subtractive primary colors. Cyan, magenta and yellow, the three colors which can be printed in specific combinations to make a great variety of other colors; used with black for press printing process color (CMYK). To broaden the range of colors that can be rendered (color gamut), some printers are now six ink systems.

thermal ink jet. A drop-on-demand ink jet technology that ejects ink droplets by applying a current to a resistor which heats rapidly to form a gas. Termed "bubblejet" by Canon. Besides Canon, the world's thermal printer manufacturers are HP, Lexmark, Xerox and Olivetti.

thermal printing. A non-impact technology which uses the heat generated by resistors to mark coated, heat-sensitive paper.

thermal transfer printing. A non-impact technology in which the image is transferred to the substrate by bringing an inked ribbon or donor sheet into contact with the substrate and selectively heating it. Three general subtechnologies are wax transfer, dye diffusion, and resistive ribbon (IBM only).

toner. The ink or marking material used by non-impact page printers. Toner commonly consists of finely ground or chemically grown plastic resin particles impregnated with pigments. Toner may be dry or suspended in a carrier liquid.

tractor feed. Mechanism for pulling or pushing continuous, pinfeed paper through a printer, normally two sets of pins mounted in plastic belt or chain loops.

trapping. Modifying color boundaries as a technique to sharpen color-to-color registration. Often synonymous with "over-trapping," in which the color boundary is slightly expanded. PostScript has automatic trapping capability.

type element. The component of a full character printer that holds the printable selection of characters. Serial printer type elements include the IBM "golf ball" and the daisywheel. Line printer type elements include type wheels or bars, drums, trains, chains, belts and bands.

typeface. A collection of fonts linked by a common design.

web. A continuous stream of paper as processed through a press or printer. Web production may be roll-to-roll, roll-to-sheet, or fanfold/flat pack.

wide format printer. Generally any printer which can handle media larger than 11 x 17 inches.

xerography. A term often used interchangeably with electrophotography, especially by Xerox.

BIBLIOGRAPHY

Books cited and/or used as sources for this history
Selected additional books
Periodicals and other sources.

BOOKS

Bashe, Charles J., et al
IBM's Early Computers
1986, MIT Press, Series in the History of Computing, 717 pages. A broad history, including processors, mag tape, system architecture. No separate chapter on printers, but excellent pages on IBM printer development from the 407 tab machine of the 1950s to the IBM 1403 and 1443 impact line printers of the 1960s.

Brown, Kenneth
Inventors at Work
1988, Tempest Books Div., Microsoft Corp.
Section on Chester Carlson and the invention of xerography.

Campbell-Kelly, M. and Williams, M., Editors
The Moore School Lectures
1985, J. P. Eckert, Jr., "Tapetypers and Printing Mechanisms"
The MIT Press
Cited by Tomash and Weiselman in *Annals of History of Computing*, 1991/Nr. 1.

Carlson, Chester, with J. J. Ermenc
The Invention and Development of Xerox Copying
1971, edited transcript of interview; hardcover; 69 pages
This interview by Ermenc, a Dartmouth professor, provides intimate insight into what inspired and drove Carlson, and the long and difficult journey from invention to commercialization. Available in Xerox archives and a few other business libraries.

Cost, Frank
Pocket Guide to Digital Printing
1997, Delmar Publishers/International Thomson, Albany, NY
272 pages, 4.5 x 7.5 inches
Impressively complete and compact overview of digital printing for the graphic arts by Cost, who is Dean of the College of Imaging Arts and Sciences at Rochester Institute of Technology. Besides covering the digital printing processes, chapters include image acquisition, image file formats, rendering type, line art and toned images, and color management. Detailed glossary of terms.

Dessauer, John H.
My Years with Xerox
1971 Doubleday, 239 pages
First person account of the author's key role in the commercialization of xerography and building of Xerox. Dessauer was with Haloid and Xerox from 1935 until 1970, serving as VP, vice-chairman of the Board, and head of Research and Engineering.

Dessauer, John H. and Harold E. Clark (both of Xerox), Editors
Xerography and Related Processes
1965, The Focal Press, 520 pages
More than you'd ever want to know about xerographic technology; excellent history of the invention; with 130 references; interesting, 12-page "skeletal outline."

Diamond, Arthur S., Editor
Handbook of Imaging Materials
1991, Marcel Dekker, Inc., 625 pages
Anthology, covering the range of digital printing system materials including photoreceptors, diazo, dielectric papers and film, ink jet inks, cyliths, etc. Heavy (but accessible) technology and science; voluminous references to books, papers and periodicals.

Durbeck, Robert C. and Sherr, Sol, Editors
Output Hardcopy Devices (An Anthology)
1988 Academic Press, Boston
Cited by Irv Weiselman in Annals of History of Computing, 1991/Nr. 1.

Eames, The Office of Charles and Ray
A Computer Perspective: Background to the Computer Age
1990, Harvard University Press, 175 pages
New edition of the 1973 book based on a 3-D exhibit created by the Eames Office and on display in New York City from 1971 until 1975. A largely pictorial history from 1890s through the 1940s, primarily computation but also a few items on printers including IBM's first demonstration of its printing calculating machine in 1919 (all the salesmen stood up and cheered).

Fenton, Howard M. and Romano, Frank J.
On-Demand Printing: The Revolution in Digital and Customized Printing
1995, Graphic Arts Technical Foundation, Pittsburgh, PA USA
192 pages, hardcover, 8 x 11 inches
The authors, industry cognoscenti affiliated respectively with GATF and RIT, have put together a readable, liberally illustrated sourcebook. Topics covered include applications and market research, a survey of the technologies (copiers, digital printers, graphic arts printing processes, networking, etc.), and a description of more than a dozen on-demand systems that were extant as of 1995. Furthermore, GATF practiced what it preached: the book was printed on-demand by Xerox DocuTech and hardcover bound off-line by ChannelBind, a Xerox company.

Foerester, Scott
Printer Bible
1990, Que Corporation

Forester, Tom
High-Tech Society
1988, MIT Press, 311 pages
One man's view of the impact of computers on business and society including the "electronic office."

Forman, Richard Lee
Tales in Peripheral Enterprise: the Rise of Dataproducts Corporation in the Computer Industry, 1962-1972
1980, Unpublished Master of Arts in History thesis, 521 Pages
A readable exercise in applying scholarly historical analysis to bear on the evolution of a business enterprise; helpful and relevant documentation of the interplay among the human, economic and technology aspects of the business.

Hiltzik, Michael A.
Dealers of Lightning: Xerox PARC and the Dawn of the Computer Age
1999, HarperCollins Books, New York, NY
450 pages (softcover edition)
Readable testimony to the contribution of the famous Xerox think-tank to computing and digital printing. A bit gushy ("America and the world are today in the grip of an unprecedented technology craze; very few are aware that most of what drives the frenzy was invented, refined, or perfected at Xerox PARC.... one of the most productive and inventive research centers ever known.") In-depth coverage of Gary Starkweather and his invention of the laser printer. Hiltzik dismisses the common notion that Xerox never had the wits to benefit from developments at PARC noting that the invention of the laser printer alone returned their PARC investment many times over.

Hinterbichler, Joan
Which Printer Should I Buy?
A Different Kind of Buyers Guide for Computer Printers
2000, E.T.C. Publishing House
44 pages, illustrated, 8 x 5.5 inches, $7.95

Jaakko Pöyry Oy (Consultants to the Pulp and Paper Industry)
Future of Paper in the Telematic World
1984, Jaakko Pöyry Oy, 231 pages
Dated, but provocative overview of the role of paper in history and speculation as to its future, liberally supported by data on reading, advertising and publishing trends, human psychology, sociology, and more.

Jacobson, Gary and Hillkirk, John
Xerox: American Samurai
1986, Collier/MacMillan, 339 pages
Lively chronicle covering how Xerox recovered from a near fatal eclipse by Japanese copier competition in the 1970s, "beating the Japanese at their own game."

Leebaert, Derek, Edited by
The Future of the Electronic Marketplace
1998, MIT Press, 383 pages
Contributions from people viewed as prime movers of the Internet revolution and its impact in various sectors of the economy including merchandising, finance, utilities; documenting the death of distance and predicting the death of bandwidth, and cautioning that the only limiting resource that may remain is peoples' time.

Levendel, Giuliana, Edited by
A Decade of Research: Xerox Palo Alto Research Center
1980, R.R. Bowker Company, 487 pages
Collection of articles by PARC staff scientists and managers on topics ranging from interactive graphics to Ethernet to various abstruse research programs in the physical sciences. Of particular interest: a 20-page chapter by Gary Starkweather entitled "High Speed Laser Printing Systems" covering the optical system for the Xerox 9700, etc. Very nuts and bolts. Also published as a chapter in *Laser Applications* by Joseph Goodman, Academic Press.

Levy, Aaron Uri and Discus, Gilles
Non-Impact Electronic Printing, The Reference Handbook
1993, Interquest, 314 Pages
Very comprehensive, liberal use of photos, line art, and diagrams; wide-ranging with chapters on hardware, PDLs, Network Printing and Connectivity; Glossary of Terms and Acronyms.

Lubar, Steven
InfoCulture/The Smithsonian Book of Information Age Inventions
1993, Houghton-Mifflin, 408 pages
Readable, illustrated segments on digital printing roots including typewriters, telegraph, copiers and printers.

Lyra Research
1996 Guide to the Printer Industry
Charles LeCompte, Publisher
369 pages
This was the last edition of an industry guide series that was published annually for a number of years. Content includes The Year in Review, Industry directory with hundreds of citations, a Technology Overview, and Glossary of Terms. The Lyra web site now contains similar, but not as complete, reference material (www. lyra.com).

Océ Printing Systems
The World of Printers
Various authors including M. Weidemer,
Dr. Gerd Goldmann, H. Brinkmeier,
G. Wächtler, and U. Mayr.
1998, English version, softcover, 327 pages
Excellent overview of laser printing as implemented by Siemens/Océ, plus chapters covering the bigger picture including a survey of non-impact printing technologies, principles of color, paper, screening techniques, print-on-demand, terminology; amazingly broad and deep.

Perrin, Timothy
Programming Laser Printers
1995, IDG Books Worldwide, Indianapolis, IN

Slater, Robert
Portraits in Silicon
1992, MIT Press, 374 pages
Biographical sketches of 31 computer pioneers from Babbage to Bill Gates; innovators of processors, memory technology, software; no printer pioneers and very little info on printers, although Donald Knuth, creator of Metafont and TeX ("one of the most important inventions in the history of typesetting") is included.

Stone, M. David
The Underground Guide to Color Printers
1996, Addison-Wesley Developers Press, 342 pages
Lively introductory guide for home and other lay users; glossary, elementary color science, "slightly askew advice on getting the best from any color printer"

Sturge, John M., Edited by
Neblette's Handbook of Photography and Reprography
Materials, Processes, and Systems
Seventh Edition
1977, Van Nostrand Reinhold Co., 641 pages
Anthology, covering the science and technology of light-based reprographics. Among the chapters of interest: History of Photographic Processes by C. B. Neblette (RIT); Silverless Imaging Systems by Richard D. Murray (Institute of Graphic Communication); Electrophotographic Processes and Systems by Donald R. Lehmbeck; Photomechanical Printing Processes by Michael Bruno. Covers graphic arts printing processes, platemaking, typesetting. Heavy science and technology; little on digital implementations. Bibliographies include hundreds of references

Urroz, Greg and Louise
The Fillmore Guide to Refilling Ink Jet Cartridges
Tips & Tricks of the Trade
1995, The Fillmore Group, Apple Valley, CA
128 pages, illustrated
Published by a company that has developed and markets refill kits for HP and other ink jet cartridges.

Webster, Edward, Edited by
Data Communications and Business Systems
1971, International Business Forms Industries, Printing Industries of America, Inc., 186 pages, hardcover.
Proceedings from the Third International Forum sponsored by IBFI, Hollywood Beach, Florida, 1970. Presentations on the state of telecommunications around the world and the CRT and hard-copy terminals of the time, the computer industry in Japan compared with Europe, OCR, the role of paper, and predictions.

OTHER SOURCES: NEWSLETTERS AND JOURNALS

Annals of the History of Computing
Marks on Paper: A Historical Survey of Computer Output Printing
Irving L. Wieselman and Erwin Tomash
1991, Volume 13, Issues #1 and #2
IEEE
A good, quick overview of computer printing technologies including technology diagrams, photos of representative printers, some industry shipment/revenue data.

Color Business Report, Mike Zeis, Editor
Blackstone Research Associates, Uxbridge, MA
Published monthly, 1990 to present

Digital Printing Report, Frank J. Romano, Editor
Digital Publishing Council, Printing Industries of America, Inc.
Alexandria, VA
Published from 1963 to present
The newsletter for Computer-to-Imposed Film, -to-Imposed Plate and -to-Digital-Press.

EDP Industry and Market Report (the "Grey Sheet")
International Data Corporation
Bi-monthly computer industry market newsletter, published since June, 1964. News plus market info on shipments of computers, and at times terminals and other related equipment.

Graphic Arts Monthly
Cahners Publishing, New York, NY
Magazine, "the Preferred Magazine of the Printing Industry

Graphic Communications World
Published bi-weekly by Green Street Communications, Inc.
Hartsdale, NY

Hard Copy Observer
Published monthly by Lyra Research, Newton, MA
1993 to present

IBM Journal of Research and Development
1981, 25th Anniversary Issue
Relevant Articles:
"IBM Typewriter Innovation," Beattie and Rahenkamp
"Printer Technology at IBM," Nickel and Fania
"Laser Electrophotographic Printing Technology," Elzinga,
  Hallmark Mattern and Woodward (IBM 3800, 6670)
1987, January issue: "Digital Halftoning on the IBM 4250
Printer," Goertzel and Thompson

Image Printing Digest
Published quarterly; "News Briefs from the Dynamic Digital
Printing Industry;" a cooperative venture of Atlantek Inc.,
Information Management Institute, and WRITE, Inc.
Information: wglover @ ibm.net

Printout Annual
Published annually in the 1980s by Datek Information Services,
Inc. Each includes "The Year in Review," other timely articles,
profiles of all major vendors, and a printer product directory.
These archived journals were particularly helpful sources for this
history.

Rip
Digital Printing and Imaging Association
newsletter published bi-monthly by DPIA, the Digital Printing
and Imaging Association, Fairfax, VA

The Seybold Reports:
Seybold Report on Desktop Publishing (monthly)
Seybold Report on Publishing Systems (bi-weekly)
Seybold Bulletin on Computer Publishing (weekly)
Seybold Publications, Media, PA USA

SPECTRUM Newsletter
Published monthly by I T Strategies since August, 1996 as a client
service.

MISCELLANEOUS

Charles Babbage Institute
c/o University of Minnesota, Digital Collections Unit, 211
Andersen Library, Minneapolis, MN, 55455
Archives include Oral Histories by various printer industry lumi-
naries including Irv Wieselman and Erwin Tomash of
Dataproducts. Also, proceedings from the conferences of Institute
for Graphic Communication, 1970s and 1980s.

IBFI, The International Association for Documents & Information
  Management Solutions, Alexandria, VA
Trade Association with the normal services plus SpecNet, a database containing basic specs for thousands of digital printing,
optical reading, and business forms handling devices

THE INTERNET

An endless web of current and historical information on printers.
Vendors web sites were a helpful supplemental source for this
book. Other helpful sites:

Deja.com. Archives postings from USENET, the non-commercial
Internet discussion forum. Knowledge sharing in diverse fields
including computers and technology. Lot's of printer lore. Mostly
anecdotes by those who used early printers, and antique computer and printer buffs and collectors looking for machines, manuals, parts, and answers to programming problems.

Anthony's History of Information Technology (IT)
Anthony J. Pennings, PhD
10-page chronology of events, 1943-1997, with emphasis on
telecommunications
http://www.academic.marist.edu/pennings/hyprhsty.htm

Home.earthlink.net/~dcrehr/
Darryl C. Rehr, "The QWERTY Connection" and miscellaneous
links to other typewriter lore. Rehr e-mail: dcrehr@earthlink.net

INDEX

PUBLISHER/CONTRIBUTOR INFORMATION

DRA of Vermont, Inc. was incorporated in 2000 to publish this history and books of similar quality that nurture understanding and lend dimension to technology, business, society, and the quality of life. Information: DRA of Vermont, Inc., 226 Handle Road, West Dover, VT 05356 USA/voice 802-464-5845/fax 802-464-6534.

Edward Webster has researched and authored dozens of studies and reports for the computer printer industry and printing industry. He founded and directed Datek Information Services, Inc., a market, technology and strategic information resource serving the digital printing and related industries. Currently he is consulting editor to I T Strategies, Inc. and manages DRA of Vermont.

I T Strategies, Inc. is a market-specific management and strategic planning consultancy serving digital printing vendors, printing press manufacturers, image processing software and silicon companies, and large user organizations worldwide. Working one-to-one, I T Strategies consultants are dedicated to helping client companies enter new markets, develop successful partnerships, capture new customers, and implement e-imaging strategies. The organization is headquartered in Massachussets with additional staff based in Tokyo, Japan. I T Strategies, Inc., 51 Mill Street, Hanover, MA 02339 USA/voice 781-826-0200/fax 781-826-0151/www.it-strategies.com.

The Océ Group, founded in 1877, offers products and systems that enable people and companies to share information effectively. In the markets for wide format printing systems and production printing systems, Océ is a world leader. The company also offers a strong assortment of document printing systems. Headquartered in Venlo, the Netherlands, the Océ Group is active in 80 countries and employs nearly 20,000 people worldwide. **Océ Printing Systems GmbH** is provider of high-end production printng solutions. For more information: Océ Printing Systems, Siemensallee 2, 85586 Poing, Germany/voice +49 (8121) 72-0/fax +49 (8121) 72-33 02/www.oce.com.

Print Unchained Advisory Board: *Marco Boer*, Consulting Partner, I T Strategies, Inc. Previously manager of printer market research at IDC and before that, senior printer analyst with BIS Strategic Decisions. *Richard A. Fotland*, President, Illuminare, Inc.; consultant to the digital printing industry; formerly Director of Advanced Development, Avery Dennison; inventor of ion printing and holder of 45 other patents. *Larry Lorah*, market research and product/strategic development consultant to the digital printing and other industries with Arthur D. Little and currently Concord Consulting Group. *John Schneider*, John M. Schneider Associates; internationally recognized pioneer in ink jet printing development over many years at Xerox and with Scitex Digital Printing as VP and Principal Scientist. *R. Hugh Van Brimer*, Founder and Chairman, I/J Printing Corporation; served as director of central research at Mead/Kodak/Scitex Digital Printing, and Founder/Chairman of Trident International via Exxon Danbury Systems Division. *Vivian Walworth*, consultant to Stereo Imaging Group, Rowland Institute for Science; long career in research and research management at Polaroid Corporation; Fellow, IS&T — The Society for Imaging Science & Technology; Editor Emeritus, *Journal of Imaging Science and Technology*. *Mike Willis*, Managing Director, Pivotal Resources Ltd., Cambridge, England. Consultant to the digital printing industry. Formerly Group Leader at Cambridge Consultants and a founder of Xaar Ltd.

THE TYPE

PRINT UNCHAINED was set in Johanna for text and Gill Sans for display, designed by Eric Gill (1882–1940). The dingbats were designed by Hermann Zapf. Eric Gill, stone carver, wood engraver, essayist and typographer, was born in Brighton, England. His father was a curate and his mother a singer. Early in life, he displayed a talent for drawing and a keen eye for proportion, inspiring his family to enroll him in an art school in Chichester.

In 1900, his father apprenticed him to the architect of the Ecclesiastical Commissioners. Shortly thereafter, he began to study lettering in evening classes with Edward Johnston. After three years, he abandoned architecture to start his own business in letter cutting and stone carving. In 1907, he joined an artists' community in Ditchling, Sussex, where he had his first experiences with printing and typography. He entered the Roman Catholic church in 1913, for which he retained a great, if somewhat unorthodox, zeal throughout his life.

In 1925, Stanley Morison approached Gill with the idea of creating a new typeface for the Monotype Corporation as the expression of a contemporary artist. He began work on what would become known as Perpetua, and not long after that a sans-serif, to be called Gill Sans. Their release three years later caused some controversy, but they became an immediate success with the public. Gill Sans has in fact become the leading British sans-serif, sometimes being described as the "national typeface of England."

Although he never considered himself to be primarily a typographer, over the course of his life he designed eleven typefaces of exceptional beauty and subtlety, and wrote a lengthy and influential *Essay on Typography*.

Book, Jacket, and binding design by Dede Cummings Designs